METHODS IN CELL BIOLOGY

VOLUME XVI

Chromatin and Chromosomal Protein Research. I

Methods in Cell Biology

Series Editor: DAVID M. PRESCOTT

DEPARTMENT OF MOLECULAR, CELLULAR
AND DEVELOPMENTAL BIOLOGY
UNIVERSITY OF COLORADO
BOULDER, COLORADO

Methods in Cell Biology

VOLUME XVI

Chromatin and Chromosomal Protein Research. I

Edited by

GARY STEIN and JANET STEIN

DEPARTMENT OF BIOCHEMISTRY AND
MOLECULAR BIOLOGY
UNIVERSITY OF FLORIDA
GAINESVILLE, FLORIDA

LEWIS J. KLEINSMITH

DIVISION OF BIOLOGICAL SCIENCES
UNIVERSITY OF MICHIGAN
ANN ARBOR, MICHIGAN

1977

ACADEMIC PRESS • New York   San Francisco   London
A Subsidiary of Harcourt Brace Jovanovich, Publishers

ACADEMIC PRESS, INC.
111 Fifth Avenue, New York, New York 10003

United Kingdom Edition published by
ACADEMIC PRESS, INC. (LONDON) LTD.
24/28 Oval Road, London NW1

LIBRARY OF CONGRESS CATALOG CARD NUMBER: 64-14220

ISBN 0–12–564116–8

PRINTED IN THE UNITED STATES OF AMERICA

CONTENTS

Part A. Isolation of Nuclei and Preparation of Chromatin. I

1. *Methods for Isolation of Nuclei and Nucleoli*
Harris Busch and Yerach Daskal

2. *Nonaqueous Isolation of Nuclei from Cultured Cells*
Theodore Gurney, Jr. and Douglas N. Foster

LIST OF CONTRIBUTORS

Numbers in parentheses indicate the pages on which the authors' contributions begin.

ROBERT PETER AARONSON, Department of Microbiology, Mt. Sinai School of Medicine, New York, New York (337)

MAX ALFERT, Department of Zoology, University of California, Berkeley, Berkeley, California (241)

LEONARD H. AUGENLICHT, Memorial Sloan-Kettering Cancer Center, New York, New York (467)

WILLIAM B. BENJAMIN, Department of Physiology and Biophysics, Health Sciences Center, School of Basic Sciences, State University of New York at Stony Brook, Stony Brook, New York (343)

ANN L. BEYER, Department of Molecular Biology, Vanderbilt University, Nashville, Tennessee (387)

E. M. BRADBURY,[1] Biophysics Laboratories, Portsmouth Polytechnic, Gun House, Hampshire Terrace, Portsmouth, Hampshire, England (179)

WOLF F. BRANDT, Department of Biochemistry, C.S.I.R.–Chromatin Research Unit, University of Cape Town, South Africa (205)

HARRIS BUSCH, Department of Pharmacology, Baylor College of Medicine, Houston, Texas (1)

JEN-FU CHIU, Department of Biochemistry, Vanderbilt University School of Medicine, Nashville, Tennessee (283)

R. DAVID COLE, Department of Biochemistry, University of California, Berkeley, Berkeley, California (113, 227)

P. MICHAEL CONN,[2] Department of Cell Biology, Baylor College of Medicine, Houston, Texas (69)

DONALD J. CUMMINGS, Department of Microbiology, University of Colorado Medical Center, Denver, Colorado (97)

NIRMAL K. DAS, Department of Cell Biology, College of Medicine, University of Kentucky, Lexington, Kentucky (241)

YERACH DASKAL, Department of Pharmacology, Baylor College of Medicine, Houston, Texas (1)

MARIE A. DiBERARDINO, Department of Anatomy, The Medical College of Pennsylvania, Philadelphia, Pennsylvania (141)

A. J. FABER, Swiss Institute for Experimental Cancer Research, Lausanne, Switzerland (447)

CARL M. FELDHERR, Department of Anatomy, University of Florida College of Medicine, Gainesville, Florida (167)

ARTHUR FORER, Biology Department, York University, Downsview, Ontario, Canada (361)

DOUGLAS N. FOSTER,[3] Department of Biology, University of Utah, Salt Lake City, Utah (45)

HIDEO FUJITANI, Department of Biochemistry, Vanderbilt University School of Medicine, Nashville, Tennessee (283)

REBA M. GOODMAN, Department of Pathology, College of Physicians and Surgeons, Columbia University, New York, New York (343)

[1] *Present address:* Biophysics Laboratories, Portsmouth Polytechnic, St. Michael's Building, White Swan Road, Portsmouth, Hampshire, England.

[2] *Present address:* Section on Hormonal Regulation, Reproduction Research Branch, National Institutes of Child Health and Human Development, National Institutes of Health, Bethesda, Maryland.

[3] *Present address:* Department of Biochemistry, University of Utah College of Medicine, Salt Lake City, Utah.

GRAHAM H. GOODWIN, Division of Molecular Biology, Chester Beatty Research Institute, Institute of Cancer Research–Royal Cancer Hospital, London, England (257)

JOEL M. GOTTESFELD, MRC Laboratory of Molecular Biology, Cambridge, England (421)

SIDNEY R. GRIMES, JR.,[4] Department of Biochemistry, Vanderbilt University School of Medicine, Nashville, Tennessee (297)

J. B. GURDON, Medical Research Council, Laboratory for Molecular Biology, Hills Road, Cambridge, England (125)

THEODORE GURNEY, JR., Department of Biology, University of Utah, Salt Lake City, Utah (45)

R. HANCOCK, Swiss Institute for Experimental Cancer Research, Lausanne, Switzerland (459)

LUBOMIR S. HNILICA, Department of Biochemistry, Vanderbilt University School of Medicine, Nashville, Tennessee (283)

NANCY J. HOFFNER, Department of Anatomy, The Medical College of Pennsylvania, Philadelphia, Pennsylvania (141)

MYRTLE HSIANG, Department of Biochemistry, University of California, Berkeley, Berkeley, California (113)

CAROLYN KATOVICH HURLEY, Laboratory of Genetics, University of Wisconsin, Madison, Wisconsin (87)

ERNEST W. JOHNS, Division of Molecular Biology, Chester Beatty Research Institute, Institute of Cancer Research–Royal Cancer Hospital, London, England (183, 257)

NINA C. KOSTRABA, Division of Cell and Molecular Biology, State University of New York at Buffalo, Buffalo, New York (317)

WALLACE M. LeSTOURGEON, Department of Molecular Biology, Vanderbilt University, Nashville, Tennessee (269, 387)

A. J. MACGILLIVRAY,[5] Beatson Institute for Cancer Research, Wolfson Laboratory for Molecular Pathology, Bearsden, Glasgow, Scotland (329, 381)

J. RICHARD MCINTOSH, Department of Molecular, Cellular, and Developmental Biology, University of Colorado, Boulder, Colorado (373)

MICHAEL B. MATILSKY, Department of Anatomy, The Medical College of Pennsylvania, Philadelphia, Pennsylvania (141)

MARVIN L. MEISTRICH, Section of Experimental Radiotherapy, The University of Texas System Cancer Center, M.D. Anderson Hospital and Tumor Institute Houston, Texas (297)

PATRICIA Z. O'FARRELL, Department of Biochemistry and Biophysics, University of California, San Francisco, San Francisco, California (407)

PATRICK H. O'FARRELL, Department of Biochemistry and Biophysics, University of California, San Francisco, San Francisco, California (407)

ROBERT D. PLATZ,[6] Section of Experimental Radiotherapy, The University of Texas System Cancer Center, M.D. Anderson Hospital and Tumor Institute, Houston, Texas (297)

[4] *Present address:* Veterans Administration Hospital, Shreveport, Louisiana.
[5] *Present address:* Biochemistry Laboratory, School of Biological Sciences, University of Sussex, Falmer, Brighton, Sussex, England.
[6] *Present address:* Frederick Cancer Research Center, Biological Markers Laboratory, Frederick, Maryland.

ELENA C. SCHMIDT, Department of Pathology, College of Physicians and Surgeons, Columbia University, New York, New York (343)

ROBERT T. SIMPSON, Developmental Biochemistry Section, Laboratory of Nutrition and Endocrinology, National Institute of Arthritis, Metabolism, and Digestive Diseases, National Institutes of Health, Bethesda, Maryland (437)

THOMAS G. SPRING,[7] Department of Biophysical Sciences, University of Houston, Houston, Texas (227)

JOHN T. STOUT, Laboratory of Genetics, University of Wisconsin, Madison, Wisconsin (87)

JOSEPH L. TURNER, Department of Molecular, Cellular, and Developmental Biology, University of Colorado, Boulder, Colorado (373)

CLAUS VON HOLT, Department of Biochemistry, C.S.I.R.–Chromatin Research Unit, University of Cape Town, South Africa (205)

TUNG YUE WANG, Division of Cell and Molecular Biology, State University of New York at Buffalo, Buffalo, New York (317)

VIRGINIA P. WRAY, Department of Cell Biology, Baylor College of Medicine, Houston, Texas (69)

WAYNE WRAY, Department of Cell Biology, Baylor College of Medicine, Houston, Texas (69)

ARTHUR M. ZIMMERMAN, Department of Zoology, University of Toronto, Toronto, Ontario, Canada (361)

SELMA ZIMMERMAN, Division of Natural Sciences, Glendon College, York University, Toronto, Ontario, Canada (361)

[7] *Present address:* Abbott Laboratories, Diagnostic Division, Abbott Park, North Chicago, Illinois.

PREFACE

During the past several years considerable attention has been focused on examining the regulation of gene expression in eukaryotic cells with emphasis on the involvement of chromatin and chromosomal proteins. The rapid progress that has been made in this area can be largely attributed to development and implementation of new, high-resolution techniques and technologies. Our increased ability to probe the eukaryotic genome has far-reaching implications, and it is reasonable to anticipate that future progress in this field will be even more dramatic.

We are attempting to present, in three volumes of *Methods in Cell Biology*, a collection of biochemical, biophysical, and histochemical procedures that constitute the principal tools for studying eukaryotic gene expression. Contained in this volume (Volume 16) are methods for isolation of nuclei, preparation and fractionation of chromatin, fractionation and characterization of histones and nonhistone chromosomal proteins, and approaches for examining the nuclear-cytoplasmic exchange of macromolecules. Volume 17 deals with further methods for fractionation and characterization of chromosomal proteins, including immunological, DNA affinity, and sequencing techniques. Also contained in Volume 17 are methods for isolation and fractionation of chromatin, nucleoli, and chromosomes. The third volume (Volume 18) focuses on approaches for examination of physical properties of chromatin, enzymic components of nuclear proteins, chromatin transcription, and chromatin reconstitution. Volume 18 also contains a section on methods for studying histone gene expression.

In compiling these three volumes we have attempted to be as inclusive as possible. However, the field is in a state of rapid growth, prohibiting us from being complete in our coverage.

The format generally followed includes a brief survey of the area, a presentation of specific techniques with emphasis on rationales for various steps, and a consideration of potential pitfalls. The articles also contain discussions of applications for the procedures. We hope that the collection of techniques presented in these volumes will be helpful to workers in the area of chromatin and chromosomal protein research, as well as to those who are just entering the field.

We want to express our sincere appreciation to the numerous investigators who have contributed to these volumes. Additionally, we are indebted to Bonnie Cooper, Linda Green, Leslie Banks-Ginn, and the staff at Academic Press for their editorial assistance.

GARY S. STEIN
JANET L. STEIN
LEWIS J. KLEINSMITH

Part A. Isolation of Nuclei and Preparation of Chromatin. I

Chapter 1

Methods for Isolation of Nuclei and Nucleoli

HARRIS BUSCH AND YERACH DASKAL

Department of Pharmacology
Baylor College of Medicine
Houston, Texas

I. Isolation of Nuclei

Procedures for the isolation of nuclei have been important in studies on histones and other nuclear proteins (*1,2*), nuclear enzymes (*2–6*), and nuclear RNA and DNA (*7,8*). Since nuclei are a starting material for isolation of nucleoli, procedures for isolation of nuclei must provide products that are satisfactory with respect to yield, morphology, and chemical composition, if isolation of nucleoli is to be satisfactory.

The objectives of procedures for isolation of nuclei are to secure a product in which (a) the nuclei are morphologically (Fig. 1) identical to those of the whole cell, (b) the contents of the nuclei as they exist in the cell are all present in the isolated product, and (c) the isolated nuclei do not contain cytoplasmic constituents. At the present time, the procedures commonly used for isolation of nuclei are (1) modification of the procedure of Chauveau *et al.* (*9*) in which the tissue is homogenized in sucrose solutions containing Ca^{2+} followed by centrifugation of the homogenate [another procedure used is the nonaqueous technique in which the tissue is subjected to rapid freezing and lyophilization followed by a milling procedure in nonaqueous solvents (*10,11*)]; (2) treatment of the tissue with citric acid following the technique of Dounce and his associates (*12–15*); and (3) techniques employing "hypotonic shock" followed by treatment with detergents and

1

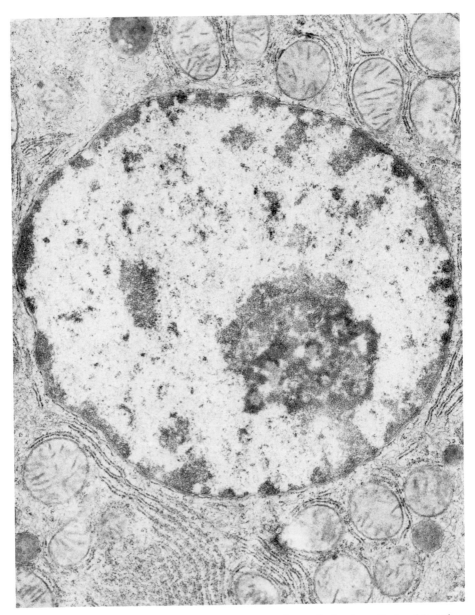

FIG.1. Electron micrograph of a nucleus within a liver cell showing mitochondria and endoplasmic reticulum in the cytoplasm and the outer layer of the nuclear envelope, which is essentially similar in structure to the endoplasmic reticulum. The inner layer of the nuclear envelope is juxtaposed to dense chromatin masses interspersed between spaces of the nuclear pores. The nucleolar stalk joins the nucleolus to the chromatin at the periphery of the nucleus.

homogenization (*16,17*). Each of the procedures has advantages from the point of view of morphology or retention of enzymic or other macromolecular constituents in the product, but each has such significant disadvantages that no ideal procedure exists for isolation of the nuclei from any kind of cell.

A. *In Vivo* Appearance and Composition of the Nuclei

The only satisfactory method for routine assessment of the appearance of the nuclei *in vivo* is the use of the light and phase microscopes. By phase microscopy of either living cells in tissue culture or smears of tissue preparations, nuclei are seen as structures that are homogeneous and less dense than the cytoplasm. The nuclear membrane is not visible as a dense, thick, defined structure although the nucleoli are usually different in density from the remainder of the nucleoplasm.

The most valuable morphological evaluation of nuclear structures is made by analysis of the electron microscopic appearance of the nuclear substructures, including the nuclear membrane, the nucleolus, and the chromatin (*7,18–20*). The nuclear membrane appears to be a double-layered structure, as suggested by light microscopists, and its appearance by electron microscopy is characteristic for both the layers of the membrane and the appearance of the "nuclear pores" (Fig. 2). The nucleoli have characteristic appearances with both granular and fibrillar components of the nucleolonemas and the light spaces between these structures. Clumped masses of chromatin are also visualized, which may be present either on the periphery of the nucleus apparently matted around the inner layer of the nuclear membrane (Fig. 2A) or randomly distributed within the nucleus. Ribonucleoproteins of varying density are seen. Some have diameters of 500 Å or greater, and these are referred to as the interchromatinic dense particles or Swift and Watson granules (*20*). Other ribonucleoprotein particles of 200–400 Å diameter and of equal density are also found in interchromatinic regions (Fig. 2). These smaller particles are apparently cross sections of the nuclear ribonucleoprotein network (*21*).

The appearance of the nucleoli constitutes one measure of the nativeness of the nuclei. The nucleoli are distinguished by their variability from tissue to tissue (*22–30*). The location of the nucleoli may be central, attached to the nuclear membrane, or random in the nucleus. In addition, the nucleoli vary in size and in number. They may be so minute as to be submicroscopic, or they may be so large as to practically fill the nuclear volume (Fig. 3) and leave only a small rim of nucleoplasm and chromatin within the nuclear membrane in cells of thioacetamide-treated animals (*31*). Their number generally varies from 1 to 10.

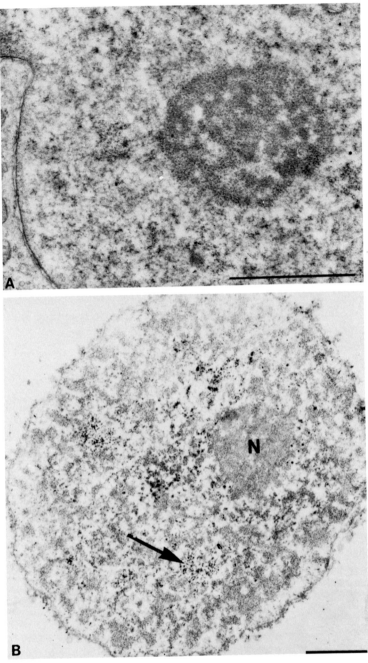

FIG.2. (A) Nucleolus within the nucleus of a normal liver cell *in situ*. Chromatin masses are composed of the faint fibrils. The marker line represents 1 μm. (b) Isolated nucleus of rat liver stained with uranyl acetate showing the nucleolus (N). ×17,000. The arrow points to the inter-chromatinic dense granules. The marker line is 1 μm.

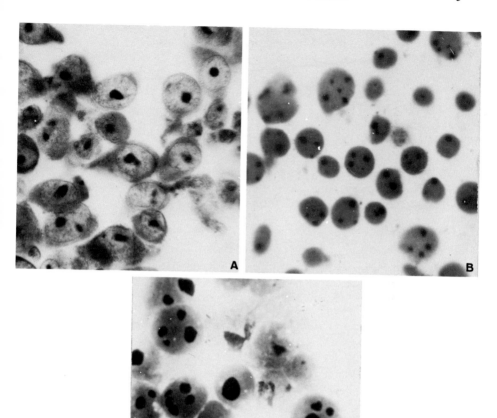

FIG. 3.   Variability in nucleolar sizes and shapes. (A) Nuclear preparation of the Walker 256 carcinosarcoma stained with azure C. (B) Nucleoli of normal liver stained with toluidine blue. (C) Nucleoli of livers of rats treated with thioacetamide (50 mg/kg) for 9 days. The stain used was toluidine blue. From Busch *et al.* (*31*).

From the biochemical point of view, certain criteria for nuclear contents now seem established. The nuclear content of DNA, RNA, and protein has now been defined within limits; isolated nuclei of normal rat liver cells contain approximately 8–11 pg of DNA, 2.2–3.0 pg of RNA, and 35–45 pg of protein per nucleus. Nuclei of tumor cells contain amounts of DNA ranging from below the diploid value for liver nuclei to as high as octaploid values. Values for RNA of Walker tumor nuclei are 1.5–2 times those of liver nuclei, but the amounts of protein present are not markedly different. The amounts of histones are approximately equivalent to the value for DNA.

B. Nuclear Enzymes

Certain enzymes have now been found to be localized to the nucleus. These include NAD pyrophosphorylase, RNA polymerase, and enzymes involved in conversion of 45 S and 55 S RNA to 28 S RNA (*2,5,32*). Since a variety of marker enzymes has been reported to be present solely in certain cytoplasmic particles, their presence in nuclear preparations is evidence of contamination of nuclear preparations (*5,32*).

II. The Sucrose–Calcium Procedure for Isolation of Nuclei

Although there are many questions about the nativeness of the nuclei isolated by procedures employing "sucrose–calcium" solutions, this method has become the one most widely used for the preparation of nuclei. Initially, this procedure was developed by Hogeboom *et al.* (*33,34*); it was later modified by Chauveau *et al.* (*9*) and brought to its present form in our laboratory (*35*). The nuclei obtained are free of cytoplasmic tags and adherent clumps of cytoplasm. By phase microscopy, the nuclear membrane is double refractile and sharp, and the nucleoli are highly refractile and readily visualized (Fig. 4). With proper purification no cytoplasmic contamination is visible by light microscopy (Fig. 5).

A. Preparation of the Tissue for Homogenization

The liver normally contains large amounts of blood which can be removed by perfusion through the portal vein. Initially, 20 ml of ice-cold 0.15 M NaCl is perfused through the liver followed by 20 ml of 0.25 M sucrose. The liver is then rapidly removed, placed in ice-cold 0.25 M sucrose, and transferred to a coldroom (2–4°) for further procedures. Inasmuch as liver tissue contains considerable arborization of the portal system as well as the hepatic blood vessels, it is necessary to separate the parenchymal cells from these more fibrous structures by passing the tissue through a mincer. In this laboratory, the Harvard tissue press is employed for this purpose. In essence, this instru-

---

FIG. 4.   Phase microscopy of a preparation of isolated nuclei of Morris hepatoma 9618. (A) ×460; (B) ×820.

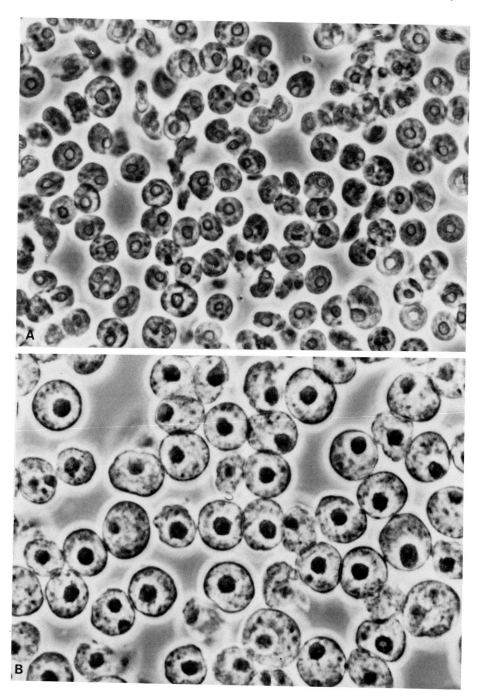

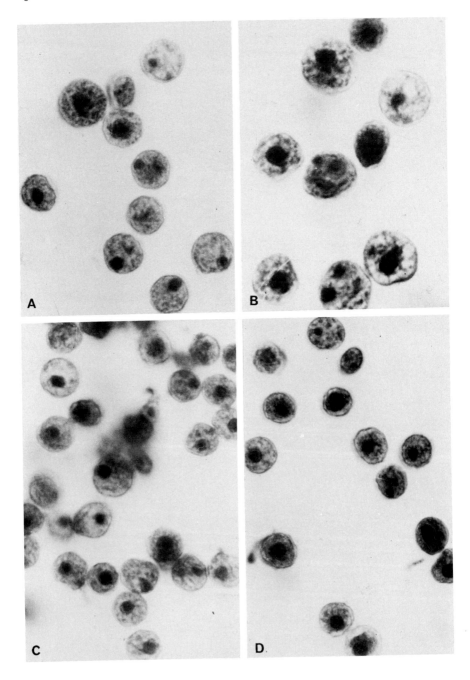

ment is a stainless steel cylinder with a perforated plate at the base and a cylinder driven by a screw at the top. Although this is a useful instrument, one modification employed includes a handle on the screw plate so as to provide a greater leverage on the handle. A more practical model is the large-size press (*36–38*), which employs the same principle except that force is applied with a rack and pinion device; this works much more rapidly and is much less tiring.

The *medium for homogenization* varies a great deal, and in this laboratory the original system of Chauveau *et al.* (*9*) has been used in which the minced or fragmented tissue is suspended in 10 volumes of 2.4 M sucrose containing 3.3 mM calcium acetate or $CaCl_2$. The concentration of the sucrose is quite critical, and care is taken that the percentage of sucrose is 60% by weight as determined either with a saccharimeter or a direct-reading refractometer (Abbe). In other laboratories, 0.25 M sucrose, 0.88 M sucrose, and 2.0 M sucrose have been employed and amounts of Ca^{2+} ranging from 0.5 mM to 10 mM (*33,39,40*) have been employed. In an endeavor to isolate nuclei with optimal enzymic activity, Widnell and Tata (*41*) used 0.32 M sucrose–3 mM $MgCl_2$ as the homogenizing medium. When large amounts of sucrose are employed as in the mass isolation of nuclei and nucleoli, it is a considerable saving to obtain sucrose from commercial sources. Preparations used for mixers with soft drinks have been obtained from the Imperial Sugar Company (Sugarland, Texas) and, in addition, granular preparations of high purity (better than is usually available from some chemical supply companies) can be obtained in 100-pound bags at low cost.

The amount of homogenizing medium has also varied, ranging from 3 to 15 ml of sucrose solutions per gram of tissue. In this laboratory, optimal preparations were obtained with 10–15 volumes volumes of 2.2 M or 2.4 M sucrose per gram of tissue.

B. Homogenization

The all-glass homogenizer used by Potter and Elvehjem (*42*) was a widely used type of homogenizing device, but with the development of polymers such as Teflon®, which are not detrimental to tissues, the use of glass homogenizers has largely been discontinued. Large test tubes made of glass pipe (1.6 in.) expanded at the top are fitted with Teflon pestles containing a stainless-steel shaft for insertion into the chuck of a drill motor (from Montgomery Ward & Co.). These motors are inexpensive and yet have the substantial

---

FIG. 5. Isolated nuclei of Morris hepatomas stained with azure C. (A) Hepatoma 9618A, ×1155; (B) hepatoma 9121, ×1540; (C) hepatoma 7787, ×1155; (D) 9108A, ×1155.

torque essential for homogenization of the liver cells normally carried out at approximately 1200 rpm. As indicated earlier (43) the pestle clearance is very critical for some purposes; for the homogenization of the liver, clearances of 0.006–0.010 in. are satisfactory. If the pestle clearance is less, considerable losses of nuclei occur inasmuch as they are fragile, particularly when "hardened" by Ca^{2+}. If the pestle clearance is too great, many whole cells remain in the homogenate.

For tissues other than the liver, such as tumors, spleen, and kidney, the conditions vary considerably because the degree of cytoplasmic adherence is much greater, especially in transplanted tumors and in some lymphocytes. For spleen, the pestle clearance may be somewhat greater and for Walker tumor it should be considerably less, i.e., 0.003 in. Even with these pestle clearances, good nuclear preparations of the Walker tumor have not been attained in this laboratory as long as there was divalent ion in the medium (43). Nuclei obtained from media in which divalent ions were omitted were not contaminated by cytoplasmic tags, but the nuclei were considerably swollen and, in addition, the nucleoli were not clearly visible. With the Morris hepatomas 9618A, 9121, and 9108A, nuclear preparations have been obtained of quality equal to those obtained with rat liver.

Other types of homogenizers have been developed that are of considerable value, although they have been used to a lesser extent in this laboratory than the Teflon–glass homogenizer. For small-scale studies, the Dounce homogenizer (15) has been utilized; the process of "Douncing" consists of several up-and-down strokes of the hand homogenizer (either of the Ten Broeck type or the Dounce ball type) in which both the pestle and the tube are ground glass with a clearance of approximately 0.006–0.010 in. The pestle of the Ten Broeck homogenizer has a crossbar at the top that is manually twisted to provide a homogenizing effect.

For large-scale homogenization, the continuous tissue homogenizer (44) is useful. This instrument (Glenco, Houston) consists of a cylinder made of either stainless-steel or glass pipe that has a screw-feed at one end and a tapered pestle at the other with very little dead space between them. The instrument is powered by a relatively low-speed motor (800 rpm) on the end containing the feeding screw, and on the other end the motor for the homogenizing pestle operates at 1700 rpm. Both the feeding screw and the homogenizing pestle are made of polyvinyl chloride (PVC) or plexiglass, and the sample flows through the instrument at a rate of 100–150 ml/minute either by gravity or under pressure of a pump.

C.   Number of Strokes or Passes for Homogenization

For homogenization with the Teflon–glass homogenizer, three strokes of the homogenizer are required to produce fragmentation of most of the cells of minced liver tissue. With the standard Potter-Elvehjem homogenizer, six to ten up-and-down strokes of the homogenizer are required, and the same is true for the Dounce procedure. With the continuous tissue homogenizer, one or two passes of the suspension suffice to produce a preparation that is equivalent to that produced by the usual homogenization in the Teflon–glass homogenizer in which approximately 90% of the nuclei are released from the surrounding cytoplasm. In its present form, the continuous homogenizer is a practical device for homogenizing 250 gm or more of tissue.

D. pH

The pH of the solutions in the sucrose–calcium procedure must be kept on the acid side, preferably in the range pH 5–6. This is necessary because at alkaline pH the Ca^{2+} or Mg^{2+} become complexed with hydroxyl groups, which results in a failure of the nucleoli to maintain their integrity. The result of this is either a poor preparation or low yield or both. In many studies no additional buffers have been added so that the pH of the homogenate approximates that of the distilled water. However, a number of buffers such as the acetate buffer (0.1–0.001 M) have been employed to maintain the pH in the range noted above.

E. Temperature Control

The shearing forces produced by homogenization may vary from very minor, when homogenizers of the Ten Broeck or Dounce type (15) are used, to very marked, when motors with high speeds and torque are employed. With hand homogenization, there is relatively little damage to the nuclei, although the products have been less satisfactory in this laboratory. With high-speed homogenization and excessive homogenization, there frequently has been very serious damage to the nuclear membranes and, in addition, the sedimentation profiles of nuclear RNA are markedly abnormal. With the high-speed, tight-fitting, motor-driven homogenizers there is also considerable friction which results in heating of the homogenate so that the temperature rises from 0–3° to 10–15°. To protect the tissue against the damaging effects of heating from internal friction as well as the external heating from the operator's hand, jacketed homogenizers have been developed. These consist of the usual glass pipe with a glass jacket through which a cooling solution of ethylene glycol–water (1:2) is pumped from a bath that is kept at −5°. Such baths can be obtained from Wilkins–Anderson or the Forma Companies and are particularly useful for experiments with small amounts of homogenates. Cracked ice is a poor medium for removing the heat from the homogenizer.

In the continuous tissue homogenizer, the sample is cooled with the aid of copper tubing wound around the stationary cylinder, by a jacket of circulating medium kept at -3 to $-5°$, or by a layer of ice slush kept around the cylinder. After the samples are homogenized, they are passed through eight layers of gauze and a 90-mesh stainless-steel wire screen prior to centrifugation.

F.   Centrifugation

The initial homogenates are subjected to a variety of centrifugation procedures depending on the initial suspending solutions and on the quality of the product that is desired. In the procedure of Chauveau *et al.* (9), the homogenate is poured into 30–ml nitrocellulose tubes and centrifuged at 40,000 g for 60 minutes. The density of the 60% sucrose solution is such that only the nuclei are sedimented under these conditions and the debris, whole cells, and cytoplasmic constituents are floated to the top of the solution or else are left on the side of the tube. Nuclear pellets prepared from normal liver using these conditions are pale amber to white in color, approaching a whitish color. Contamination is apparent when the surface of the pellet is either coated with a brownish layer or streaks. In the procedure generally used in this laboratory, the pellet is resuspended in 1.0 M sucrose–1 mM calcium acetate in a volume of 1 ml per gram of original tissue. The sample is rehomogenized with a loosely fitting Teflon pestle (0.010–0.012 in. clearance) attached to a motor operating at 500 rpm. This homogenate is centrifuged at 3000 g for 5 minutes, and the precipitate consists of highly purified nuclei (Figs. 1–5).

Although it was a common practice at one time to refer to the "600 g precipitate" of the homogenates prepared in 0.25 M sucrose or 0.88 M sucrose as "nuclear preparations," this terminology has now been largely abandoned. The precipitates obtained by centrifugation of homogenates for 10 minutes at 600 g contain whole cells, much debris, and considerable cytoplasmic contamination. This centrifugation procedure is of value only if the mitochondria or other more slowly sedimenting cellular components are desired and the supernatant fraction is to be saved.

G.   Large-scale Preparation of Liver Nuclei (45)

The weighed (about 1 kg) liver tissue is disrupted in a Hobart model 4612 meat chopper (grinder) fitted with a specially made stainless-steel chopper plate (1-mm diameter holes). A tissue–sucrose slurry (8 to 12 liters) is prepared and maintained at 2–4° by continuous stirring of the macerated tissue into 1.6 M sucrose (1:9 w/v)-3.3 mM $CaCl_2$. Homogenization is effected by pumping this slurry (150 ml/minute) through a continuous tissue homogenizer equipped with a lucite pestle having a total clearance of 0.006 in. A low temperature (8–12°) is maintained during homogenization by circulating a water-antifreeze mixture at −2° through a copper cooling coil wrapped around the stainless-steel tube of the homogenizer. At this temperature virtually no degradation of nuclear RNA occurs. Fibrous tissue and incompletely homogenized liver tissue are removed from the homogenate by filtration through two layers of 20-mesh stainless-steel wire cloth. The filtered homogenate is cooled and maintained at 2° prior to isolation of nuclei.

One homogenization with 1.6 M sucrose releases 90% of the liver nuclei from cytoplasm in 12 liters of tissue-sucrose slurry in approximately 90 minutes. Less concentrated sucrose is not as effective in stripping the cytoplasm from the nuclei; more concentrated sucrose produces excess heat.

H. Isolation of Nuclei

Nuclei are isolated by means of two air turbine drive Sharples T-1 open-type laboratory supercentrifuges equipped with clarifier bowl assemblies and cooling coils in the frame assembly. Heating of the bowls is prevented by circulation of water–antifreeze (−2°) through the cooling coils and free passage of air into the centrifuge chamber through holes ($\frac{3}{8}$ in. each) drilled at the top and bottom of the frame assembly. Sucrose solutions and homogenate are passed into the two centrifuge bowls at equal rates with the use of a pair of 45-drill feed nozzles connected with $\frac{1}{2}$-in. Tygon tubing and a Y tube to a single variable-speed pump. The clarifier bowls are filled with 2.4 M sucrose during acceleration of the centrifuges to 50,000 rpm (60,000 g). The combination of the high density of 2.4 M sucrose and the high g force produces a large layer of 2.4 M sucrose between the walls of the bowl and the flow-through core as the less dense homogenate passes through the bowl. This standing discontinuous gradient is effective in producing morphologically satisfactory nuclei. At 180 ml/minute, the homogenate passes through the bowl as a thin stream confined to the core of the bowl with little if any mixing of the two sucrose layers. Nuclei without cytoplasmic tags are sufficiently dense to penetrate the 2.4 M sucrose and sediment against the walls while cytoplasmic materials and whole cells pass out with the 1.6 M sucrose homogenate as a supernatant fraction. Nuclei are collected from 12 liters of homogenate with the use of two instruments in 35 to 40 minutes. At the end of each run, fresh 1.6 M sucrose is used to flush the remaining homogenate from the tubing, pump, and clarifier bowls before deceleration and collection of the nuclei from the walls of the bowls. Under these conditions, approximately 40% of the total nuclei is collected during one centrifugation run. A second batch of nuclei is collected by centrifugation of the first supernatant fraction under identical conditions.

Approximately 80% of the free nuclei are collected from 12 liters of homogenate by two such centrifugations in two instruments in a total time of 90 to 100 minutes. Centrifugation is usually started when the homogenization is two-thirds complete. In experiments with 120 animals, it is possible to kill the animals, perfuse and dissect the livers, disrupt and homogenize the tissue, and collect the nuclei in good yields and high quality in a total time of 4 to $4\frac{1}{2}$ hours.

I. Criteria of Purity

The most satisfactory initial criterion of morphological purity of the preparations is direct examination of the product by phase microscopy. Under the conditions of the phase examination in which the pellet is suspended either in 0.25 M sucrose alone or in 0.25 M sucrose–3.3 mM $CaCl_2$, the optimal preparations contain only rounded nuclei with no visible mitochondrial particles and clear, sharp nucleoli. For counting the nuclei in a hemacytometer, they may be stained with either azure C (7,46) or Giemsa stain (47). Direct determinations of the recovery of nuclei can be made by light microscopic analysis.

When the morphological criteria of purity are satisfactory (Figs. 1–5), i.e., less than one whole cell per 4000 nuclei and no visible cytoplasmic particles, enzyme analyses for contaminating cytoplasm may be useful. Widnell and Tata (41) utilized glucose-6-phosphatase, cytochrome oxidase, glucose-6-phosphate dehydrogenase, and NADH-cytochrome c reductase as cytoplasmic enzyme markers. In nuclear preparations obtained in this laboratory, Siebert et al. (32) found that the activity of the cytoplasmic marker enzymes, glutamate dehydrogenase and adenylate kinase, were 0.4% and 0.5% of the whole homogenate, respectively.

Electron microscopic analysis of nuclear preparations is difficult to evaluate on a quantitative basis inasmuch as the procedures for preparation of the samples on the grids and the techniques for fixation and staining may produce some damage to the samples (Fig. 6). Nonetheless, the purity can be ascertained on the basis of either total or recognizable particle counts. With the optimal preparations, contamination of the nuclear preparations with either mitochondria or lysosomal particles is not found, or these particles are present in amounts of less than 0.1% of the total particles in the pre-

FIG. 6. Electron micrographs of isolated nuclei of Morris hepatoma 9618A. (A) Little or no cytoplasm is apparent, but outer nuclear membranes are still attached to the nuclei. The large nucleoli of these cells are visible in several of the sections. ×6170. (B) Focusing on the largest nucleolus. This is seen to consist mainly of granular elements, and one ribonucleoprotein branch of the nuclear ribonucleoprotein network is seen emerging from the nucleolus toward the nuclear membrane. ×11,600.

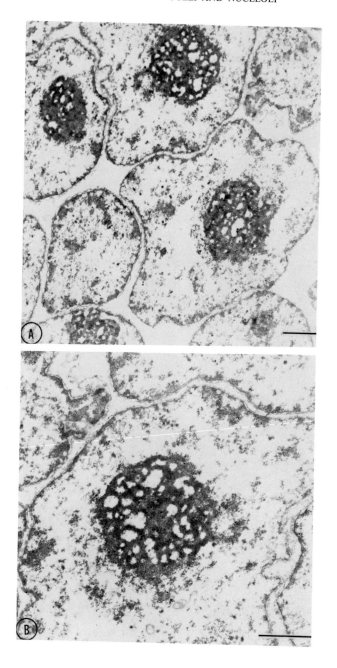

paration. However, small amorphous components are present, generally as a very small percentage of the total nuclear area in amounts ranging from 1 to 8% of the total particles in the preparation.

The yields of nuclei with the Chauveau procedure (9) are approximately 25–35% of the total in the homogenate, and with the continuous homogenization procedure they approximate 50–60% of the total in the homogenate. The former, however, have a higher degree of purity. Improved yields up to 90% are obtained by the method of Blobel and Potter (48, 49).

Analysis of the constituents of isolated liver nuclei provides the following values for DNA, RNA, and proteins as picograms per nucleus: 11, 3, and 44. The ratios of RNA to DNA and protein to DNA are approximately 0.25 and 4, respectively.

III.   The Citric Acid Procedure for Isolation of Nuclei

The use of citric acid as a suspending medium for preparation of nuclei either from whole homogenates or in the initial extraction of cells was popularized by Dounce (12–14) after Marshak (50) had reported that homogenization in citric acid removed the cytoplasm from the nuclear membrane. This method has been particularly used in preparation of nuclei from cells that represented difficult problems with the sucrose procedure, such as tumor cells, thymocytes, and cells of skin. Although the precise properties that make citric acid valuable in removal of cytoplasm are not established, reduction of pH, chelation of calcium ion, and extraction of acid-soluble proteins and lipid may be involved as factors reducing the adherence of the cytoplasm to the nuclear membrane.

In this laboratory, the citric acid procedure has been used for preparation of the nuclei of the Walker tumor cells in both small-scale and mass preparations (51). With this procedure, RNA of high molecular weight, particularly the 45 S RNA, can readily be isolated from tumor nuclei and, by comparison with the sucrose–calcium procedure, the amount of associated ribosomal RNA is reduced to a minimum. The preservation of the 45 S RNA may occur because ribonuclease, a basic protein, is removed by citric acid or is inactivated at the low pH or because Ca^{2+} is chelated by the citric acid. In any event, the method is of particular value for preservation of nuclear RNA.

A.   Morphology of Nuclei Isolated with Citric Acid Procedure

The morphology of these nuclei is quite different from that which is seen by phase microscopy of the nuclei in whole cells in that the nuclear membrane is pale and the nucleolus is enlarged, less refractile, and apparently flattened; the chromatin appears to be clumped (Fig. 7A). The cytoplasm is largely removed, but many clumps of apparently denatured cytoplasm are found in the preparations; these clumps can be removed by centrifugation of the preparation after it is layered over 0.88 M sucrose–1.5% citric acid.

Electron microscopy of these preparations reveals that the organization of the chromatin and the subcomponents of the nucleoli are different from those seen with the sucrose–calcium procedure (Fig. 7B). The borders of the nucleoli are considerably more hazy than those observed after isolation with the sucrose–calcium procedure and the nucleoli appear to be larger. The chromatin is clumped. These changes may be due to the considerable extraction of protein with citric acid in the concentrations used and to the low pH necessary to effect removal of the cytoplasm. Since protein exerts an important effect on the stainability of the citric acid preparations for electron microscopy, it is not surprising that there are changes in the electron micrographs of preparations treated with citric acid (Fig. 7B).

B. Solutions for Homogenization

The homogenizing medium employed has varied only in the concentration of citric acid employed and in other components added. Marshak (50), who first used the method for liver and tumor tissue, utilized 5% citric acid solution, although Stoneburg and Haven (52) used the method in conjunction with a pepsin digestion in 1940. The pH of the homogenizing solution is dependent upon the concentration of the citric acid employed so that 1.5% citric acid solutions have a pH of 2.3 and 4% citric acid solutions have a pH of 2.2. Since there is variation in the tenacity of the cytoplasmic adherence to the nuclear membrane in various tissues, for some tissues lower concentrations of citric acid may be employed than for others; e.g., for the Walker tumor a concentration of 2.5% was found to be optimal (15,51).

C. Temperature

As in the case of isolation of nuclei with either the sucrose–calcium procedure or the nonaqueous media, it is generally believed to be optimal to keep the temperature as close to 0° as possible throughout the homogenization and centrifugation. When high molecular weight RNA is isolated, it is very important to minimize RNase activity, and cold conditions are an absolute requirement.

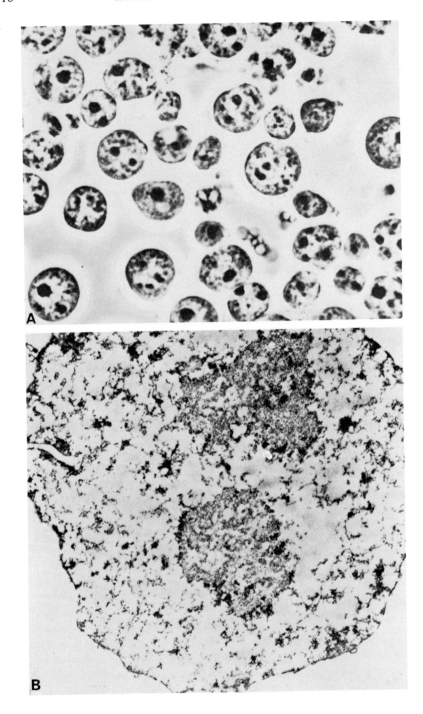

D. Homogenization

In this laboratory, tumor tissue is homogenized in 2.5% citric acid solutions in a weight to volume ratio of 1:10. Liver preparations are made in 1.5% citric acid in the same ratio of weight to volume. There are no special problems involved in the citric acid procedure that have not already been dealt with in connection with the sucrose–calcium method. The same Teflon–glass system is used for small-scale preparations in this laboratory. For larger preparations, the continuous homogenizers are used, and satisfactory preparations are obtained with either stainless-steel or glass cylinders.

Although years ago Schneider and Hogeboom (53) pointed out the unsuitability of the Waring blender and similar instruments for most biochemical preparations, such instruments as well as some oscillators are still being used. These instruments are unsuited to the isolation of nuclei by any method because they destroy not only nuclei but also high molecular weight RNA.

E. Centrifugation

Initially, the homogenates of tumor preparations are centrifuged at 600 g for 10 minutes and then are rehomogenized for short periods in 0.25 M sucrose–1.5% citric acid; these suspensions are then layered over 0.88 M sucrose–1.5% citric acid. The samples are centrifuged at 900 g for 10 minutes after which the supernatant solutions are discarded and the sediment which contains the nuclear preparation is saved. For liver, the procedure may be simpler, depending upon the starting preparation, i.e., the layering procedure is not necessary if a fairly good preparation is obtained initially.

F. Yield and Purity

The yields of nuclei of high purity obtained with the citric acid procedure are somewhat higher than those obtained with the sucrose–calcium procedure, i.e., they are approximately 43%, which is slightly more than the 35% usually obtained with the Chauveau technique (9). Whether this difference is statistically significant is not yet defined. Although the purity of these preparations is reasonable, it has not yet been possible to obtain a preparation that is completely free of cytoplasmic contamination as detected by light microscopy. The separation of the cytoplasm from the nucleus occurs by a

---

FIG. 7. (A) Nuclei of Walker tumor cells isolated by the citric and procedure. The nucleoli are large and appear to be intact. Little cytoplasmic contamination is seen. The nuclei are pale and appear to be extracted. (B) Electron micrograph of a nucleus of a Walker tumor cell isolated by the citric acid procedure. ×10,200. From Busch and Smetana (35.)

process that resembles the shelling of a seed or hard-boiled egg yolk from the surface shell of denatured protein. Since this shelling process is not always complete, an occasional nucleus is found with some of the cytoplasm partially adhered to it. These cytoplasmic tags differ from those found in the Chauveau procedure (9) in that they are wedge shaped with the larger end of the wedge farther from the nuclear membranes; in the Chauveau procedure, the opposite is generally true, i.e., the larger part of the wedge is closer to the nuclear membrane.

Electron microscopic analysis shows that there are generally a number of small segments of the denatured cytoplasm associated with these preparations so that the purity index by particle count (nuclear particles/total particles) is not as satisfactory as would be desirable. These particles are generally bits of cytoplasm that apparently become as dense as the nuclei by dehydration or denaturation. With nuclear preparations obtained by the Chauveau procedure (9), the purity of the preparation may be as high as 99% by particle count in the phase microscope and as high as 92% by particle count in the electron micrographs. The best purity of the nuclear preparations obtained with the citric acid procedure in this laboratory was about 97% by particle count with the phase microscope and about 85% by particle count of electron micrographs.

For further assays of purity, chemical methods have been employed and assays of DNA, RNA, and protein have been reported. However, no assays have been presented of enzymes in the nuclei obtained by this procedure, probably because of the assumption that they are denatured.

In this laboratory, values for DNA, RNA, and protein have been obtained in comparison with the sucrose–calcium procedure (see Table I). Dounce (15) has reported that in nuclei isolated by the citric acid procedure, DNA composed 20–23% of the dry weight, a value that agrees well with the results obtained in studies in this laboratory. Citric acid–sucrose solutions at pH

TABLE I

NUCLEAR CONTENTS[a]

Sample	Sucrose–calcium (46)	Citric acid (51)
Liver		
DNA	10.7	8.8
RNA	2.9	2.2
Protein	43.9	35.1
Tumor		
DNA	15.4	11.9
RNA	23.5	5.7
Protein	171.5	34.8

[a] In picograms per nucleus.

5.8 were then used and the DNA was markedly reduced in amount. Dounce (15) has indicated that the procedure employing buffers at pH 5.8 results in a more satisfactory preparation from the point of view of the protein content and minimal proteolytic activity in the preparation.

IV. The Use of Detergents for Nuclear Isolation

The objectives of the detergent procedures for the isolation of nuclei are similar to those of other methods. Frequently, these are parts of other cell fractionation procedures. However, it has been found that detergents enable isolation of nuclei from either tumor tissue, ascites cells, tissue culture cells, or nontumor tissues. Nuclei have been isolated that have good RNA polymerase activity as well as morphological characteristics.

A. Nonidet P-40 (NP-40) Isolation Procedure

This method has been explored by Muramatsu *et al.* (54) and further modified. The procedure was useful for isolation of nuclei from Novikoff hepatoma ascites cells, normal liver (54), Ehrlich ascites, HeLa, and L cells (54). Detergent concentrations used for nuclear isolation varied from 0.3 to 0.5%, depending on the cell lines. The procedures are carried out at 4 °C.

For Novikoff hepatoma, the ascites cells are diluted with ice-cold TNKM buffer [50 mM Tris-HCl (pH 7.0)–0.13 M NaCl–25 mM KCl–2.5 mM Mg acetate] and washed once. The washed cell pellet is taken up in 6 volumes (of the original cell volume) of ice-cold RSB buffer [10 mM Tris-HCl (pH 7.4)–10 mM NaCl–1.5 mM $MgCl_2$] and centrifuged at 10,000 g for 10 minutes. The pellet is resuspended in 8 volumes of fresh RSB and gently swirled or stirred with a Tissuemizer ® (55), at a setting of 3.5–4.0, for 20–30 seconds, or until the cell suspension is homogenous.

As the cells are stirred, a 10% solution of NP-40 (Shell Oil Co.) is slowly added drop by drop to a final concentration of 0.5%. The cell suspension is stirred for an additional 30–45 seconds with 15–seconds intervals. This procedure is monitored continuously by phase microscopy; it is terminated when nuclei are free of cytoplasmic contaminants.

The homogenate is centrifuged at 10,000 g for 12 minutes. The resulting fluffy pellet is gently resuspended in 10 volumes (of original cell volume) of 0.25 M sucrose–10 mM $MgCl_2$ and overlayered on top of an equal volume of 0.88 M sucrose and centrifuged at 1200 g for 10 minutes.

The resulting pellet contains the purified nuclei. If after microscopic examination the nuclear pellet still contains some debris or cytoplasmic contaminants, the final purification step should be repeated. The nuclear yields of this procedure are in excess of 60% (Ballal, *57a*).

1. MORPHOLOGY OF THE NUCLEI

When examined by phase microscopy, the isolated nuclei appear rounded (probably due to the hypotonic treatments), with conspicuous nucleoli, and a very thick dense rim of perinuclear chromatin.

Examination with the electron microscope shows that the morphology of NP-40 isolated nuclei differs from nuclei *in situ*. No chromocenters or dense perinuclear chromatin clumps are seen. Although the nucleoli contain both fibrillar and granular elements, their compactness is decreased. The nucleoplasm lost some electron density and the perichromatin granules were not identified. Frequently, the internuclear spaces are seen to contain partially fragmented and lysed cytoplasmic membranes, detergent-induced micelles, and ribosomes.

2. BIOLOGICAL ACTIVITY OF NP-40 NUCLEI

Chromatin can be easily isolated from "NP-40 nuclei" by routine procedures (*56,57*). There is no evidence for the degradation of either histone, nonhistone proteins, or DNA. Moreover, nucleoli isolated from these nuclei contain intact rRNA. Assays for template activity of these nucleoli *in vitro* showed they are capable of synthesizing high molecular weight rRNA (*57a*), with or without addition of exogenous polymerase I.

B. Ivory ® Isolation Procedure for Nuclei

This method provides a means for rapid cell fractionation into nuclear and cytoplasmic fractions. As detailed below, the nuclear and polysomal RNAs retain their biological activity.

This method may be used for ascites cells and liver, as well as for cells in regular tissue culture. The only modification to be introduced is the final concentration of the detergent (*58*).

1. PROCEDURE

Novikoff hepatoma ascites cells are washed twice in ice-cold TKM buffer [50 mM Tris-HCl (pH 7.5)–50 mM KCl–5 mM MgCl$_2$]. The washed cells are resuspended in 10 volumes of ice-cold hypotonic buffer [10 mM Tris-HCl(pH 7.4)–5 mM MgCl$_2$] and allowed to swell in the cold for 10–15 minutes (monitored by phase microscopy). As soon as cells are completely swollen, the cell suspension is brought to room temperature (within 5 minutes) and a concentrated solution of commercial Ivory[®] detergent (Procter and Gamble, Cincinnati, Ohio) is added to a final concentration of 0.7–1%. The suspension is continuously and gently swirled to prevent cell clumping. Aliquots of the suspension are monitored by phase microscopy. When 90–95% of the cells are lysed, the suspension is chilled and centrifuged at 10,000 g for 10 minutes. The resulting pellet contains crude nuclei. The nuclear pellet is resuspended in 10 volumes of 0.25 M sucrose–10 mM MgCl$_2$, overlayered on top of an equal volume of 0.88 M sucrose and centrifuged at 1200 g for 10 minutes. The resulting pellet contains purified nuclei.

2. MORPHOLOGY

On phase microscopy, nuclei appear to contain large clumps of chromatin. The nucleoli are barely visible (Fig. 8). The condensed perinucleolar chromatin is also observed by electron microscopy (Fig. 9). Only a few chromocenters are visible within the nucleoplasm. Nucleoli appear dispersed and are recognized only by the presence of nucleolar chromatin and nucleolar granular components (Fig. 10). At higher magnification (Fig. 11), it can be seen that the outer portion of the nuclear envelope is largely removed along with the membrane-bound polyribosomes. Other nuclear organelles such as the inter- and perichromatin granules are not altered by this detergent treatment (Fig. 12 and 13).

3. BIOLOGICAL ACTIVITY

The integrity and biological activity of the isolated nuclei was assayed (58) and was comparable to other isolation procedures. Nuclear and nucleolar polymerase activities were preserved as well as the interactions of nuclear histone and nonhistone proteins. Nuclear and rRNAs were not degraded, as analyzed by sucrose density gradient sedimentation pattern.

V.   Isolation and Purification of Nucleoli

Morphologically satisfactory nucleoli have been isolated by microdissection under phase microscopy (59,60). Such techniques are both laborious and demanding of special skills; they are unsuitable for biochemical analysis because only a few nucleoli can be isolated in any given time. For studies on nucleoli of *Drosophila*, however, microdissection has been valuable in conjunction with ultramicromethods (59) for analysis of RNA.

The first satisfactory method for isolation of nucleoli from nonmammalian species came from the studies of Vincent (61,62) on nucleoli of starfish

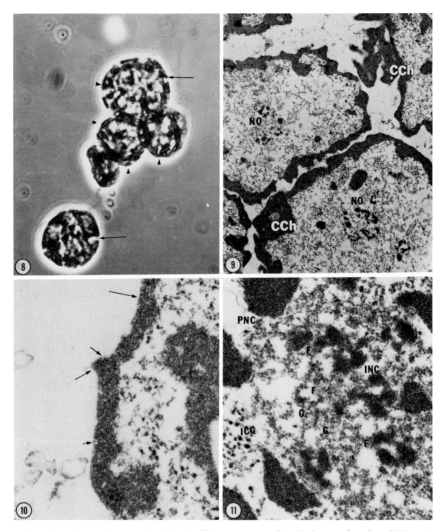

FIG. 8. Phase micrograph of Novikoff hepatoma nuclei, 4 minutes after the addition of detergent to whole-cell suspension. Nucleoli are inconspicuous due to chromatin clumping (pointers). The apparent gaps along nuclear perimeters (arrows) are a refractile artifact of phase microscopy. ×1300. From Daskal *et al. (58)*.

FIG. 9. Electron micrograph of the nuclear preparation from Fig. 8. Condensed chromatin (CCh) around nuclear peripheries gives the reticulated appearance of nuclei in light microscopy (Fig. 8). Nucleoli (NO) appear to have lost their compact structure. ×5850. From Daskal *et al. (58)*.

FIG. 10. Higher-magnification electron micrograph of nuclear peripheries for the evaluation of the presence of cytoplasmic tags and nuclear envelope components. Arrows note regions where osmiophylic membranelike structures are seen attached to peripherally condensed chromatin. ×32,000. From Daskal *et al. (58)*.

eggs. Monty *et al.* (*63*) in Dounce's laboratory announced the first isolation of nucleoli from mammalian cells using sonication to break up the cells and cell nuclei.[1]

Maggio *et al.* (*39,40*) reinvestigated the procedure of Monty *et al.* (*63*) and found that the sonication procedure could be markedly improved provided the container was cooled and Ca^{2+} was added to the medium.[2] Details of their procedure were provided to us at a time when efforts for isolation of nucleoli from tumors and precancerous livers were underway in the authors' laboratory. A reinvestigation of the variables of the procedure employing sonication led to the method (*31,46,67–69*) now successfully employed in many laboratories for isolation of nucleoli of normal and tumor tissues.

VI. The Sonication Procedure for Isolation of Nucleoli

A. Isolation and Sonication of Nuclei

For the purposes of isolation of nucleoli, it is essential that a divalent ion be present in the medium for isolation of the nuclei; hence, the sucrose–calcium procedure (see above) for isolation of nuclei is commonly employed.

For sonication of the nuclei, the Branson Sonifier is now employed; it has a power of 0–150 W and operates at 20 kHz. The Raytheon Sonic Oscillator, Model DF–101, which has a power of 200 W and operates at 10 kHz, may also be used. The containers for these two instruments differ, as well as the size of the plate from which the sonic vibrations emanate. The Raytheon instrument contains a vibrating reed which oscillates in an alternating elec-

---

[1] Among the early and less successful procedures tested for isolation of nucleoli were such methods as treatment of nuclei with citric acid and with 50% acetic acid (*7*), a method useful for preparation of chromosomes (*64*). A "panker" was also developed that consists of an upper rotating head which is kept at a defined distance from a lower stationary plate by means of a micrometer screw (*65,66*). For isolation of nuclei, this head was set at a greater distance from the plate. The reported isolation of nucleoli by this method has not been verified.

[2] Although methods have been reported for isolation of nuclei and nucleoli from plant cells, at present such preparations are in relatively little use. Apparently the methods either lacked adequate specificities or do not readily provide as satisfactory a product as is obtainable from mammalian systems.

FIG. 11. Although nucleoli appear dispersed at low magnifications (Fig. 9), at high magnifications all nucleolar components can be identified. Perinucleolar chromatin (PNC) and intranucleolar chromatin (INC) are highly condensed. Fibrillar (F) and granular (G) components of the nucleolus are dispersed but are still confined to a nucleolar "region." ICG, interchromatin granules. ×32,000. From Daskal *et al.* (*58*).

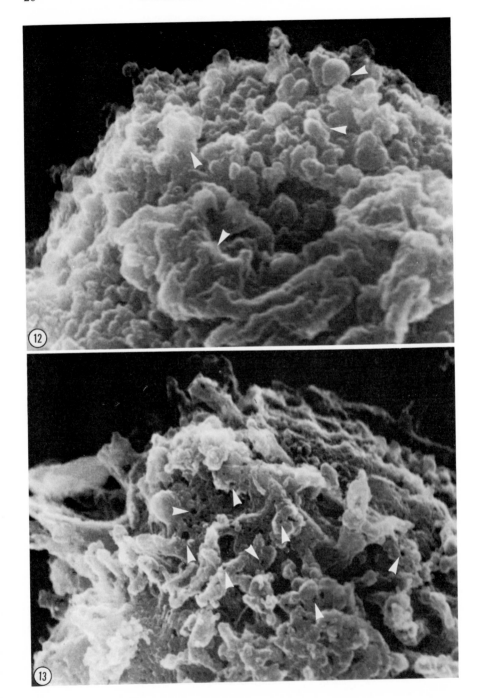

tromagnetic field. The reed must be screwed tightly into the bottom of a stainless-steel diaphragm approximately 2.5 cm in diameter which forms the base of a stainless-steel cup with a volume of approximately 50 ml. This container is surrounded by a stainless-steel jacket which can be connected conveniently to a pump carrying cooling fluid which froms part of a recirculating cooling system. The cup of a Branson Sonifier is an all-glass tapered vessel which has several glass semicircular tubes emerging from its base in the form of a rosette. This rosette must be placed into a circulating ice or cooling bath which operates at a temperature of -5 to $-10°$. For the Raytheon sonicator, the temperature of the cooling bath (Wilkens-Anderson Low Temp) is maintained at $2°$: each of these baths contains a mixture of ethylene glycol and water in a 1:3 ratio. With the Branson Sonifier, the tip must be examined periodically for micro to macro pitting. It should be burnished frequently to provide maximum output.

The Raytheon Sonic Oscillator is adjusted in this laboratory to provide 1.0–1.1 A of output current for periods of 15–30 seconds. At each time interval, a drop of the solution is used to determine the extent of destruction of the nuclei and the sonic oscillation is discontinued when the nuclei are virtually completely destroyed. The destruction of the nuclei does not ordinarily require longer than 45–120 seconds in the Raytheon Sonic Oscillator, and with the Branson Sonifier, it generally requires from 25 to 120 seconds. Although the Branson Sonifier operates at higher amperages, i.e. 7.5–8.0 kHz, and more, the sonic oscillations emanate from what is virtually a point source. The sonic oscillations in the Branson Sonifier are of sufficient strength to set up waves throughout the rosette which circulates the contained suspension. This circulation permits cooling of the entire contents of the sample.

B. Staining of the Nuclei and Nucleoli

It is necessary to follow the stages of the destruction of the nuclei quite closely inasmuch as excessive sonic oscillations result in destruction of the nucleoli. The stain most commonly used is azure C which is dissolved in 0.25 M sucrose to a final concentration of 0.1%. This strains permits rapid visualization of the nucleoli as well as the ribonucleoprotein components of the cytoplasm. The pH of the stain is 3.4 following direct addition of azure

---

FIGS. 12–13. Scanning electron micrographs of nuclei obtained by routine homogenization (Fig. 12) and the detergent procedure (Fig. 13). Nuclei from homogenized cells (Fig. 12) are covered by patches of vesicular membranous structures (pointers). In detergent-treated nuclei, the membranous vesicles appear less flocculent, and conspicuous pores (pointers) ranging in size from 50 to 150 nm are apparent. Fig. 12, ×17,500; Fig. 13, ×13,475. From Daskal et al. (58).

C to the sucrose solution. Other basic dyes including toluidine blue, Giemsa stain, and methylene blue, may also be used.

C.   Effect of Increased Temperature

The rate of destruction of the nucleoli and nuclei is dependent upon the temperature of the sonication medium. At 2°, the rate of destruction of the nuclei was markedly reduced by comparison with that at 15°, i.e., 50% nuclear destruction was found approximately 10 seconds earlier and 90% nuclear destruction occurred 20 seconds earlier. However, the number of nucleoli recovered was reduced and the rate of nucleolar destruction was markedly accelerated at 15° so that the higher temperature is not utilized.

D.   Volume of the Solutions for Sonication

The optimal volume is 20 ml. In the system employing the Branson Sonifier, the rosette supplied is used for sonication of 35 ml of homogenate, or a beaker could be substituted in which the sample was stirred with the aid of a magnetic stirrer; the temperature is maintained below 4°. Under the conditions of the experiments, there was a greater time requirement for destruction of the nuclei with a greater volume of suspending medium. However, within 3–4 minutes, the nuclei in a total volume of 150–200 ml could be destroyed with this instrument. With the Raytheon Sonic Oscillator, it was found that the time for destruction of liver nuclei was $\frac{1}{2}$ minute in a volume of 10 ml, 1 minute in a volume of 20 ml, and 5 minutes in a volume of 40 ml.

E.   Purification of the Nucleoli from the Sonicate

The whole sonicate in a volume of 20 ml is layered over 20–25 ml of 0.88 M sucrose, and the mixture is centrifuged for 20 minutes at 2000 g in a swinging bucket rotor (68, 69). The sedimented nucleoli may be resuspended in 0.25 M or 0.88 M sucrose and centrifuged once more to remove coprecipitated nucleoprotein. This preparation constitutes the type of preparation shown in Fig. 14, which is a satisfactory nucleolar preparation.

F.   Criteria of Purity

Criteria for isolated nucleoli of liver and Walker tumor include morphological, chemical, and enzymic analyses. In view of the substructure of the nucleoli, which includes the well-defined nucleolonemas as well as the light spaces which lend to the whole structure a spongelike appearance, it is possible to identify the nucleoli more exactly by electron microscopy than by

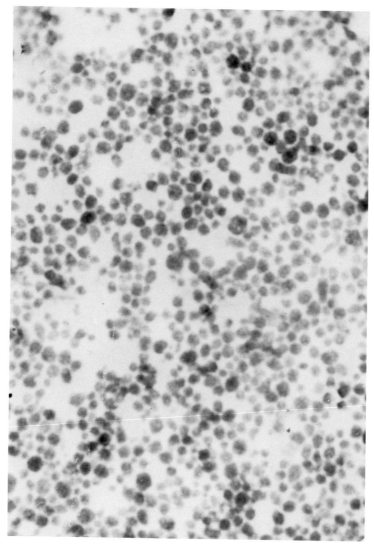

FIG. 14.   Isolated nucleoli of normal rat liver stained with toluidine blue. ×1200.

light microscopy. In addition, the nucleolonemas contain ribonucleoprotein aggregates that appear granular in electron micrographs (Fig. 15A–C).

By light microscopy, nucleoli have a reasonably characteristic appearance as seen by their spherical or irregular shapes when stained with basic dyes. By phase microscopy, however, it is possible that ribonucleoprotein aggre-

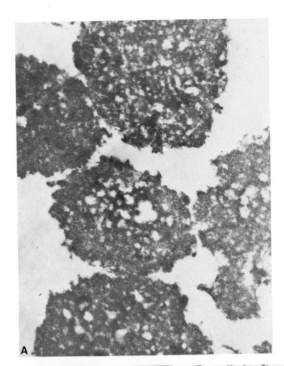

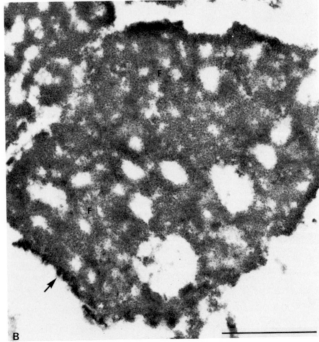

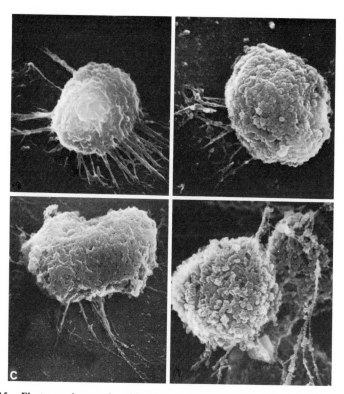

FIG. 15. Electron micrographs of isolated nuclei of Morris hepatoma 9618A. (A) Little or no cytoplasm is apparent, but outer nuclear membranes are still attached to the nuclei. The large nucleoli of these cells are visible in several of the nuclei. ×3000. (B) Focusing on the largest nucleolus. This is seen to consist mainly of granular elements, and one ribonucleoprotein branch of the nuclear ribonucleoprotein network is seen emerging from the nucleolus toward the nuclear membrane. × 5500. From Busch and Smetana (35). (C) Scanning electron micrographs of isolated nucleoli. (*Top left*) Nucleolus isolated by the sucrose–calcium procedure. The surfaces of these nucleoli are compact. Multiple lateral chromatin fibers are clearly visible. × 7000. (*Top right*) Nucleoli isolated from Novikoff cells after sonication in sucrose containing 12 mM MgCl$_2$. The nucleolar surface contains numerous microglobular components. The prolonged sonication used for this isolation procedure results in removal of considerable nucleolus-associated chromatin. ×13,000. (*Bottom left*) Novikoff hepatoma nucleoli isolated by the original procedure of Muramatsu *et al.* (*54*) using the NP-40 detergent. Although nucleoli appear compact and have some associated chromatin, the surface is microglobular. ×10,500 (*Bottom right*) Novikoff hepatoma cell nucleoli isolated by modification of the procedure of Muramatsu *et al.* (*54*) by sequential reduction of the Mg^{2+} ion concentration to 0.05 mM. The nucleoli appear to contain some fibrous strands similar in structure to isolated nucleolar fibrillar component. ×10,500.

gates from the cytoplasm may be aggregated and may produce a similar appearance. For this reason, electron microscopy is a more satisfactory index of the type of product isolated. From the standpoint of purity as indicated by the presence of whole nuclei. the minimal standard of preparation in this laboratory is a maximum of one nucleus per 4000 nucleoli. However, in many preparations no nuclei are found and electron microscopic analysis is utilized to make particle counts of contaminants. Since both large and small contaminants are counted, purities of 85% are considered satisfactory, i.e., of all particles present, 85% are nucleoli. In some instances, purities of 92% have been obtained.

Chemical analyses of nucleoli have shown that the major nucleolar components are RNA, DNA, and protein in the approximate ratio of 1:1:10. The high DNA content may reflect the presence in the nucleolar preparations of either perinucleolar or intranucleolar chromatin. Evidence for the presence of intranucleolar DNA strands has been provided by electron microscopic studies. Perinucleolar DNA and associated proteins appear to be essential for the maintenance of the nucleolar structure during the course of isolation of the nucleoli. Differences have not yet been found between the intra- and extranucleolar DNA or nuclear proteins on the basis of elementary base or amino acid analysis.

Nucleolar RNA differs in sedimentation characteristics from whole nuclear RNA (70,71) as indicated by the sucrose density sedimentation diagrams (35). RNA of isolated nucleoli has much larger amounts of both 35 S and 45 S RNA than whole nuclear RNA, but the relative amount of 18 S RNA is much less. In fact, in the isolated liver nucleoli, there is little if any RNA in the 18 S region. In nucleoli isolated from the Walker tumor a similar pattern was observed. The larger amounts of higher molecular weight RNA and the limited amount of 18 S RNA provide a distinguishing feature of nucleolar RNA. Differences between nucleolar and extranucleolar proteins have been extensively studied (72,73).

G.   Enzymic Assays for Nucleolar Purity

There are no specific nucleolar enzymes that are readily distinguishable from the nuclear enzymes as a whole. However, by comparison with nuclei, purified nucleoli from liver cells contain lower specific activities and disproportionately smaller amounts of the enzymes glutamate dehydrogenase, adenylate kinase, catalase, and glucose-6-phosphate (32). In addition, they contain larger amounts of RNA polymerase, ribonuclease, DPN pyrophosphorylase, and, to a lesser extent, ATPase A than whole nuclei. The activity of nucleoli in conversion of GC-rich 45 S to 28 S RNA is remarkable, but the specific activities of the responsible enzymes have not yet been evaluated (74,75).

H. Mass Isolation of Nucleoli Using the Sonication Procedure

There are no special requirements for the mass isolation of nucleoli from the nuclei obtained by large-scale techniques since batch processes are used in this laboratory for this purpose (36,37,44,45). However, the sonic oscillation is carried out in a Branson Sonifier because it can hold a greater volume than the Raytheon Sonic Oscillator. Moreover, centrifugation of the sonicate which is layered over 0.88 M sucrose is carried out in the rotor of the RC–3 centrifuge that contains four swinging buckets, each of which has a volume of 650 ml. The total amount of nucleoli that has been sedimented in the course of any one centrifugation is 7 gm. If the degree of nucleolar purification is not quite satisfactory, the sediment is resuspended in 0.25 M sucrose and either the sonic oscillation, the centrifugation, or both may be repeated.

VII. Compression-Decompression Procedure for Isolation of Nucleoli

The basis for the compression-decompression procedure for isolation of nucleoli was the concept of Somers et al. (64) that such delicate structures as chromosomes can be obtained from nuclei without serious damage to their morphology by alternate aspiration and ejection of nuclear suspensions from a syringe. It was then found (47) that when either nuclei or whole cells were incubated with Ca^{2+} in concentrations of 5–10 mM for periods of approximately 60 minutes, the nuclei or whole cells became selectively "hardened," which permits them to rupture on release from a French pressure cell. Under these conditions, the nucleoli were not destroyed and could then be isolated from the "pressate" by differential centrifugations.

A. Homogenization

The tissues are homogenized with teflon–glass homogenizers as described for isolation of nuclei. The homogenization medium consists of 0.25 M sucrose containing 5–10 mM $CaCl_2$. The pestle clearance is 0.005–0.006 in. regardless of whether Walker tumor or liver is studied. The homogenate is filtered through eight layers of gauze over a 90-mesh stainless-steel screen. The degree of purification varies considerably, but highly purified nuclear preparations are not essential for continuation of the procedure.

B.   Incubation

All homogenates are kept for periods of 40–60 minutes at 0–2 ° in the cold laboratory or in the refrigerator. If the samples are kept for longer or shorter times, the ratio of nuclei to nucleoli in the final product is increased considerably.

C.   Disruption of the Nuclei

Forty milliliters of the homogenate are placed in a French pressure cell, which is then set into a Carver hydraulic press. The pressure is increased to 7000–9000 lb/in.² The needle valve of the pressure cell is opened a crack, and pressure is then adjusted to 5000–7000 lb/in.² as the sample is permitted to emerge from the pressure cell. A variety of pressures ranging from 3000 to 9000 lb/in.² were utilized in this step. At the lower end of the pressure scale, the destruction of the nuclei was not satisfactory, and large numbers of whole or partially broken nuclei were found. At the higher pressures, there was an alteration in the color of the preparations—they developed a grayish hue. The duration of the release of the sample from the French pressure cell varies from 5 to 15 minutes.

D.   Differential Centrifugation

A discontinuous gradient of sucrose utilized for the preparation of the nucleoli from the pressed product consisted of an upper layer of 5 ml of 1.5 M sucrose and a lower layer of 3 ml of 2.2 M sucrose in a 30-ml centrifuge tube of the Beckman 25.1 swinging bucket rotor; 20 ml of pressed product were layered over the gradient. After centrifugation of the discontinuous gradient at 25,000 rpm for 10–20 minutes, the supernatant solutions were removed by aspiration with a water pump and the pellet that remained at the bottom of the tube constituted the nucleolar preparation (Fig. 16).

The optimal conditions for centrifugation of the pressed product were studied for the Walker tumor and the liver. Inasmuch as the nucleoli of the Walker tumor are larger and probably more dense than those of the liver,

FIG. 16.   (A) Smear of isolated nuclei from Walker tumor stained with toluidine blue. The nucleoli and the cytoplasmic basophilic structures are stained intensely. ×1800. (B) Smear of isolated nuclei from Walker tumor stained with toluidine blue after hydrolysis with 1 N HCl. Chromatin-containing structures are stained. The nucleolus is not stained, but the perinucleolar nucleolus-associated chromatin is stained. ×1800. (C) Smear of isolated nucleoli from Walker tumor, stained with toluidine blue at pH 5.0. Note the similar morphology and staining properties of isolated nucleoli and nucleoli *in situ.* ×1800. (D) Isolated nucleoli from the Walker tumor stained with toluidine blue after hydrolysis with 1 N HCl. The nucleolus-associated chromatin is stained, whereas the rest of the nucleolus is not stained. Septa penetrating from the nucleolus-associated chromatin into the nucleolus are visible in both the isolated nucleoli and the nucleoli *in situ.* × 1800.

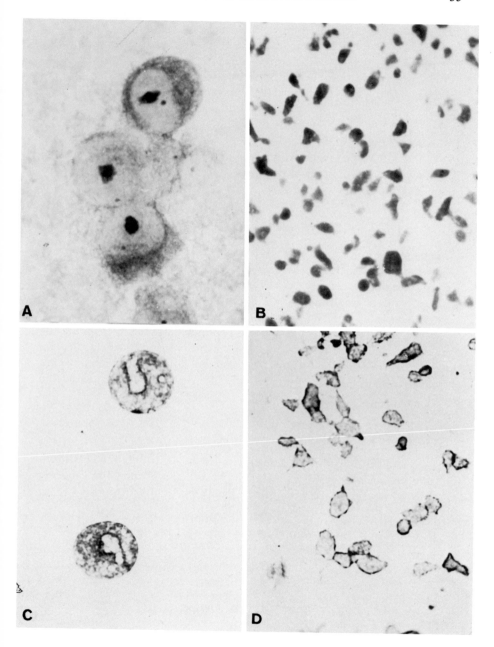

50% of the total nucleoli are sedimented in approximately $7\frac{1}{2}$ minutes in a force field of 64,000 g. For the nucleoli of the liver, the time for sedimentation of 50% of the nucleoli is approximately 20 minutes. The type of product varies inasmuch as the longer the period of sedimentation, the more chromatin tends to sediment with the nucleoli. At the optimal time points indicated above, the percentage of the total nucleic acid and proteins in nucleoli obtained from the Walker tumor are 72, 16, and 12 for protein, RNA, and DNA, respectively. In the nucleoli obtained from the liver, these values are 89, 7, and 3%, respectively. At longer times more contamination is visible by direct particle counts of the nucleolar preparations in the light microscope and in electron micrographs.

E.   Composition of Nucleoli Isolated by the Press Procedure

As indicated in Table II, some differences were found in the composition of the individual nucleoli isolated by the sonication or press (compression-decompression) procedure. Less DNA was found in the nucleoli of the liver and less protein was found in the nucleoli of the Walker tumor with the compression-decompression procedure than with the sonication procedure. However, it is likely that different-size populations of the nucleoli may be studied with these methods and, in addition, the long duration of the incubation in the course of the compression-decompression procedure may have produced some changes due to enzymic activity. The similarities of the results for the different preparations are striking.

F.   Recovery

The overall recovery of the nucleoli under the centrifugation conditions indicated above was approximately 50% that of the whole homogenate in studies on both the Walker tumor and the liver.

TABLE II

COMPOSITION[a] OF INDIVIDUAL NUCLEOLI ISOLATED
BY THE SONICATION OR THE PRESS PROCEDURE

Sample	Protein	RNA	DNA
Liver			
Press (47)	4.0	0.34	0.14
Sonicate (46)	3.8	0.38	0.38
Walker tumor			
Press (47)	6.0	1.4	1.1
Sonicate (46)	21.6	1.9	1.7

[a] In picograms per nucleolus.

VIII. "Nucleolar" Preparations Obtained by Chemical Extractions

After nuclei are extracted with 0.15 M NaCl and 1 or 2 M NaCl, a residue fraction is left in which the labeling of RNA is significantly greater than that of the extract (76). A similar result is obtained when the nuclei are initially extracted with 0.1–0.2 M phosphate buffer. The residue fraction has been referred to as the "nucleolar fraction" because of the presence of structures that stain with density of nucleoli when the samples are stained with pyronine. The basis of the chemical technique for preparation of the nucleolar fraction is, then, successive extraction of the nuclei with dilute and concentrated saline solution (77, 78).

A. Extraction of Liver Nuclei

All operations are carried out in the cold laboratory at 2–4°. The Tris-NaCl extract is obtained by treatment of the isolated nuclei twice with 0.14 M NaCl–50 mM Tris-HCl (pH 7.6)–1 mM Mg acetate; 0.5 ml of the solution is used per gram of original liver (78). The nuclei are homogenized in this solution for a period of 1 minute with a loosely fitting Teflon–glass homogenizer (pestle clearance 0.009–0.012 in.). Addition of sodium polyvinyl sulfate (PVS) at this step produces a much more satisfactory yield of high molecular weight RNA in the product (78) than is obtained when no RNase inhibitor is added; the final concentration of PVS is 0.2 mg/ml of solution. The extracted nuclei are centrifugated at 4000 g for 10 minutes.

The nuclear residues are then suspended in a small volume of the Tris-saline solution, 0.2 ml/gm of original liver, after which 2 M NaCl is added in the amount of 4.0 ml/gm of liver. This mixture is homogenized in a Teflon–glass homogenizer. This suspension is then shaken for 30 minutes in an equipoise shaker. To prevent foaming, a few drops of octanol are added. The suspension is then centrifuged for 30 minutes at 40,000 g. The supernatant solution contains the deoxyribonucleoproteins and ribonucleoproteins, and the pellet is the "nucleolar" or residue fraction.

B. Extraction of Calf Thymus Nuclei

In the studies of Sibatani et al. (77), the nuclei were suspended in 10 volumes of 0.1 M Tris-HCl (pH 7.6)–3 mM CaCl$_2$. This mixture was blended at 8000 rpm for 1 minute and centrifuged at 1000 g for 15 minutes.

The extraction of DNA was effected by stirring with 18–20 volumes of 1 M NaCl. The gel was stirred at high speed, and extraction was carried out for 16 hours at 2°, which was the temperature of the entire procedure. Sibatani *et al.* (*77*) noted that the Waring blender should not be used at this point because the organization of the nucleoli is destroyed. The sample was then centrifuged at 40,000 g in order to sediment the "nucleolar" fraction. In the case of calf thymus, this fraction accounted for about 2% of the dry weight of the nucleus, but the specific activity of the RNA was high by comparison with that extracted with the dilute or concentrated saline solutions.

C. Nature of the Product

Morphological studies on the product obtained after saline extraction revealed that a number of ribonucleoproteins remained in the residue and that most of them were components of the nuclear ribonucleoprotein network (*21*). This network apparently consists of the nucleolus and ribonucleoprotein strands that emanate from it (Fig. 17).

The sucrose density sedimentation diagrams of the RNA extracted from isolated nucleoli are substantially different from those of the "nucleolar residues" obtained in the chemical extraction procedures. However, some similarities are apparent, such as the small amount of 18 S RNA and the predominant amounts of higher molecular weight RNA. Altough Sibatani *et al.* (*77*) were able to isolate RNA from the fraction obtained from calf thymus nuclei that had virtually the same composition as DNA, the "nucleolar preparations" from rat liver have not been found to contain such a fraction. Evidence obtained in this laboratory has indicated that some AU-rich RNA is present in the "nucleolar residues" (*80*).

A number of experiments have been made with the "nucleoli" (Fig. 18) prepared by the procedure of Penman (*81*). Initial reports indicated that this method might be of value for studies on the components of the nucleus. For isolation of HeLa cell nucleoli by the procedure of Penman (*81*) the cells were washed with Earle's saline swollen in RSB [10 mM NaCl–10 mM Tris-HCl (pH 7.4)–1.5 mM MgCl$_2$] for 10 minutes and then homogenized. The nuclear pellet (1600 g) was resuspended in a mixture of one part of 10% deoxycholate and two parts of 10% Tween 40. The pellet was mixed on a Vortex mixer and washed. The crude nuclear pellet was resuspended in HSB buffer [0.5 M NaCl–50 mM MgCl$_2$–10 mM Tris-HCl (pH 7.4)] and

---

FIG. 17. (A) The nuclear RNP network (arrows) radiating from nucleoli (pointer) in a liver cell extracted with 0.14 M NaCl prior to smearing and staining with toluidine blue for RNA; nucleus (N). ×2100. (B) The RNP network in a Walker tumor cell after treatment with 0.14 and 2 M NaCl prior to fixation in osmium tetroxide and embedding. The chromatin is extracted and the RNP network apparently corresponds to the various filaments and granules of the interchromatin areas (I) to which the nucleolus (No) seems to be attached. ×22,000. From Busch and Smetana (*79*).

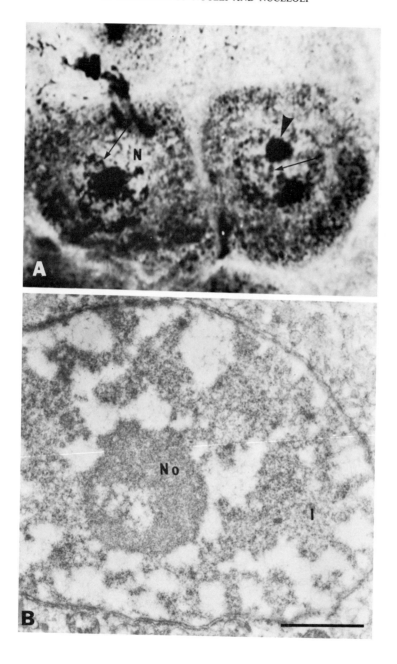

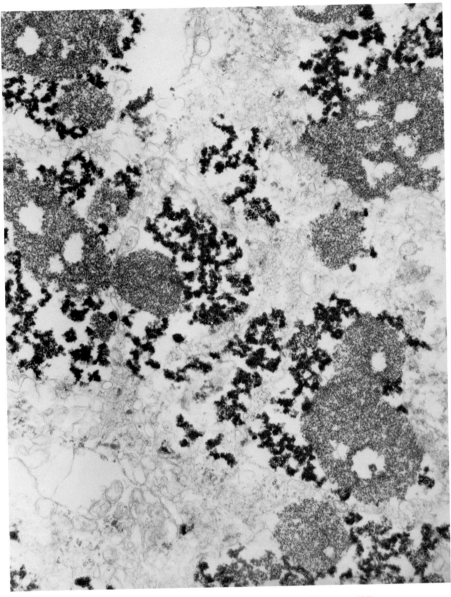

FIG. 18.   "Nucleoli" isolated by the procedure of Penman (81).

digested with DNase. The nuclear lysate was sedimented and washed with HSB buffer twice and processed for electron microscopy. Electron micrographs obtained in this laboratory with this method have not shown a morphology similar to those of the nucleoli *in situ* or to those obtained with the sucrose–Mg^{2+} or sucrose–Ca^{2+} methods. In addition to the lack of nucleolar granular elements, there is a large amount of the nuclear ribonucleoprotein network and membrane components. Accordingly, this procedure is like the successive salt extraction method noted above rather than the methods used for isolation of nuclear products for satisfactory morphology.

ACKNOWLEDGMENTS

These studies were supported by the Cancer Research Center Grant CA–10893, awarded by the National Cancer Institute, Department of Health, Education, and Welfare, the Pauline Sterne Wolff Memorial Foundation, the Davidson Fund, and a generous gift by Mrs. Jack Hutchins.

REFERENCES

1. Allfrey, V. G., and Mirsky, A. E., *in* "The Nucleohistones" (J. Bonner and P. Ts'o, eds.), p. 267. Holden-Day, San Francisco, California, 1964.

2. Busch, H., "Histones and Other Nuclear Proteins." Academic Press, New York, 1965.

3. Reid, B. R., and Cole, R. D., *Proc. Natl. Acad. Sci. U.S.A.* **51**, 1044 (1964).

4. Schneider, W. C., *Exp. Cell Res. Suppl.* **9**, 430 (1963).

5. Siebert, G., and Humphrey, G. B., *Adv. Enzymol.* **27**, 239 (1965).

6. Stern, H., Allfrey, V. G., Mirsky, A. E., and Saetren, H., *J. Gen. Physiol.* **35** 559 (1951).

7. Busch, H., Byvoet, P., and Smetana, K., *Cancer Res.* **23**, 313 (1963).

8. Okamura, N., and Busch, H., *Cancer Res.* **25**, 693 (1965).

9. Chauveau, J., Moule, Y., and Rouiller, C., *Exp. Cell Res.* **11**, 317 (1956).

10. Allfrey, V. G., Stern, H., Mirsky, A. E., and Saetren, H., *J. Gen. Physiol.* **35**, 529 (1951).

11. Siebert, G., *Exp. Cell Res. Suppl.* **9**, 389 (1963).

12. Dounce, A. L., *Ann. N. Y. Acad. Sci.* **50**, 982 (1950).

13. Dounce, A. L., *Int. Rev. Cytol.* **3**, 199 (1954).

14. Dounce, A. L., *in* "The Nucleic Acids" (E. Chargaff and J. N. Davidson, eds.), Vol. 2, p. 93. Academic Press, New York, 1955.

15. Dounce, A. L., *Exp. Cell Res. Suppl.* **9**, 126 (1963).

16. Lazarus, H. M., and Sporn, M. B., *Proc. Natl. Acad. Sci. U.S.A.* **57**, 1386 (1967).

17. Penman, S., Smith, I., and Holtzman, E., *Science* **154**, 786 (1966).

18. Granboulan, N., and Granboulan, N. P., *Exp. Cell Res.* **34**, 71 (1964).

19. Smetana, K., and Busch, H., *Cancer Res.* **24**, 537 (1964).

20. Swift, H., *Exp. Cell Res. Suppl.* **9**, 54 (1963).

21. Smetana, K., Steele, W. J., and Busch, H., *Exp. Cell Res.* **31**, 198 (1963).

22. Brachet, J., *C. R. Seances Soc. Biol. Ses Fil.* **133**, 88 (1940).

23. Brachet, J., *Arch. Biol.* **53**, 207 (1942).

24. Caspersson, T. O., "Cell Growth and Cell Function. A Cytochemical Study." Norton, New York.

25. Caspersson, T. O., and Santesson, L., *Acta Radiol. Suppl.* **46**, 1 (1942).

26. Caspersson, T. O., and Schultz, J., *Proc. Natl. Acad. Sci. U.S.A.* **26** 507 (1940).
27. MacCarty, W. C., *Am. J. Cancer* **26**, 529 (1936).
28. MacCarty. W. C., *Am. J. Cancer* **31**, 104 (1937).
29. MacCarty, W. C., and Haumeder, E., *Am. J. Cancer* **20**, 403 (1934).
30. Montgomery, T. H., *J. Morphol.* **15**, 265 (1898).
31. Busch, H., Muramatsu, M., Adams, H. R., Smetana, K., Steele, W. J., and Liau, M. C. *Exp. Cell Res. Suppl.* **9**, 150 (1963).
32. Siebert, G., Villalobos, J., Jr., Ro, T. S., Steele, W. J., Lindenmayer, G., Adams, H. R., and Busch, H., *J. Biol. Chem.* **241**, 71 (1966).
33. Hogeboom, G. H., Schneider, W. C., and Striebich, M. J., *J. Biol. Chem.* **196**, 111 (1952).
34. Hogeboom, G. H., Schneider, W. C., and Striebich, M. J., *Cancer Res.* **13**, 617 (1953).
35. Busch, H., and Smetana, K., "The Nucleolus." Academic Press, New York, 1970.
36. Desjardins, R., Smetana, K., and Busch, H., *Exp. Cell Res.* **40**, 127 (1965).
37. Desjardins, R., Smetana, K., Grogan, D., Higashi, K., and Busch, H., *Cancer Res.* **26**, 97 (1966).
38. Desjardins, R., Grogan, D. E., and Busch, H., *Cancer Res.* **27**, 159 (1967).
39. Maggio, R., Siekevitz, P., and Palade, G. E., *J. Cell Biol.* **18**, 267 (1963).
40. Maggio, R., Siekevitz, P., and Palade, G. E., *J. Cell Biol.* **12**, 293 (1963).
41. Widnell, C. C., and Tata, J. R., *Biochem. J.* **92**, 313 (1964).
42. Potter, V. R., and Elvehjem, C. A., *J. Biol. Chem.* **114**, 495 (1936).
43. Busch, H., Starbuck, W. C., and Davis, J. R., *Cancer Res.* **19**, 684 (1959).
44. Busch, H., and Desjardins, R., *Exp. Cell Res.* **40**, 353 (1965).
45. Hodnett, J. L., and Busch, H., *J. Biol. Chem.* **243**, 6336 (1968).
46. Muramatsu, M., Smetana, K., and Busch, H., *Cancer Res.* **23**, 510 (1963).
47. Desjardins, R., Smetana, K., Steele, W. J., and Busch, H., *Cancer Res.* **23**, 1819 (1963).
48. Blobel, G., and Potter, V. R., *Science* **154**, 1662 (1966).
49. Blobel, G., and Potter, V. R., *J. Mol. Biol.* **26**, 279 (1967).
50. Marshak, A., *J. Gen. Physiol.* **25**, 275 (1941).
51. Higashi, K., Shankarnarayanan, K., Adams, H. R., and Busch, H., *Cancer Res.* **26**, 1582 (1966).
52. Stoneburg, C. A., and Haven, F. L., *Am. J. Cancer* **38**, 377 (1940).
53. Schneider, W. C., and Hogeboom, G. H., *Annu. Rev. Biochem.* **25**, 201 (1956).
54. Muramatsu, M., Hayashi, Y., Onishi, P., Sakai, M., Takai, K., and Kashiyama, T., *Exp. Cell Res.* **25**, 693 (1974).
55. Taylor, C. W., Yeoman, L. C., Daskal, I., and Busch, H., *Exp. Cell Res.* **82**, 215 (1973).
56. Ballal, N. R., Goldberg. D. A., and Busch, H., *Biochem. Biophys. Res. Commun.* **62**, 972 (1975).
57. Yeoman, L. C., Jordan, J. J., Busch, R. K., Taylor, C. W., Savage, H. E., and Busch, H., *Proc. Natl. Acad. Sci. U.S.A.* **73**, 3254 (1976).
57a. Ballal, N. R., Daskal, Y, Choi, Y. C., and Busch, H., submitted for publication.
58. Daskal, Y., Ramirez, S. A., Ballal, N. R., Spohn, W. H., Wu, B., and Busch, H., *Cancer Res.* **36**, 1026 (1976).
59. Edström, J-E., *in* "The Role of Chromosomes in Development" (M. Locke, ed.), p. 137. Academic Press, New York, (1964).
60. Kopac, M. J., and Mateyko, G. M., *Adv. Cancer Res.* **8**, 121 (1964).
61. Vincent, W. C., *Proc. Natl. Acad. Sci. U.S.A.* **38**, 139 (1952).
62. Vincent, W. S., *Int. Rev. Cytol.* **5**, 269 (1955).
63. Monty, K. J., Litt, M., Kay, E. R. M., and Dounce, A. L., *J. Biophys. Biochem. Cytol.* **2**, 127 (1956).
64. Somers, C. E., Cole, A., and Hsu, T. C., *Exp. Cell Res. Suppl.* **9**, 220 (1963).

65. Poort, C., *Biochim. Biophys. Acta* **46**, 373 (1961).
66. Poort, C., *Biochim. Biophys. Acta* **51**, 236 (1961).
67. Busch, H., Lane, M., Adams, H. R., DeBakey, M. E., and Muramatsu, M., *Cancer Res.* **25**, 225 (1965).
68. Ro, T. S., and Busch, H., *Cancer Res.* **24**, 1630 (1964).
69. Ro, T. S., Muramatsu, M., and Busch, H., *Biochem. Biophys. Res. Commun.* **14**, 149 (1964).
70. Muramatsu, M., and Busch, H., *J. Biol. Chem.* **240**, 3960 (1965).
71. Muramatsu, M., Hodnett, H. L., Steele, W. J., and Busch, H., *Biochim. Biophys. Acta* **123**, 116 (1966).
72. Olson, M. O. J., and Busch, H., *in* "The Cell Nucleus" (H. Busch, ed.), Vol. III, p. 212. Academic Press, New York, 1974.
73. Olson, M. O. J., Prestayko, A. W., Jones, C. E., and Busch, H., *J. Mol. Biol.* **90**, 161 (1974).
74. Liau, M. C., Craig, N. C., and Perry, R. P., *Biochim. Biophys. Acta* **169**, 196 (1968).
75. Vesco, C., and Penman, S., *Biochim. Biophys. Acta* **169**, 188 (1968).
76. Allfrey, V. G., and Mirsky, A. E., *Proc. Natl. Acad. Sci. U.S.A.* **43**, 821 (1957).
77. Sibatani, A., deKloet, S. R., Allfrey, V. G., and Mirsky, A. E., *Proc. Natl. Acad. Sci. U.S.A.* **48**, 471 (1962).
78. Steele, W. J., Okamura, N., and Busch, H., *J. Biol. Chem.* **240**, 1742 (1965).
79. Busch, H., and Smetana, K., *in* "The Molecular Biology of Cancer" (H. Busch, ed.), p. 41. Academic Press, New York, (1974).
80. Choi, Y. C., and Busch, H., *Biochim. Biophys. Acta* **174**, 766 (1969).
81. Penman, S., *J. Mol. Biol.* **17**, 117 (1966).

Chapter 2

Nonaqueous Isolation of Nuclei from Cultured Cells

THEODORE GURNEY, JR. AND DOUGLAS N. FOSTER[1]

Department of Biology,
University of Utah,
Salt Lake City, Utah

I. Introduction

The nonaqueous method of cell fractionation, devised by Behrens (*1,2*) and modified by others (*3–5*) should make possible bulk separation of cell components while avoiding the extraction or redistribution of water-soluble material encountered in conventional aqueous fractionation. Nonaqueous isolation of nuclei can be used to determine the species and concentrations of soluble nuclear components, including ions and soluble nuclear proteins. The concentrations of many of these substances and their distributions between the nucleus and cytoplasm of living cells are largely unknown today despite extensive study (*6*).

Nonaqueous nuclear isolation consists of lyophilization of tissues or dispersed cells, homogenization of the dry powder in a nonaqueous liquid, and then sedimentation of nuclei, again in a nonaqueous liquid. The method has not been used widely because of its technical difficulty and because of reports that nonaqueous procedures introduced morphological artifacts and gave incomplete fractionation (*7*). Despite these frequently encountered difficulties, Siebert and co-workers (*8,9*) have recovered several enzyme activities in nonaqueous fractionation and have reported clear partition of some activities between cell fractions. Nevertheless, the possibilities of structural damage introduced during freezing and loss of enzyme activity from exposure to organic solvents remain with all nonaqueous methods.

Kirsch *et al.* (*5*) have made three important refinements of the nonaqueous

[1]*Present address*: Department of Biochemistry, University of Utah College of Medicine, Salt Lake City, Utah 84132.

method by substituting dichlorodifluoromethane ("Freon-12") for isopentane in a histochemical freezing method (*10*), by chilling frozen samples during the drying, and by using glycerol as the nonaqueous homogenization medium, replacing the petroleum ether, benzene, and cyclohexane of older methods. Clean nuclei without apparent structural damage were reported to be obtained in good yield from tissues or from cultured cells.

We have found that the method of Kirsch *et al.* (*5*) can be applied with varying degrees of success to several types of cultured cells, although the procedure is not straightforward and has not yet been published in detail sufficient for routine use. As described below, we have found that several simplifications make the procedure quite easy without noticeably compromising the quality of isolated nuclei. We have used the methods given here to show that DNA polymerase-α, previously thought to be a cytoplasmic activity, can be quantitatively recovered from nuclei of cultured mouse and human cells (*11,12*).

II. Lyophilization Apparatus

The borosilicate glass manifold, lyophilization tubes, and cold traps used in the work reported here are shown in Fig. 1. The dimensions of this apparatus permitted the isolation of nuclei from 0.05 to 2.0 ml of wet packed cells in each of three samples prepared simultaneously. The large diameters of the tubing (2–3 cm) and cold traps (7 cm) gave rapid cryogenic pumping. The two cold traps were refrigerated at approximately $-70\,^{\circ}\text{C}$ with a mixture of acetone and solid CO_2. The glass portion of the apparatus was attached to a Welch model 1402 mechanical vacuum pump through rubber vacuum tubing. During lyophilization the vacuum was measured with a Hastings thermocouple vacuum gauge which had been calibrated with a McLeod gauge.

Detachable lyophilization test tubes (3 $\times$ 22 cm) with female ground-glass joints were used first to freeze the cells and then to dry them after attachment to the vacuum manifold. During drying, the tubes containing frozen cells were held at $-25\,^{\circ}$ to $-30\,^{\circ}\text{C}$ by partially immersing tubes in methanol refrigerated with a "Multicool MC-4-40" apparatus (FTS Systems, Stone Ridge, New York). We have also used a constant-temperature freezing mixture of approximately 25% (w/v) anhydrous $CaCl_2$ in ice water to replace the refrigerated methanol. The $CaCl_2$ freezing mixture required frequent replenishment to maintain the correct temperature, however, and we do not recommend its use.

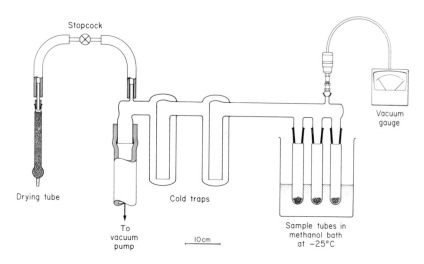

FIG. 1.   Schematic drawing of apparatus used to dry frozen cells. Construction was of boro-silicate glass except for rubber tubing to the vacuum pump and to the stopcock and a metal fitting to the Hastings thermocouple vacuum gauge tube. The lyophilization sample tubes were attached to the vacuum manifold by greased 29/42 ground-glass joints. The cold traps were emptied of their water after use by aspiration through thin rubber catheters. The glass parts of the apparatus could be dismantled and cleaned by washing with detergents, organic solvents, or acid.

III.   Cell Culture

A.   Monolayer Culture

The mouse fibroblast cell line BALB 3T3 (*13*) and a line derived from BALB 3T3, 12 A3 cl 10 (*14*), were cultured in monolayers in Dulbecco's modification of Eagle's medium (*15*) plus 36 mM sodium bicarbonate buffer plus 5% or 20% calf serum in 10-cm plastic Petri dishes in a humid 7% CO_2 atmosphere at 37°C. Medium, serum, and dishes were purchased from Flow Laboratories. The same procedures were used to culture the monkey kidney cell line BSC-1 (*16*), secondary cultures of fibroblasts from mouse or chicken embryo, and the human cell lines KB (*17*) and HeLa S-3 (*18*). The SV40 virus-transformed diploid BALB/c mouse cell line, SVT2 (*19*), was cultured with only 2% or 5% calf serum. Each cultured cell line could be grown to 0.10–0.25 ml wet packed cells per culture by daily changes of medium. Cell lines were examined for contamination with mycoplasmas by making auto-radiographs of [³H]thymidine-labeled cultures (*20*). We found no evidence of cytoplasmic labeling which would have indicated contamination.

B. Suspension Culture

HeLa S-3 cells were grown for 24 hours in the formulation of Eagle's medium for suspension culture (21) plus 10% fetal calf serum at 37 °C to a cell density of 5–6 × 10^5 cells/ml.

IV. Preparation of Cells for Freezing

A. Monolayer Culture

Six or fewer Petri dish cultures were removed from the incubator and rushed to a cold room (2 °C) where the medium was decanted and replaced within 3 seconds with ice-cold PBS.[2] The cell sheets were then washed twice by pouring with cold PBS. To suspend cells from culture dishes, all cells except transformed cultures (SVT2, KB, HeLa S-3, and Rous sarcoma virus transformed chick fibroblasts) had to be partially loosened with trypsin at 2 °C (22, 23) followed by neutralization of the trypsin and further washing. Transformed cells could be dislodged by vigorous pipetting in cold PBS without trypsin treatment, although we usually used trypsin with all cultures. Trypsin treatment consisted of exposing washed monolayers to 100 μg/ml of crystallized trypsin (Worthington) in PBS for 2 to 10 minutes at 2 °C. Transformed cells required less trypsinization time than normal cells; the length of treatment required was determined empirically for each cell line. The correct trypsin treatment left the cell sheets attached to the culture dishes firmly enough for washing after treatment but loosely enough to allow the cell sheet to be removed and dispersed by gentle pipetting after washing. Therefore, after the correct time of trypsin treatment at 2 °C (2–10 minutes), the trypsin solution was decanted, and the tenuously attached cell sheet on the Petri dish was washed quickly with 20 μg/ml of soybean trypsin inhibitor (Sigma) in 2 °C PBS and then twice in 2 °C PBS without additions. The cells were then suspended from each dish in 2 to 3 ml of PBS by pipetting and sedimented (1000 g, 30 seconds, 2 °C). The supernatant fraction was discarded. The pellet of cells was then resuspended after adding twice the pellet volume of PBS or hypotonic sodium phosphate [5 mM NaH$_2$PO$_4$– 5 mM Na$_2$HPO$_4$, pH 7.2 at 25 °C] at 2 °C. The cells were frozen within 5 minutes after resuspending. A maximum pellet volume of 0.5 ml of cells and 1.0 ml of buffer (1.5 ml the total volume) could be frozen at one time. Four

---

[2] Phosphate-buffered saline (PBS) contained 140 mM NaCl–2.7 mM KCl–8 mM Na$_2$HPO$_4$– 1.5 mM KH$_2$PO$_4$–0.9 mM CaCl$_2$–0.5 mM MgCl$_2$(pH 7.2 at 25 °C).

such batches as slurries of frozen cells, prepared as described below in Section V, could be combined into one lyophilization tube after freezing, and three tubes could be lyophilized at one time. Hence, the capacity of our apparatus was 6.0 ml of packed cells.

B. Suspension Culture

Cultured cells at 37 °C were sedimented (1000 g, 30 seconds, 25 °C) and resuspended in warm (37 °C) complete growth medium by adding one pellet volume of growth medium and stirring briefly with a Vortex mixer. The cells were frozen within 1 minute after resuspending the pellet. The limitations on the volumes of cells frozen were the same as given for monolayer cells.

C. Viability Test

Cells prepared for freezing were tested for viability by mixing the concentrated suspension with 100 volumes of monolayer culture medium with serum at 2 °C. Cells and medium were then poured into Petri dishes, incubated at 37 °C, and examined at 2 hours, 6 hours, and 30 hours after planting with an inverted phase-contrast microscope. In most cases, cells adhered and began to spread by 2 hours with a doubling in cell number by 30 hours. By this test, suspended cells were over 90% viable, except mouse cells suspended in hypotonic buffer, which were 20–50% viable. KB cells were found to be >80% viable after hypotonic treatment. Of all the procedures used to prepare cells for freezing, only hypotonic treatment reduced viability significantly.

In other work, we found that our method of dispersing monolayer cultures preserved polysome patterns (23) and quantitatively retained the soluble DNA polymerase-α within the cell nucleus (12). Scraping cells with a rubber policeman, or very vigorous pipetting without trypsin treatment, reduced viability and caused some loss of DNA polymerase-α.

V. Freezing Cells

Cells were frozen in 3 × 22 cm glass tubes with female ground-glass fittings (Fig. 1). Before preparing cells for freezing, 15 ml of liquid Freon-12 (Union Carbide Corp.) were poured or condensed into each tube which was chilled in liquid nitrogen. The Freon-12 was then allowed to freeze in the tube at liquid nitrogen temperature. Five minutes before freezing the cells, the tube containing the solid Freon-12 was removed from the liquid nitrogen, the liquid nitrogen overlying the solid Freon-12 was decanted, and the tube was placed in a straight-sided Dewar flask, which was then covered. The Freon-12 partially melted during the 5 minutes after being removed from liquid nitrogen, with 2–4 mm of liquid Freon-12 appearing over the solid Freon-12. At this point, the cell suspension was drawn into a Pasteur pipette and dripped into the thawing Freon-12 from a height of 20 cm. The tube was shaken while the cells were being frozen. Monolayer cells were frozen in the cold room (2°C) while suspension cells were frozen at 23°C. The result of these procedures was to freeze the concentrated cell suspension into flakes at the melting point of Freon-12, −158°C. The high heat capacity of melting Freon-12 froze the cells very rapidly, and the resulting frozen thin flakes presented a large surface area to speed lyophilization. The temperature of the freezing cells could be maintained at −158°C by making sure that solid Freon-12 was present throughout the freezing process. This situation was obtained by using 15 ml of Freon-12, which was about 10% thawed at the beginning of dripping the concentrated cell suspension, and by freezing no more than 1.5 ml of cell suspension at once. It was also important that the slurry of freezing cells be shaken continuously and vigorously during the freezing process and that the drops of cell suspension fall directly into the melting Freon-12 and not hit the sides of the lyophilization tube. The procedure required careful timing and some practice, but was by no means impossibly difficult.

After the cells were frozen in liquid Freon-12, the tubes were chilled to −70°C in a mixture of solid CO_2 and acetone. Several frozen preparations could be combined into one tube by shaking and then pouring the slurries of frozen cells, giving up to 6 ml of frozen cell suspension (2.0 ml packed cells) in one tube. The liquid Freon-12 was then removed with a chilled (−70°C) pipette and saved to be reused. The frozen cells in tubes minus Freon-12 were then either attached to the lyophilization apparatus immediately or were stored indefinitely in a liquid nitrogen refrigerator and lyophilized later. It was advisable to remove nearly all of the liquid Freon-12 from each tube before placing it for storage in liquid nitrogen in order to avoid the potentially explosive condition caused by liquid nitrogen becoming trapped under solid Freon-12

VI.   Drying Cells

Lyophilization tubes containing frozen cell flakes and traces of liquid Freon-12 were attached to the glass vacuum manifold (Fig. 1) and were half submerged in methanol or a $CaCl_2$ freezing mixture at $-25°$ to $-30°C$. The residual Freon-12 (bp_{760} $-30°C$) was allowed to boil off; escaping Freon-12 gas was detected by venting the vacuum system through water. When the rate of escaping Freon-12 had dropped below 0.5 ml/second, the cold traps were chilled in acetone plus solid CO_2 and the stopcock was closed. After waiting 3 minutes for the last of the Freon-12 to condense in the cold traps, the pump was started. With the system being used at its maximum capacity, the pressure dropped to 10 torr after 5–10 hours and to 3 torr by 36 hours. Samples containing less than 1.0 ml of water per tube could be lyophilized to 3 torr in 18 hours. After the pressure reached 3 torr, the pump was stopped and air was admitted to the system through the two chilled cold traps. The lyophilization tubes were then detached, stoppered, and removed to a cold room ($2°C$).

VII. Homogenization

All operations were conducted in a cold room at $0-2°C$. We worked rapidly to avoid contamination of nonaqueous material with atmospheric water.

The dried cells were poured from the lyophilization tube through a glassine paper funnel into the 10-ml fluted glass flask of a Virtis model 23 microhomogenizer (Virtis Co., Gardiner, New York). The cells were then mixed with 2–4 ml of chilled spectral grade glycerol (Mallinckrodt or Matheson, Coleman, and Bell) by adding glycerol with a syringe (minus needle) and stirring with a glass rod. [Reagent grade glycerol usually contains about 5% water. If spectral grade glycerol (less than 0.1% water) is not available, anyhydrous glycerol can be prepared by vacuum distillation.] The homogenizer flask was chilled in ice water and the cell suspension was homogenized for 10 to 15 minutes at 4000 to 6000 rpm (25 to 35 V ac applied to the homogenizer motor) using a single stainless-steel blade 1 cm in diameter. New cooling ice was added every 5 minutes.

Immediately after suspension in glycerol, but before homogenization, lyophilized preparations appeared as single cells or as small clumps in phase-contrast microscopy (Fig. 2). The cells retained the flat or spindle shapes seen in living cultures despite dispersion of the cell sheets, lyophilization, and mixing dried cells with glycerol.

Homogenization was monitored by examining samples of the homogenate with a phase-contrast microscope at $400 \times$. The times and speeds of homogenization required to strip most of the visible cytoplasm from the nuclei

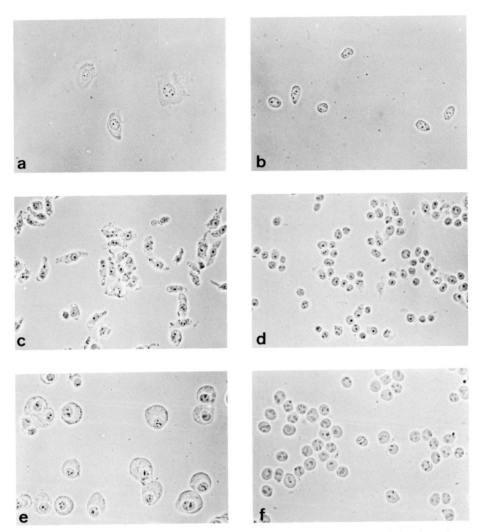

FIG. 2.   Phase-contrast light micrographs of (a) dried whole 3T3 cells suspended in glycerol; (b) the same 3T3 cells homogenized 10 minutes at 30 V; (c) dried whole SVT2 cells; (d) homogenized SVT2 cells, 12 minutes at 32 V; (e) dried whole KB cells; and (f) purified KB nuclei. The magnification in all six micrographs is the same; KB nuclei are approximately 10 μm in diameter.

were reproducible but differed among the several cell types. Small cells such as chick embryo fibroblasts, SVT2 mouse cells, and HeLa S-3 cells required 15–30 minutes at 35 V whereas larger cells such as 3T3 mouse cells required only 10 minutes at 30 V. Figure 2 shows cell suspensions in glycerol before and after homogenization.

Methods of preparing cells for freezing had pronounced effects on the success of homogenization. Trypsinization of monolayers at 23° or 37°C, instead of the usual temperature of 2°C, gave preparations requiring higher speeds of homogenization to break cells and even then yielded nuclei visibly contaminated with cytoplasm. Cells frozen in isotonic buffer, PBS, often required more homogenization to produce clean nuclei than did cells frozen in hypotonic buffer, 10 mM sodium phosphate. (It should be remembered, however, that treatment with hypotonic buffer often reduced cell viability.) HeLa S-3 cells grown in suspension culture required more homogenization than the same cells grown in monolayer culture. Mitotic cells were not visibly disrupted by homogenization.

Nuclei appeared to remain whole, by microscopic examination, if the homogenization speed was kept at 6000 rpm (35 V) or below. Above this speed, considerable fragmentation of nuclei occurred. The homogenization speed required to strip cytoplasm from nuclei of the larger mammalian monolayer cells, such as 3T3 or KB, was well below the 35 V limit, and in these cases clean nuclei could be obtained in good yield. By contrast, smaller, more rounded cells, such as Rous sarcoma virus transformed chick embryo fibroblasts and HeLa S-3 cells grown in suspension culture, were more difficult to homogenize successfully since the homogenizer speed required to strip cytoplasm from over 90% of the nuclei also fragmented more than 20% of the nuclei. Clean nuclei could be obtained even in these cases, although in relatively low yield, by homogenizing until all nuclei were visibly clean (e.g., 37 V, 20 minutes) and then centrifuging the homogenate, as described below, just long enough to sediment whole nuclei, while leaving nuclear fragments in the supernatant.

Recently we have found that both RNA and DNA polymerase activities were unstable in glycerol suspensions of lyophilized cells or purified nuclei at temperatures above −25°C. The approximate half-lives of DNA polymerase-α and RNA polymerase II in glycerol at +4°C were 4 hours (12; S. Harmon and T. Gurney, unpublished). The instability of enzyme activities made rapid homogenization necessary in some cases.

VIII.   Separation of Cell Fractions

Homogenized cells in glycerol were centrifuged to sediment nuclei. We used a Beckman Spinco SW 50.1 swinging-bucket centrifuge rotor and the matching Beckman 5-ml nitrocellulose centrifuge tubes. The centrifuge chamber, rotor, and tubes were chilled to 0 °C, or to −25 °C in studies of polymerase activities. Before pouring the homogenate in the tubes, a layer of 1–2 ml of pure glycerol was placed in the bottom of each tube. The homogenate was then poured over the layer of glycerol from the homogenization flask, the flask was rinsed once or twice with pure glycerol by stirring with a clean glass rod, and the rinses were also added to the centrifuge tube. The viscosity of glycerol at 0 °C prevented the bottom layer of pure glycerol from mixing with the homogenized cells. During subsequent centrifugation, the nuclei therefore sedimented through the layer of clean glycerol which separated nuclei from cytoplasm. Because of our experience with unstable polymerase activities, we kept the glycerol homogenate just warm enough to pour, about 0 °C, and worked without delay. The filled tubes were placed in prechilled buckets, the rotor was placed in the chamber which was then evacuated to 100 torr or below, and the rotor was accelerated to 40,000 rpm (average centrifugal force of 130,000 g). We centrifuged the homogenate for 2 hours at 0 °C or for 12 hours at −25 °C. The rotor temperature could be maintained at −25 °C in the older Beckman Model L ultracentrifuges; the new Beckman L-5 instruments could not be refrigerated below 0 °C without special modifications.

After sedimentation of nuclei, the cytoplasmic supernatant fraction was decanted at 0 °C and the inside of the tube above the pellet of nuclei was wiped four times with lint-free tissue paper, as suggested by Kirsch et al. (5), to remove traces of the viscous supernatant glycerol. The isolated nuclei were stored at −28 °C or at −70 °C in the nitrocellulose centrifuge tube, which was placed inside a larger glass test tube over anhydrous $CaSO_4$. The glass tube was capped with a rubber serum stopper.

Kirsch et al. (5) used a layer of denser nonaqueous liquid, 15% (v/v) 3-chloro-1,2-propanediol plus 85% glycerol, in the bottom of the centrifuge tube to aid separation of dense nuclei from lighter cytoplasmic particles. This general procedure has been used since the work of Behrens (1); Kirsch et al. (5) substituted their mixture for the carbon tetrachloride solutions used earlier. We have compared fractionation of 3T3 cells, by criteria given below, with and without the use of 15% 3-chloro-1,2-propanediol and could find no significant differences. We have concluded, therefore, that the use of the denser solvent in the fractionation of undifferentiated cultured cells was an unnecessary complication and could be omitted. Furthermore, we found that glycerol solutions containing 3-chloro-1,2-propanediol changed the appearance of nuclei in the phase-contrast microscope to featureless gray spheres and sometimes lysed the nuclei unless commercial preparations of 3-chloro-1,2 propanediol (Eastman or Aldrich) were vacuum-distilled before use. The use of denser liquids might be reconsidered, however, in fractionation of other more complex cells.

IX.   Methods of Handling Glycerol Suspensions

A.   Supercooling

Anhydrous glycerol (mp + 17.8 °C) formed supercooled liquid suspensions of cells or cell fractions at 0°, −28°, and even at −70 °C. Freezing of glycerol occurred if suspensions were stirred vigorously below 0 °C, and for this reason we recommend homogenization at 0 °C. Samples stored at −28 °C remained in liquid form indefinitely, but we observed occasional spontaneous freezing of suspensions stored longer than a few days at −70 °C.

B.   Pipetting

Quantitative transfers of whole cells or cell fractions suspended in glycerol at 0 °C could be made with an Eppendorf (or similar) micropipette of capacity 10 μl or larger, if the polypropylene pipette tip was truncated to give an internal diameter of at least 1.5 mm. The pipette could be used either in the "to contain" or "to deliver" mode of operation. In the latter case, the inside of the pipette tip was first coated with the glycerol suspension and then the desired amount was delivered with one quick stroke.

C.   Suspension in Ethanol

In biochemical analysis of nucleic acids and triphosphates, glycerol suspensions were often diluted in 95% ethanol because ethanol gave homogeneous stable suspensions without the clumping of nuclei, and resulting sampling errors, frequently encountered in aqueous suspensions. Ethanol was the diluent of choice when counting nuclei with a hemocytometer or for assaying DNA or ATP. Nucleoside triphosphates were stable in ethanol–glycerol mixtures.

D.   Concentration of Macromolecules from Dilute Glycerol Suspensions

Dilute whole-cell homogenates or the cytoplasmic supernatant fraction could be precipitated to concentrate macromolecules before extraction of nucleic acids or analysis of proteins by polyacrylamide gel electrophoresis.

To concentrate nucleic acids, the glycerol suspension was mixed with four volumes of 150 mM sodium acetate–1mM EDTA–0.5% (w/v) sodium lauryl sulfate (SLS) (pH 7.0). Then ethanol was added to 70% (v/v), the mixture was chilled overnight at −28 °C, and then it was centrifuged (10,000 g, 30 minutes, 0 °C). The precipitate was resuspended in low salt buffer [10 mM

Tris-Cl–1 mM EDTA–0.5% (w/v) SLS (pH8.0)] and could be extracted with phenol-chloroform by standard procedures.

To concentrate (denatured) proteins, 0.5 ml of glycerol suspension containing < 300 μg of protein was mixed with 2.0 ml of the SLS-containing buffer given above and then chilled on ice. Trichloroacetic acid (TCA) was added to 20% (w/v), and the mixture was allowed to precipitate on ice 30 minutes. The high concentration of TCA was necessary to precipitate histones. The precipitate was collected by sedimentation (10,000 g, 10 minutes, 0 °C) and dissolved in 1 ml 1 N NaOH at 0 °C. A sample of 50 or 100 μl was then removed for protein determination. The remaining material was neutralized with 1 ml 1N HCl, reprecipitated with 20% TCA, and centrifuged. The precipitate was then washed by three 5-ml rinses of the centrifuge tube with a mixture of 25% (v/v) ethanol and 75% ethyl ether at room temperature to remove the TCA. The precipitate was then dissolved in the gel electrophoresis sample buffer which contained 50 mM Tris-Cl–5.4 M urea–18% glycerol–1.6% (w/v) SLS (pH6.8) (24). Electrophoresis profiles of proteins from whole cells before and after precipitation and resuspension were identical.

E. Measurement of DNA Concentrations

Glycerol suspensions were diluted in aqueous buffers or in 95% ethanol as described above and then precipitated in 10% trichloroacetic acid. Dilute samples were mixed with 50 μg/ml purified yeast RNA (Sigma) as a precipitation carrier. Precipitates were collected on 2.4-cm Whatman GF/C glass fiber filters, washed twice with 10 ml of 1 N HCl at 0 °C, once with 2 ml of 95% ethanol at 23 °C, and air dried. DNA on the filter was then assayed by a fluorometric method based on that of Kissane and Robins (25). Each filter was placed in a glass scintillation counter vial, wetted with 0.25 ml of a 2 M solution of purified 3,5-diaminobenzoic acid dihydrochloride, and hydrolyzed by heating the closed vial at 60 °C for 40 minutes. The filter was then extracted for 10 minutes at room temperature with 5 ml of 1 N NCl, and the fluorescence of the extract was read at 510 nm with excitation at 410 nm. The assay was useful between 0.1 and 50 μg of DNA per sample.

F. Measurement of ATP Concentrations

Samples were diluted in 95% ethanol as above. ATP contents of ethanol mixtures were assayed using Sigma firefly lantern extract and a scintillation spectrometer according to methods of Addanki, Sotos, and Rearick (26) as modified by Emerson and Humphreys (27). The final mixture contained 0.5% ethanol. The assay was useful between 1 and 50 pmol of ATP per sample.

X. Appearance of Purified Nuclei

The appearance of purified nuclei in the phase-contrast microscope was similar to that obtained using an aqueous isolation procedure in which cells were homogenized with a tight-fitting Dounce homogenizer in hypotonic detergent-containing buffer [15 mM KCl–3 mM MgCl$_2$–10 mM Tris-HCl–0.5% (v/v) Triton X-100[3] (pH 7.4)]. The size of the nucleus was the same, as measured using an ocular grid, as that in cells fixed in monolayer culture. The perimeter of the nucleus was often slightly fluted when isolated by nonaqueous procedures, as opposed to detergent-isolated nuclei in aqueous buffer which always had smooth perimeters.

In electron microscopy, thin sections showed typical nuclear morphology (Fig. 3A). The inner nuclear membrane was continuous, but the outer nuclear membrane was nearly absent in every section, in agreement with Kirsch et al. (5). Occasional nuclei, fewer than 5% in most preparations, showed gross clumping of chromatin (Fig. 3B). Clumped chromatin occurred more frequently when freezing more than 1.5 ml of cell suspension in one tube. Clumping could be minimized by freezing no more than 1.5 ml of suspension and by vigorous shaking of the melting Freon-12 during freezing.

In the scanning electron microscope, the nuclear surface was irregular and convoluted (Fig. 4), in agreement with the fluted appearance seen in the light microscope.

XI. Recovery of Nuclear DNA

DNA concentrations were measured in nuclear and cytoplasmic cell fractions using the fluorometric assay given in Section IX, E. Relative amounts of DNA in the two fractions were also estimated from recovery of acid-insoluble radioactivity obtained from cells labeled with [14C]thymidine. Mitochondrial DNA, less than 1% of cellular DNA in our cell cultures, was not considered to be a significant source of DNA, as measured by either means.

The results of these measurements, in seven experiments on four cell types, are shown in Table I. Recovery of DNA in the nuclear fraction from the cell homogenate was 91–94%. DNA remaining in the cytoplasmic fraction varied between 5 and 10%. These values were typical of many experiments, although cytoplasmic DNA was often well below 5% in occasional experiments with KB cells. A doubling of the homogenization time, in one

[3] Triton X-100 is a nonionic detergent, trademark of Rohm and Haas Corporation, New York.

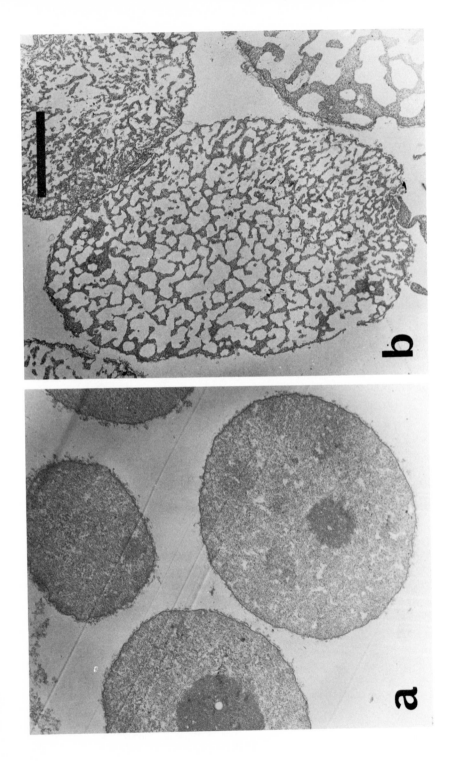

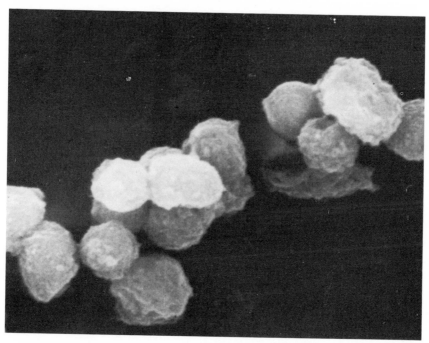

FIG. 4. Scanning electron micrograph (kindly prepared by Dr. Jack Risius, Donner Laboratory, University of California, Berkeley) of purified SVT2 nuclei.

experiment, did not increase the nonsedimenting proportion of DNA. Our standard sedimentation conditions in these studies were 130,000g average, 2 hours, and 0 °C. Additional centrifugation of the supernatant (130,000 g average, 5 hours, 0 °C) sedimented about half of the acid-insoluble radioactivity of the supernatant, in one experiment. An interpretation of these results is that a minority of the nuclei, 5 to 10% in these experiments, fragmented into nonsedimenting pieces at homogenizer speeds which did not affect the remaining nuclei. Perhaps a freezing artifact made a minority of nuclei especially fragile. The issue of presumptive nuclear material in the cytoplasmic fraction was examined further, as described below, by studying distributions of RNA and proteins.

FIG. 3. Thin-section electron micrographs [kindly prepared by Dr. T. D. Dixon, Dept. of Molecular Biology, University of California, Berkeley, according to methods of Cook and Aikawa (28) of (a) purified SVT2 nuclei and (b) clumped nuclei from 12A3 cl 10 cells prepared after freezing more than 1.5 ml of cell suspension at one time. The bar in (b) represents 2.5 μm in both micrographs.

TABLE I

DISTRIBUTION OF DNA IN NONAQUEOUS CELL FRACTIONS[a]

Cell line	Counts per minute incorporated in vivo from [14C]dT			Chemical assay for DNA (µg)		
	Homogenate	Nuclei	Cytoplasm	Homogenate	Nuclei	Cytoplasm
SVT2	406,191 ± 17,194 (100.0 ± 4.2%)	371,826 ± 11,322 (91.4 ± 2.8%)	28,857 ± 703 (7.3 ± 2.5%)	883 ± 35 (100.0 ± 4%)	812 ± 24 (92.1 ± 3.1%)	58 ± 1 (6.6 ± 2%)
KB	399,467 ± 16,790 (100.0 ± 4.3%)	366,867 ± 35,329 (91.6 ± 9.8%)	37,693 ± 760 (9.4 ± 2.0%)	3348 ± 201 (100.0 ± 6.1%)	3039 ± 152 (90.8 ± 5.1%)	288 ± 12 (8.6 ± 3.9%)
12A3 cl 10 G1	ND[b]	ND	ND	374 ± 7.5 (100.0 ± 2.1%)	346 ± 6.9 (92.1 ± 2.0%)	34 ± 0.7 (9.1 ± 2.1%)
12A3 cl 10 S	668,544 ± 59,560 (100.0 ± 8.9%)	632,469 ± 13,510 (94.6 ± 2.1%)	43,107 ± 1385 (6.4 ± 3.2%)	544 ± 16 (100.0 ± 2.9%)	508 ± 10 (93.2 ± 1.8%)	29 ± 0.6 (5.3 ± 1.9%)

[a] Distributions of DNA between nuclear and cytoplasmic cell fractions were determined by two methods: (1) Radioactivity incorporated into acid-insoluble material from cells labeled 48 or 72 hours with [2-14C]thymidine, 50 mCi/mmol, 0.002 µCi/ml, 15 ml per culture. (2) DNA was measured by a fluorometric assay for acid-insoluble purine deoxyribonucleotides. The data obtained by either method represent in each instance one sedimented cell homogenate and were obtained by four or five determinations on aliquots of each preparation.

[b] ND = not determined.

XII. Distributions of RNA Species

The extent of cell fractionation was measured from the distributions of cellular RNA molecules between nuclear and cytoplasmic cell fractions, which were determined according to methods of Penman (29). Mouse 12A3 cl 10 cells were labeled 90 minutes with [³H]uridine, labeled cells were fractionated by our procedures, and RNA was extracted from the homogenate and from the two cell fractions by the methods of Bandman and Gurney (23). It was necessary to remove glycerol by ethanol precipitation as described above (Section IX, D) or else to dilute the glycerol to 5% (v/v) in order to get the phenol emulsion to separate after extraction. Purified labeled RNA was then analyzed by sedimentation in nondenaturing sucrose gradients. Five labeled species of RNA were resolved, as shown in Fig. 5. From the work of Penman (29) and Darnell (30) we know that nuclear species are the > 45 S HnRNA, 45 S rRNA, and 32 S rRNA, while predominantly cytoplasmic species are 18 S rRNA and 4 S tRNA. The 28 S rRNA is found in both cell fractions.

Nonaqueous fractionation of the five species 12A3 cl 10 cells (Fig. 5) was considerable but not complete. We estimated that more than 90% of the 18 S and 4 S RNA of the homogenate was recovered in the cytoplasmic fraction, while about 90% of the 45 S and HnRNA was recovered in the nuclear fraction. Resedimentation of either fraction alone (130,000 g average, 5 hours, 0 °C), did not result in further purification. These results also were typical of monolayer cultures of KB and HeLa cells. Assuming that HnRNA and 45 S RNA were exclusively nuclear and that 18 S RNA and 4 S RNA were exclusively cytoplasmic, we concluded that our method of cell fractionation gave 10% or less cross-contamination of the two fractions for monolayer mammalian cells. The extent of fractionation of smaller cells, such as transformed chick cells or HeLa S-3 cells grown in spinner culture, was not as complete, for reasons discussed above in Section VII.

XIII. Electrophoresis of Nuclear and Cytoplasmic Proteins

In another demonstration of cell fractionation, proteins from whole cells and from the two cell fractions were analyzed by SLS polyacrylamide slab gel electrophoresis according to methods of Laemmli and Favre (31). Proteins were concentrated from the dilute cytoplasmic glycerol suspension according to methods given above in Section IX, D. Results from an experiment with monolayer cultures of HeLa S-3 cells are shown in Fig. 6.

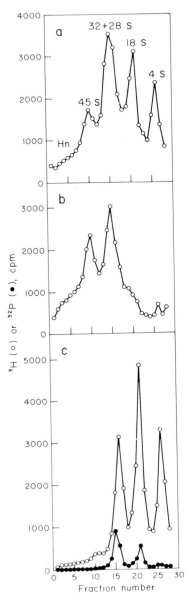

FIG. 5.   Sucrose gradient sedimentation analysis of labeled RNA extracted from (a) whole cells, (b) nonaqueous nuclei, and (c) nonaqueous cytoplasm. 12A3 cl 10 cells were labeled for 90 minutes with 5 μCi/ml [³H]uridine (20 Ci/mmol), cells were fractionated, and RNA was purified from whole cells and each fraction. RNA was analyzed by sedimentation in 15–30% SLS sucrose gradients in the Beckman SW-36 rotor (5 hours, 25°C, 35,000 rpm). Sedimentation was from right to left. Open circles: cpm from [³H]RNA; solid circles: cpm from [³²P]RNA extracted from purified mouse ribosomes.

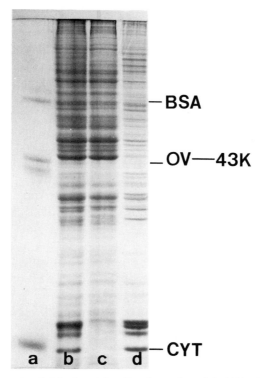

FIG. 6. Photographs of stained proteins after electrophoresis in 11% polyacrylamide SLS gels (*31*) of (a) protein reference markers: BSA bovine serum albumin 65,000 daltons, OV ovalbumin 40,000 d, and CYT cytochrome C 13,000 daltons; (b) proteins from whole HeLa S-3 cells; (c) proteins from the nonaqueous cytoplasmic fraction of HeLa; (d) proteins from HeLa nuclei.

Known prominent nuclear proteins are the four small histones H2A, H2B, H3, and H4 seen as a cluster of four bands at the bottom of Fig. 6. Several other nuclear proteins can be discerned in Fig. 6C, including the H1 histones seen as three or four bands of apparent molecular weights near 30,000. Several characteristic cytoplasmic proteins are also seen, including a prominent band running slightly slower than the ovalbumin reference (40,000 daltons), most probably actin subunits (43,000 daltons) (*32*).

Histones were confined almost exclusively to the nuclear fraction, while the 43,000 dalton protein was confined almost exclusively to the cytoplasmic fraction, thus confirming the successful fractionation observed for nucleic acids in Sections XI and XII above. We did not attempt to quantitate the fractionation of the protein species, although quantitation should be straightforward using densitometry and normalizing data to the histones or the 43,000 dalton protein.

If the 43,000 dalton cytoplasmic protein can be assumed to represent actin, as seems most likely, then the location of actin, determined by nonaqueous fractionation of HeLa cells, is almost exclusively cytoplasmic. We have observed the same distribution of a 43,000 dalton protein also in mouse 3T3 cells, both in G1 phase contact-inhibited cells and in synchronized, serum-stimulated S phase cells (N. Neff and T. Gurney, unpublished data). From these observations, it seems unlikely that actin can be a major nonhistone chromosomal protein as proposed recently by Douvas *et al.* (*33*) from observations on rat liver chromatin prepared by aqueous procedures. It also seems unlikely to us that the apparent cytoplasmic location of actin arose from selective degradation by endogenous proteases during cell fractionation as Douvas *et al.* (*33*) have suggested, because hydrolases should not have been active in the absence of water. The finding of actin in the nuclei of *Physarum polycephalum* (*34,35*) may represent a special case quite distinct from mammalian cells because in *Physarum*, unlike mammalian cells, the nuclear membrane encloses the mitotic apparatus during mitosis. Comings and Harris (*36*), using conventional aqueous cell fractionation, have independently concluded that actin, and other contractile proteins as well, are not likely to be major chromosomal proteins in mammals. The presence or absence of actin should therefore serve as a convenient criterion for cell fractionation and for the purity of nuclear proteins from mammalian cells.

XIV.   ATP Content of Whole Cells and of Isolated Nuclei

The concentration of ATP was determined in isolated nuclei to demonstrate retention of a water-soluble molecule in nuclei prepared by the nonaqueous method. Nuclei and whole cells were diluted in ethanol as described in Section IX, C above. The DNA concentrations of the ethanol suspensions were determined by methods given in Section IX, E and the ATP concentrations by methods given in Section IX, F.

In three determinations, the ATP content of glycerol homogenates from SVT2 cells was 220 ± 50 pmol of ATP per μg of DNA. In eight determinations the ATP content of isolated SVT2 nuclei was 25 ± 10 pmol of ATP per μg of DNA. The ATP content of nuclei, approximately one-tenth that of the whole cell, was also approximately equal to the level of cytoplasmic contamination observed by the distribution of 18 S and 4 S RNA, assuming that the RNA species were exclusively cytoplasmic (see Section XII above). Hence the apparent nuclear ATP could have been derived substantially from cosedimenting cytoplasmic material. Kirsch *et al.* (5) also observed rather low nuclear ATP content.

ATP could be quantitatively eluted from nuclear preparations by diluting nuclei in aqueous buffers or in ethanol and resedimenting the nuclei (3,000 g, 15 minutes, 0 °C). By contrast, nuclei diluted in glycerol at 2 °C and resedimented (130,000 g, 2 hours, 0 °C) retained 60 to 100% of the ATP measured after the first sedimentation in glycerol (three experiments) if the second sedimentation was performed within 3 hours of the first. Nuclei stored in glycerol for 24 hours or longer at +2 °C or at −28 °C retained all of the original ATP in the glycerol suspension, but less than half of this ATP sedimented again with nuclei (three experiments). Hence, ATP was retained with rapidly sedimenting structures, including nuclei, long enough to be measured approximately by these procedures although it was eventually eluted. These experiments also showed that sedimentation of nuclei for 2 hours at 0 °C was preferable to sedimentation for 12 hours at −25 °C for the measurement of ATP and other nuclear constituents of small molecular weight.

XV.   Conclusions

A.   Advantages of the Method

Aside from the time required to lyophilize cells and to centrifuge the glycerol homogenate, the time required to prepare nuclei by the nonaqueous method was not much greater than that required for aqueous cell fractionation. We estimate here the times required for one person to process 1.0 ml of wet packed cells (3–4 × 10⁹ cells) from monolayers of HeLa or SVT2 cells.

Preparation and freezing cells	30 minutes
Setting up lyophilization equipment	20 minutes
Lyophilization	(24 hours)
Homogenization	30 minutes
Centrifugation	(2 hours)
Storage of cell fractions	10 minutes
Total time minus lyophilization and centrifugation:	90 minutes

The procedure, as we have developed it, fractionated the cells with about 10% cross-contamination of the two fractions, using one homogenization step and one centrifugation step. The cell homogenate was processed in rather concentrated form, up to about 1 × 10⁹ cells/ml. The nuclei were sedimented to a pellet and the cytoplasmic fraction was recovered after dilution of the homogenate by only 30–50% from the layer of pure glycerol in the bottom of the centrifuge tube and from rinsing the homogenization flask. Recoveries depended on the extent of rinsing of the homogenization flask; with care, losses due to the viscous glycerol adhering to cold glass could be kept to 10%.

Once the cells had been fractionated, cell fractions could be stored indefinitely in stable form at −28° or −70°C. Polymerase activities were stable at these temperatures, and molecular weights of macromolecules were unchanged after a year of storage.

Glycerol was a convenient homogenization medium since it was stable, rather inert, and freely miscible in water.

B.   Disadvantages of the Method

In our present development of the nonaqueous procedure, we have not fractionated suspension cultured cells successfully. Clean nuclei in rather low yield could be prepared after prolonged high-speed homogenization, but in that case broken nuclei extensively contaminated the cytoplasmic fraction. We encountered the same trouble, to a lesser extent, with chick monolayer cultures. We are continuing to develop techniques for suspension cells.

A potential disadvantage of the techniques required to fractionate monolayer cells is the necessary chilling of the cultures before freezing. During trypsinization and the associated washing, cells were chilled from 37°C to 0–2°C for about 10 minutes before being frozen in melting Freon-12. It is quite conceivable that some soluble molecules became redistributed within the cell during this time, despite the fact that the cells were shown to be viable. This reservation about the technique for monolayer cultures makes the perfection of techniques for suspension cultures doubly important.

The extent of fractionation, 10% or less cross-contamination, places a limit on the precision of measurement of soluble nuclear components. This point was well illustrated in our measurement of nuclear ATP. We could not distinguish between low levels of nuclear ATP and adventitiously bound cytoplasmic ATP.

We have found that some polymerase activities were unstable in 100% glycerol at temperatures above −25°C although DNA polymerase activities quantitatively survived freezing and lyophilization (*12*). The demonstrated instability of two enzymes points to the likelihood that other activities will also be found to be unstable, perhaps even more unstable, thus making quantitation difficult.

The homogenization medium, glycerol, cannot be removed easily after fractionation without denaturing macromolecules of the nucleus or cytoplasm. Removal of the medium by evaporation would make possible concentrations of active cell components. For this purpose, other more volatile nonaqueous solvents might be surveyed for applicability in nonaqueous cell fractionation.

ACKNOWLEDGMENTS

This investigation was supported by U.S. National Institutes of Health Grants CA 12407 and CA 17504 from the National Cancer Institute. Part of this research was carried out at the Department of Molecular Biology, University of California, Berkeley, where Douglas N. Foster was a graduate student and predoctoral trainee of the U.S. Public Health Service, Grant GM 01389. We wish to thank Elizabeth G. Gurney, Ellen F. Hughes, Bruce R. Gordon, Richard Junghans, and Richard D. Macdonald for help in developing techniques.

REFERENCES

1. Behrens, M., *Hoppe-Seyler's Z. Physiol. Chem.* **209**, 59 (1932).
2. Behrens, M., *in* "Biochemische Taschenbuch" (H. Rauen, ed.), p. 910. Springer-Verlag, Berlin and New York, 1956.
3. Allfrey, V. G., Stern, H., Mirsky, A. E., and Saetren, H., *J. Gen. Physiol.* **35**, 529 (1952).
4. Siebert, G., *Methods Cancer Res.* **2**, 287 (1967).
5. Kirsch, W. M., Leitner, J. W., Gainey, M., Schultz, D., Lasher, R., and Nakane, P., *Science* **168**, 1592 (1970).
6. Kato, T., and Lowry, O. H., *J. Biol. Chem.* **248**, 2044 (1973).
7. Busch, H., *in* "Methods in Enzymology," Vol. 12: Nucleic Acids (L. Grossman and K. Moldave, eds.), Part A, p. 421. Academic Press, New York, 1967.
8. Siebert, G., *Exp. Cell Res. Suppl.* **9**, 389 (1963).
9. Siebert, G., Furlong, N. B., Roman, W., Schlatterer, B., and Jaus, H., *in* "Methodological Developments in Biochemistry" (E. Reid, ed.), Vol. 4, p. 13. Longmans Group, London.
10. Anfinsen, C. B., Lowry, O. H., and Hastings, A. B., *J. Cell. Comp. Physiol.* **20**, 231 (1942).
11. Foster, D. N., and Gurney, T., Jr., *J. Cell Biol.* **63**, 103a (1974).
12. Foster, D. N., and Gurney, T., Jr., *J. Biol. Chem.*, in press (1976).
13. Aaronson, S. A., and Todaro, G. J., *J. Cell. Physiol.* **72**, 141 (1968).
14. Smith, H. S., Gelb, L. D., and Martin, M. A., *Proc. Natl. Acad. Sci. U.S.A.* **69**, 152 (1972).
15. Dulbecco, R., and Freeman, G., *Virology* **8**, 396 (1959).
16. Meyer, H. M., Jr., Hopps, H. E., Rogers, N. G., Brooks, B. E., Bernheim, B. C., Jones, W. P., Nisalak, A., and Douglas, R. D., *J. Immunol.* **88**, 796 (1962).
17. Eagle, H., *Proc. Soc. Exp. Biol. Med.* **89**, 362 (1955).
18. Puck, T. T., Marcus, P. I., and Cieciura, S. J., *J. Exp. Med.* **103**, 273 (1956).
19. Aaronson, S. A., and Todaro, G. J., *Science* **162**, 1024 (1968).
20. Gurney, T., Jr., *Proc. Natl. Acad. Sci. U.S.A.* **62**, 906 (1969).
21. Eagle, H., *Science* **130**, 432 (1959).
22. Levine, E. M., Becker, Y., Boone, C. W., and Eagle, H., *Proc. Natl. Acad. Sci. U.S.A.* **53**, 350 (1965).
23. Bandman, E., and Gurney, T., Jr., *Exp. Cell Res.* **90**, 159 (1975).
24. Hughes, S. H., Wahl, G. M., and Capecchi, M. R., *J. Biol. Chem.* **250**, 120 (1975).
25. Kissane, J. M., and Robins, E., *J. Biol. Chem.* **233**, 184 (1958).
26. Addanki, S., Sotos, J. F., and Rearick, P. D., *Anal. Biochem.* **14**, 261 (1966).

27. Emerson, C. P., Jr., and Humphreys, T., *Anal. Biochem.* **40**, 254 (1971).
28. Cook, R. T., and Aikawa, M., *Exp. Cell Res.* **78**, 257 (1973).
29. Penman, S., *J. Mol. Biol.* **17**, 117 (1966).
30. Darnell, J. E., *Bacteriol. Revs.* **32**, 262 (1968).
31. Laemmli, U. K., and Favre, M., *J. Mol. Biol.* **80**, 575 (1973).
32. Pollard, T. D., and Weihing, R. R., *CRC Crit. Rev. Biochem.* **2**, 1 (1974).
33. Douvas, A. S., Harrington, C. A., and Bonner, J., *Proc. Natl. Acad. Sci. U.S.A.* **72**, 3902 (1975).
34. Jockusch, B. M., Becker, M., Hindennach, I., and Jockusch, H., *Exp. Cell Res.* **89**, 241 (1974).
35. LeStourgeon, W. M., Forer, A., Yang, Y-Z, Bertram, J. S., and Rusch, H. P., *Biochim. Biophys. Acta* **379**, 529 (1975).
36. Comings, D. E., and Harris, D. C., *J. Cell Biol.* **70**, 44 (1976).

Chapter 3

Isolation of Nuclei Using Hexylene Glycol

WAYNE WRAY, P. MICHAEL CONN,[1] AND
VIRGINIA P. WRAY

Department of Cell Biology,
Baylor College of Medicine,
Houston, Texas

I. Introduction

The isolation of nuclei of defined purity with an adequate yield has established guidelines. There should be maximal stabilization of nuclear structure with minimal degradative effects on nuclear components. Concomitantly there should be maximal elimination of extranuclear material with minimal loss of intranuclear material. To this end, there must be a controlled disruption of tissue and cells to obtain a nuclear suspension in a solvent system that stabilizes nuclear structure and component integrity. Finally, after a physical separation of extranuclear material from the nuclei, a product is obtained which has properties that are a consequence of the method used.

The common nuclear isolation methods use either aqueous or nonaqueous homogenization media; the choice depends upon the nature of the end products desired. The nonaqueous organic solvent techniques unavoidably damage the nuclear membrane, whereas the aqueous methods allow a leakage of soluble nuclear components. The established criteria for the purity of isolated nuclei are light and electron microscopic analysis for cytoplasmic contamination and nuclear morphology, localization of enzymic activity, chemical composition, and the integrity of nuclear components (*1–4*). Problems which may affect these criteria adversely include the extraction of intranuclear components and lysis of fragile nuclei. Contamination from extranuclear material in the form of cytoplasmic tags and adsorbed or cose-

[1] *Present address*: Section on Hormonal Regulation, Reproduction Research Branch, National Institutes of Child Health and Human Development, National Institutes of Health, Bethesda, Maryland 20014.

dimented material also may cause serious misinterpretations of data. Most techniques satisfy only one or a few of these criteria of purity, and in fact, nuclear purity has been rigorously demonstrated in only a few cases (*3*). As yet, no isolation method described meets all of the theoretical considerations. Excellent reviews by Wang (*1*), Busch (*2*), Muramatsu (*3*), and Roodyn (*4*) discuss and evaluate in detail several isolation procedures and their relative advantages.

The selection of a nuclear isolation procedure has classically depended on both the cell type and the nature of the investigation. An excellent technique for one tissue may not be suitable for another (*5*). Furthermore, some procedures yield products that are acceptable for further studies in some areas, but are unsuitable for investigation in others. For example, a citric acid procedure is useful for the subsequent isolation of RNA, but it is definitely unacceptable for the isolation of histone. In addition to requirements for purity, it would be useful if a method were applicable to a large variety of cell and tissue types.

This chapter describes a general procedure for the isolation of nuclei in buffered 2-methyl-2,4-pentanediol (hexylene glycol) solutions which meets the requirements of nuclear purity and general applicability. The method is based on original observations by Kane (*6*) and later Sisken *et al.* (*7*), who used hexylene glycol for the isolation of the mitotic apparatus. Wray and Stubblefield (*8*) extended these observations to the isolation of metaphase chromosomes and showed the method to be applicable to the isolation of interphase nuclei from cultured mammalian cells. The efficacy of the hexylene glycol method prompted us to extend these original observations to a number of varied cell and tissue types, including several that are refractory to standard procedures. We have chosen tissues from mouse brain, chicken liver, rat liver, rat uterus, chick oviduct, hen oviduct, rabbit oviduct, and Novikoff hepatoma ascites cells and from the tissue culture cell lines Chinese hamster ovary and HeLa to demonstrate the minor modifications necessary for general applicability of the isolation method.

II.   General Methods

A.   Tissue Culture and Cell Lines

Some of the experiments presented were performed using Chinese hamster ovary (CHO) and HeLa tissue culture cell lines. The cells were cultured in McCoy's medium 5A supplemented with 10% fetal calf serum. The pH of the medium was buffered by a bicarbonate system requiring 5%

CO_2 in the atmosphere. The CHO cell line has a cell cycle of approximately 12 hours and the HeLa cell cycle is about 22 hours in length. Cultures were subdivided every 24 hours.

B. Photography and Electron Microscopy

A Leitz Orthoplan microscope equipped with an Orthomat camera was used for phase-contrast microscopy. Pictures were made on High Contrast Copy Film 5069 (Eastman Kodak, Rochester, New York) rated at an ASA of 12 and were developed in Acufine (Acufine, Inc., Chicago, Illinois) diluted with water 1:1 at 20 °C for 6 minutes.

Thin sections were prepared from aliquots of nuclear fractions suspended in 0.1 mM $MgCl_2$–10 mM Tris-HCl (pH 7.0)–10% sucrose. About 500 μl of nuclear suspension were placed in size 00 cylindrical-tip BEEM capsules and centrifuged at 1000 g for 10 minutes. The capsule was sliced at both ends of the constriction forming a cylinder around the nuclear pellet. The cylinder was further sliced to 5-mm lengths and fixed for 2 hours in 5% v/v glutaral-dehyde in 0.2 M sodium cacodylate buffer (pH 7.4), washed in buffer for 0.5 hours, and treated with 1% OsO_4 for 2 hours. The nuclear pellet was gently removed from the cylinder with a toothpick, washed in distilled water, stained *en bloc* with 1% uranyl acetate, and washed again. The stained material was dehydrated progressively in 25%, 50%, 70%, and 100% ethanol in the cold, followed by two washes with 100% ethanol at room temperature. After infiltration with propylene oxide and 1:1 propylene oxide–Epon mixture, the pellet was infiltrated and mounted in Epon (corrected for Epoxide index), polymerized at 70 °C, and sectioned with a Dupont diamond knife on a Porter-Blum microtome. The sections thus prepared were examined in a Siemens 102 transmission electron microscope after contrast staining in lead and uranium.

C. Biochemical Analysis and Polyacrylamide Slab Gel Electrophoresis

DNA, RNA, and protein were measured by the methods of Burton (9) Fleck and Munro (10), and Lowry et al. (11), respectively. Chromosomal proteins were prepared for polyacrylamide gel electrophoresis by dissolving chromosomes in 3% sodium dodecyl sulfate (SDS) (BDH Chemicals, Ltd., Poole, England)–0.062 M Tris (pH 6.8) (Sigma, St. Louis, Missouri) at 100 °C for 10 minutes. The amount of protein in the sample was assayed by spectrophotometric assay (12), after which β-mercaptoethanol (Eastman Organic) was added and the samples reheated. Chromosomal proteins were analyzed on 9.5% polyacrylamide slab gels using a Tris-glycine buffered

SDS system (*13*). Gels were stained with 0.05% Coomassie blue (Colab, Glenwood, Illinois) in methanol:acetic acid:water 40:7.5:52.2 (v/v/v) and destained by diffusion in 7.5% acetic acid. Gels were photographed on Contrast Process Pan 4155 (Eastman Kodak) 4 × 5 in. sheet film and developed in D-11 for 5 minutes at 20 °C. Gels were scanned on a Helena Quick Scan Jr. (Helena Laboratories, Beaumont, Texas) which had been modified to accept slab gels.

III.   Specific Nuclear Isolation Methods

A.   Buffer System

The isolation buffer (*8*) contains hexylene glycol (2-methyl-2,4-pentanediol) (Eastman Organic, Rochester, New York), piperazine-*N*,*N* '-bis (2-ethane sulfonic acid) monosodium monohydrate ("PIPES") (Calbiochem, La Jolla, California), and $CaCl_2$. The buffer is best made by dilution of stock solutions of $CaCl_2$ and PIPES, adjusted to the desired pH with 1.0 *N* NaOH. A final check of the pH is made before addition of the hexylene glycol. It is imperative that the pH be adjusted before the addition of the hexylene glycol because some pH electrodes respond sluggishly or incorrectly if the solution contains an appreciable amount of this organic compound.

Different tissues and preparations or conditions require slight modifications of the concentrations of the buffer components in order to yield nuclei of the high standards we require. Buffer-to-tissue ratios, temperature, and rapidity of preparation are also relevant factors in the success of the procedure. The following sections give specific nuclei isolation procedures and buffer recipes to give the researcher a specific isolation procedure from which to begin his investigation. In reading this section one should recognize that in no two laboratories does the same technique always give identical results, and therefore a slight modification of these basic procedures might be required.

The rationale for the composition of this buffer system will be discussed in detail in Section V.

B.   Tissue Culture Cells

Solution:
   Buffer HPN:

0.5 M hexylene glycol
1.0 mM CaCl$_2$
0.05 mM PIPES (pH 6.8)

Pellets of harvested tissue culture cells are washed once in 120 volumes of buffer HPN to remove culture medium. The washed cells are sedimented by centrifugation at 1100 g for 3 minutes at 4 °C. After the wash the hypotonic cells are suspended gently in nuclei buffer at 25 °C. The cells are lysed by nitrogen cavitation at 200 lb/in.^{2} or by homogenization with a Dounce homogenizer. Degree of cell lysis is monitored by phase-contrast microscopy. The nuclei from the broken cells are sedimented by centrifugation at 1100 g for 3 minutes at 4 °C. Residual cellular debris is removed by subsequent washing of the nuclei with buffer HPN.

Nuclei can be isolated directly from cell monolayers. It is necessary to purge the CO_2 from the culture atmosphere to prevent a drop in pH, which prevents cell lysis. Fibroblasts are washed with the nuclear isolation buffer. A second wash is left on the cells for about 10 minutes at room temperature, after which the cells spontaneously detach from the glass. The suspension is then lysed by homogenization.

The nuclei may be further purified using standard sucrose density solutions prepared using buffer HPN as the solvent. Adjustments in final sucrose concentration and centrifugation conditions are necessary for each specific cell type.

C. Novikoff Ascites Tumor Cells

Solutions:
1. Buffer HP-125:
   0.5 M hexylene glycol
   0.125 mM CaCl$_2$
   0.05 mM PIPES (pH 7.5)
2. 50% sucrose–HP-125 (pH 7.5) ($n = 1.4200$)
3. 60% sucrose–HP-125 (pH 7.5) ($n = 1.4420$)

All solutions and samples are kept on ice throughout the procedure.

Harvested Novikoff ascites tumor cells in the ascites fluid are dispensed as 5 to 10 ml aliquots into 50-ml conical plastic centrifuge tubes. Each aliquot of cells is diluted to 45 ml with buffer HP-125 and centrifuged at 1000 g for 3 minutes at 4 °C. This wash is repeated three times, as rapidly as possible.

Each final cell pellet is then suspended in 40 ml of buffer HP-125, transferred to a 40-ml Dounce homogenizer, and homogenized vigorously with 8 to 12 passes of an "A" pestle. Efficiency of cell breakage versus integrity of the freed nuclei is determined by phase microscopy. Nuclear pellets are

obtained by centrifugation of the homogenate at 1500 *g* for 5 minutes at
4 °C. The pellets are individually resuspended in 40 ml of buffer HP-125 and
homogenized gently in a 40-ml Dounce homogenizer with an "A" pestle
until an even suspension is obtained. Nuclei are repelleted by the same
centrifugation procedure.

Ten milliliters of solution 2 are pipetted into 30-ml Corex centrifuge
tubes. The nuclear pellets are suspended in solution 3 and the density is
adjusted to 46.1% sucrose ($n = 1.412$) using buffer HP-125 and solution 3.
This nuclear suspension is layered into the centrifuge tubes. The interphase is mixed slightly and the samples are centrifuged at 19,600 *g* for 20
minutes at 4 °C in a Sorvall HB-4 rotor. The supernatant is discarded and
the pellets are resuspended in 30 ml of the buffer HP-125. The samples are
centrifuged at 1500 *g* for 5 minutes at 4 °C and the supernatant is discarded.
This wash is repeated twice with the resulting pellet being clean nuclei.

D.   Normal Rat Liver

Solutions:
1. Buffer HPN (pH 7.5):
   0.5 *M* hexylene glycol
   1.0 m*M* CaCl$_2$
   0.05 m*M* PIPES (pH 7.5)
2. 60% sucrose–HPN (pH 7.5) ($n = 1.442$)
3. 54.6% sucrose–HPN (pH 7.5) ($n = 1.430$)

All solutions and tissue samples are maintained at ice-bath temperature.
As rapidly as possible the livers are removed, trimmed of connective
tissue, and weighed. A normal preparation uses 15–20 gm of tissue. The
livers are rinsed briefly with ice-cold distilled water and then minced with
scissors. The minced tissue is taken in approximately 3-gm lots, mixed with
40 ml of buffer HPN (pH 7.5), and homogenized vigorously using a 40-ml
Dounce homogenizer and an "A" pestle. Degree of breakage is monitored
using phase microscopy. The pooled homogenate is filtered through one
layer of nylon organza. A crude nuclear pellet is obtained by centrifuging
the homogenate in 50-ml plastic conical centrifuge tubes at 1500 *g* for
5 minutes at 4 °C. The supernatant is discarded and the 50-ml wash is
repeated.

Ten milliliters of solution 3 are pipetted into 30-ml Corex centrifuge tubes.
The pellets are suspended in approximately 20 ml of buffer HPN (pH 7.5),
and an equal volume of solution 2 is added. After mixing, this suspension is
layered into the centrifuge tubes. The interface is mixed slightly and then
centrifuged at 16,900 *g* for 30 minutes at 4 °C in a Sorvall HB-4 rotor. The
pellet is saved, resuspended in buffer HPN (pH 7.5), and centrifuged at

1500 g for 5 minutes at 4 °C. This wash step is repeated yielding the final nuclear pellet.

E. Chick Oviduct, Hen Oviduct, Mouse Brain, and Other Tissues

Solutions: The homogenization buffers used all contained 0.5 M hexylene glycol and 1 mM PIPES (pH 7.0). Final concentrations of $CaCl_2$ varied according to the tissue.

1. Buffer HPC-1   (0.001 mM $CaCl_2$)
2. Buffer HPC-10   (0.010 mM $CaCl_2$)
3. Buffer HPC-100   (0.1 mM $CaCl_2$)
4. Buffer HPC-1000   (1.0 mM $CaCl_2$)

1. CHICK OVIDUCT

Oviducts (1–5 gm), pooled from 14-day diethylstilbesterol (DES) treated chicks, are cleaned of vascular and connective tissue. The oviducts are diced and homogenized at room temperature in buffer HPC-1 in a Dounce homogenizer (10–12 passes of the "A" pestle). The homogenate is filtered through four layers of cheesecloth and a single layer of organza; then it is centrifuged for 5 minutes at 900 g in a clinical centrifuge. The pellets are washed three times in buffer HPC-1, suspended in 30 ml of cold 2.0 M sucrose–0.1 mM $MgCl_2$–10 mM Tris-HCl (pH 7.5), and centrifuged at 4 °C for 40 minutes at 25,000 g in a Sorvall SS-34 rotor. The resulting pellet is the purified chick oviduct nuclear fraction.

2. HEN OVIDUCT

Oviducts are cleaned as described for chick. Diced oviducts (20–30 gm) are homogenized at room temperature in Buffer HPC-10 in a 600-ml beaker using four 30-second bursts of a Tekmar Tissumizer (SDT-182E). All subsequent steps are carried out at 4 °C. After filtration as described above, the homogenate is centrifuged at 1200 g for 5 minutes in a Sorvall HS-4 rotor. Following three washes with buffer HPC-10, the nuclear pellet is suspended by Teflon–glass homogenization in 2.3 M sucrose–0.1 mM $MgCl_2$–10 mM Tris-HCL (pH 7.5). The suspension is adjusted to 2.1 M sucrose in a total volume of approximately 200 ml and centrifuged at 25,000 g for 60 minutes in a Sorvall SS-34 rotor. The pellet is the purified hen oviduct nuclear fraction.

3. MOUSE BRAIN

Brains (0.5 gm) from C57BL/6$_m$ mice are diced and homogenized in buffer HPC-1000 with three 5-second bursts in a Tissumizer (SDT-100 EN). After three washes with buffer HPC-1000, the nuclei are pelleted through cold 2.1 M sucrose buffer as described for hen oviduct.

4. OTHER TISSUES

Nuclei were prepared from other tissues as for hen and chick using the modifications described in Table I.

TABLE I

PROCEDURE MODIFICATIONS FOR NUCLEAR ISOLATION
OF OTHER TISSUES

Tissue	Method of homogenization	Homogenization buffer	Sucrose buffer[a]
Rat uterus	Tissumizer	HPC-10	2.1
Chicken liver	Dounce	HPC-100	2.1
Rabbit oviduct	Tissumizer	HPC-100	2.0

[a] M sucrose in 0.1 mM MgCl$_2$–10 mM Tris-HCl (pH 7.5).

F.   Modification of Buffer System for Any Tissue

The concentration of 0.5 M hexylene glycol has consistently given the best preparations of isolated nuclei, no matter what cells or tissues were used. When adapting this buffer system to your own needs, the variables which should be considered are the concentration of PIPES, the concentration of CaCl$_2$, the pH of the buffer, and the temperature. In our laboratory the concentration of PIPES used varies between 0.05 mM and 1.0 mM, and the pH varies between 6.5 and 7.5. The temperature is normally kept at 4°C, but may be raised for certain applications. The concentration of CaCl$_2$ is probably the most important consideration since it may vary from 1.0 mM down to 0.001 mM. For cell lines that are particularly recalcitrant to initial disruption the Ca^{2+} concentration should be lowered. Care should be exercised, however, since the nuclear membrane as well as the plasma membrane becomes easier to break when the Ca^{2+} concentration is lowered. The total amount of cytoplasmic contamination of the nuclei is decreased with lower Ca^{2+} concentrations, so another application that may be used successfully is to break the nuclei at one concentration of Ca^{2+} and then to wash the cells or centrifuge them through heavy sucrose containing a different Ca^{2+} content.

IV.   Nuclear Characterization

A.   Morphology

Phase micrographs of nuclei isolated from chick and hen oviduct, mouse brain, and Novikoff hepatoma ascites cells are shown in Fig. 1. The chick

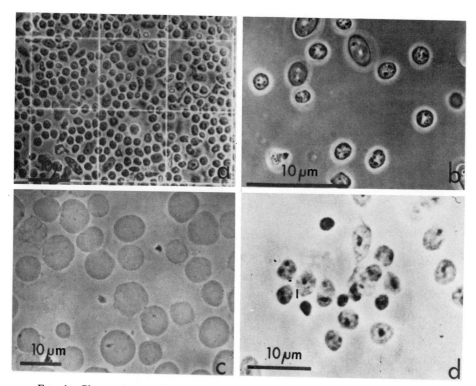

Fig. 1. Phase micrographs of isolated nuclei. (a) Chick nuclei in a hemacytometer chamber. (b) Hen oviduct nuclei showing perinuclear chromatin. (c) Mouse brain nuclei isolated in HPC-1000. (d) Novikoff ascites tumor cells isolated in HP-125.

nuclei (Fig. 1a) which are viewed on a hemocytometer have no apparent cytoplasmic tags and contain visible nucleoli and heterochromatin. In hen oviduct nuclei (Fig. 1b) perinuclear chromatin is easily seen, and cyto-plasmic contamination is not evident. Micrographs of nuclear preparations from mouse brain (Fig. 1c) and Novikoff hepatoma ascites cells (Fig. 1d) show that they also are free of major contaminants. Figure 2 is a more highly concentrated and enlarged phase-contrast micrograph of Novikoff ascites nuclei, which gives a clear indication of the presence of nucleoli and hetero-chromatic material and the absence of cytoplasmic contamination.

Electron micrographs (Fig. 3) of chick and hen oviduct nuclei confirm and extend conclusions from light microscopy. The integrity of the nuclear membrane as well as condensed and extended chromatin can be seen. A small amount of membranous material is, in some places, continuous with the outer nuclear membrane. Thus, phase and electron microscopic observa-

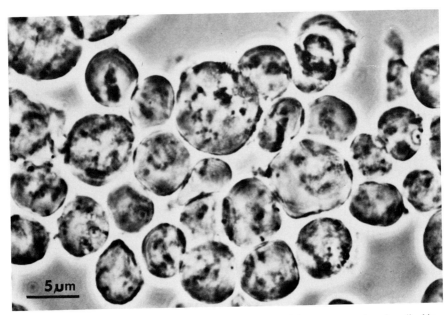

FIG. 2.   Isolated Novikoff ascites tumor cell nuclei. Nuclei were prepared as described in Section III, C. Phase-contrast microscopy.

tions indicate that our preparations are not contaminated with cytoplasmic structures, and electron microscopy shows that the nuclear fine structure is well preserved.

B.   Biochemical Content

For fresh and frozen rat liver, mouse brain, and Novikoff hepatoma cell nuclei, the ratio of acid-soluble proteins to DNA is near unity, a typical ratio for histone to DNA (Table II). Nuclei prepared from chick or hen oviduct have a higher ratio of acid-soluble proteins of DNA than nuclear preparations from other tissues. This increased value reflects the contribution of nuclear lysozyme, a basic protein that is present in isolated chick oviduct nuclei (*14*). Nonhistone protein/DNA and phospholipid/DNA ratios are typical. The slightly elevated protein/DNA ratio of hen oviduct nuclei compared to chick oviduct nuclei may reflect increased articulation of the outer nuclear membrane with the endoplasmic reticulum. The yields of nuclei

FIG. 3.   Electron micrographs of nuclear fractions. (a) Chick oviduct nuclei. (b) Hen oviduct nuclei. Arrows indicate blebbing of the outer nuclear membrane (B), ribosomes (R), heterochromatin (H), euchromatin (E), and pore structure (P). The unit character of the outer (0) and inner (I) nuclear membrane is apparent.

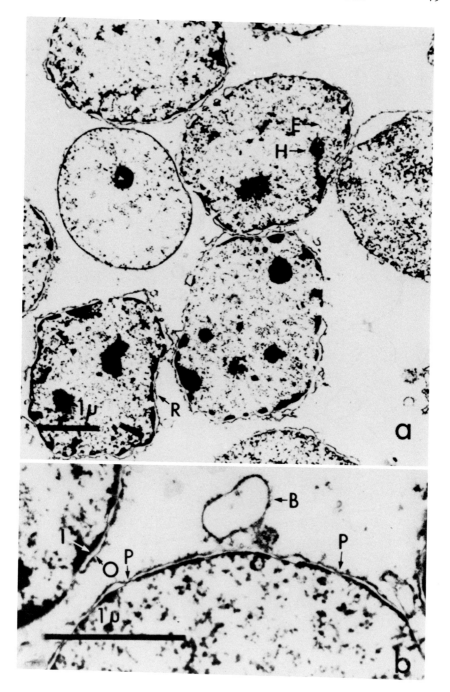

TABLE II

CHEMICAL COMPOSITION AND YIELD OF NUCLEAR FRACTIONS[a]

Nuclei	Acid-soluble protein/DNA	RNA/DNA	Protein/DNA	NHP/DNA	Phospholipid/DNA	Yield
Mouse brain	0.95	0.24	4.6	3.65	0.04	35
Chick oviduct	1.24	0.22	4.4	3.16	0.04	40
Hen oviduct	1.30	0.24	4.8	3.50	0.04	40
Rat liver (fresh)	0.97	0.149	4.4	3.43	0.03	40
Rat liver (frozen)	1.05	0.23	4.4	3.35	0.04	45
Novikoff hepatoma	0.85	0.19	4.5	3.65	0.09	48

[a] Chemical composition is w/w and yield is in percent. Protein (11), RNA (10), and DNA (9) were measured by standard methods. The latter was determined in nuclear pellets which were extracted twice in ice-cold 0.4 N perchloric acid to remove sucrose and hexylene glycol, both of which interfere with assay color formation. Acid-soluble proteins (mostly histone) were extracted as previously described in 0.4 N H_2SO_4 (14). The extract was titrated to pH 7.0 with NaOH, lyophilized, and extracted protein was determined (11). Nonhistone proteins (NHP) were determined by subtracting acid-soluble from total proteins. Lipid was extracted in chloroform: methanol (2:1), washed in NaCl (15), and phosphate determined (16). The yield was determined by measuring recovery of filtered homogenate DNA in the nuclear pellet.

range from 35 to 48% for all tissues. The yields reported are for preparations that began with 3 gm of tissue (or 20 ml packed cell volume in the case of Novikoff hepatoma cells). In larger preparations of oviduct tissue (15–50 gm), the yields are 50–60% (*14a*).

The major peptides of nuclei isolated by these methods are consistent from preparation to preparation as demonstrated in Fig. 4. The clarity and sharpness of the individual bands further substantiate the low degree of proteolysis which occurs in the preparations.

C. Enzymic Characterization

Enzymic studies reveal that isolated Chinese hamster nuclei contain DNA polymerase, RNA polymerase, and adenylate kinase. Table III shows

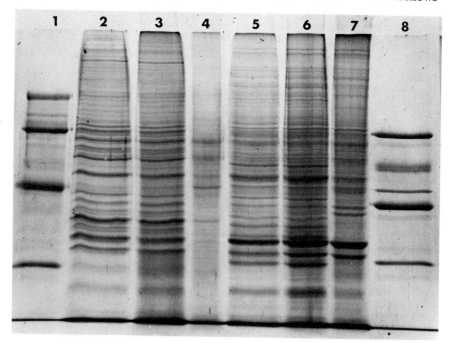

FIG. 4. Peptide composition of isolated nuclei from Novikoff ascites tumor cells and rat liver tissue. Samples were electrophoresed on 9.5% polyacrylamide gels according to the procedure of Laemmli (*13*). *Slot 1*—Molecular weight standards: phosphorylase A, 94,000; bovine serum albumin, 68,000; ovalbumin, 45,000; and α-chymotrypsinogen A, 25,700. *Slots 2,3*—Peptides from different preparations of isolated Novikoff ascites tumor cell nuclei. *Slots 5,6*—Peptides from different preparations of rat liver nuclei. *Slot 8*—Molecular weight standards: bovine serum albumin, 68,000; glutamate dehydrogenase, 53,000; actin 45,000; liver alcohol dehydrogenase 41,600; and carbonic anhydrase 29,000. Nuclei samples contained 40 μg of protein per slot. *Slots 4,7*—Samples do not apply to this report.

these activities along with many cytoplasmic enzymes which are found associated only with the cytoplasmic fraction. It is obvious, also, that exposure to the buffer solution used does not destroy the activity of most enzymes. Determination of enzyme kinetic rates after isolation has not been attempted.

Studies of DNase activity in the hamster cytoplasmic lysate indicate that this enzyme is not released when the cell is broken open (Table IV). The DNA in nuclei prelabeled with thymidine-2-^{14}C appears to remain stable when incubated at 37°C in the chromosome isolation buffer. Added Mg^{2+} (1.6 mM) enhances stability, whereas the presence of added DNase (670 μg/ml) greatly accelerated degradation. The nuclei are somewhat resistant to added DNase, and the kinetics suggest that about two-thirds of the nuclei may be totally resistant, perhaps because of an intact nuclear membrane. At any rate, it seems clear that the isolated preparations do not contain appreciable DNase activity.

The results of enzyme assays on isolated chick and hen nuclei and total homogenate are shown in Table V. Two forms of malate dehydrogenase are present in most cell types. One is present in a soluble cytoplasmic form and the other is particulate and associated with the mitochondria. Thus, with

TABLE III

ENZYMES ASSOCIATED WITH CELL FRACTIONS[a,b]

Enzyme	Nuclei	Cytoplasm
DNA polymerase	+	+
RNA polymerase	+	
Adenylate kinase	+	+
Catalase	−	+
Alkaline phosphatase	−	+
Acid phosphatase	−	+
Lactic dehydrogenase	−	+
Alcohol dehydrogenase	−	+
Phosphoglucomutase	−	+
Creatine kinase	−	+
6-Phosphogluconate dehydrogenase	−	+
Glucose-6P dehydrogenase	−	+
Hexokinase	−	+
Isocitric dehydrogenase	−	+
α-Glycerol phosphate dehydrogenase	−	+

[a] From Wray and Stubblefield (8).

[b] After isolation of Chinese hamster nuclei starch gel electrophoresis was performed on samples of nuclei and cytoplasm. Enzyme activity was then determined by appropriate staining reactions (17). The presence (+) or absence (−) of enzyme is noted in the table.

TABLE IV

DNase Digestion of Nuclei[a]

		Percent DNA hydrolyzed		
Substrate[b]	Additions	In 0 minutes	In 15 minutes	In 30 minutes
Nuclei	—	0.3	1.1	1.5
Nuclei	$+Mg^{2+}$	—	0.5	0.6
Nuclei	$+DNase$	—	27.5	26.0
Nuclei	$+Mg^{2+} + DNase$	—	37.8	36.5

[a] From Wray and Stubblefield. (8).

[b] The reaction system contained 0.3 ml of Chinese hamster nuclei (labelled with thymidine-2-^{14}C) in buffer. Where indicated, $MgCl_2$ was added to a concentration of 1.6 mM and DNase was added to a concentration of 670 μg/ml. The hydrolysis of DNA was measured as increase of trichloroacetic acid–nonsedimentable isotope into the supernatant. After centrifuging for 10 minutes at 10,000 rpm at 4°C, 0.1-ml samples were counted in an NCS-toluene scintillating fluid in a Beckman Liquid Scintillation Counter.

TABLE V

Enzyme Activites of Chick and Hen Nuclear Fractions[a]

Enzyme	Homogenate		Nuclei	
	Chick	Hen	Chick	Hen
Malate dehydrogenase				
Relative activity (%)	100	100	0.8	0.42
Specific activity (mU/mg protein)	480	430	140	140
Mg^{2+}-ATPase				
Relative activity (%)	100	100	1.56	1.60
Specific activity (mU/mg protein)	33.1	67.9	35.2	84.3
Na^+, K^+-ATPase				
Relative activity (%)	100	100	0.9	3.18
Specific activity (mU/mg protein)	18.0	22.9	11.3	56.3
RNA polymerase				
Specific activity (pmol [^{3}H]UTP				
incorporated/mg DNA/10 minutes)				
I	—	—	195	54
II	—	—	1140	583

[a] Malate dehydrogenase (EC 1.1.1.37) was assayed by monitoring the oxidation of NADH spectrophotometrically at 340 nm. Mg^{2+}-dependent ATPase was determined by including ouabain (Sigma) in a final concentration of 0.4 mM. Na^+,K^+-dependent activity was determined by subtracting Mg^{2+} dependent from total activity (no ouabain). Endogenous DNA-dependent RNA polymerase was assayed according to the methods of Schwartz et al. (18).

respect to nuclei, the presence of either indicates contamination. While malate dehydrogenase is active in filtered homogenates of oviduct, the relative activities shown in Table V demonstrate that malate dehydrogenase was essentially absent ($<1\%$) from the nuclear fraction.

ATPase activity is found in plasma, mitochondrial, microsomal, and nuclear membranes (19). The nuclear membrane enzyme is Mg^{2+}-dependent and not activated by Na^+ or K^+, whereas the plasma membrane contains both forms. Table V shows that the relative activities of Na^+, K^+-ATPase were considerably reduced in the nuclear fraction. The specific activities are similar to those reported for nuclei isolated from a number of tissues (20,21).

The endogenous activities of nucleolar and nucleoplasmic RNA polymerase are retained in the purified nuclear fraction (Table V). In both hen and chick oviduct nuclei, the specific activity of nucleoplasmic enzyme is higher than the nucleolar enzyme. Although endogenous RNA polymerases remain very active in the nuclear preparation, quantitative comparisons relative to other reported values are difficult to make, due to variations in optimal assay conditions.

V. Discussion

Established criteria dictate that isolated nuclei should contain all of the subnuclear morphological components and all of the functional activities initially present. They should also be free of contamination by cytoplasmic material. A definite advantage of nonaqueous nuclear isolation methods is the prevention of leakage of soluble nuclear components. The difficulty of preparation plus the degree of cytoplasmic organelle contamination limits the use of nonaqueous isolation methods to specific purposes. Aqueous isolations are much easier to accomplish and generally include the use of components that either physiologically or chemically stabilize nuclear structure and protect labile components against degradative effects; however, certain soluble components are inevitably lost.

The hexylene glycol–PIPES buffer system presented here probably has advantages of both the nonaqueous and aqueous techniques. The hexylene glycol imparts many hydrophobic properties to the nuclei, while the general aqueous nature of the buffer allows the isolation procedure to be easily accomplished. In addition, the hexylene glycol is utilized as a stabilizing agent for the nuclei. Kane (6) stated: "the addition of glycol to water reduces the dielectric constant and increases the electrostatic free energy of the protein molecules, thus reducing their solubility. The addition of glycol to water also influences the hydrophobic interactions between the nonpolar groups of the protein molecules, which play an important role in controlling the conformation of protein molecules in solution." The nonmetal-binding property and $pK_a = 6.8$ of PIPES buffer is appropriate for careful control of divalent cations in a near-neutral buffer (22).

The hexylene glycol–PIPES buffer is efficacious both *in vivo* and *in vitro*. The relative nontoxicity of the buffer has been demonstrated by growth of Chinese hamster fibroblasts in media containing 10% isolation buffer (8). Using the rigorous guides of localization of enzyme activity, morphologic stability of nuclear components, and enzyme functionality, the buffer appears to act primarily as a support medium and prevents subcellular degradation. This is substantiated by the observation that in this buffer system cellular organelles such as ribosomes, polysomes, lysosomes, mitochondria, and chromosomes have been isolated and, in some cases, biochemically and morphologically characterized to be equal to or better than those prepared by other isolation methods (23).

The purity and morphology of the hexylene glycol–isolated nuclei have been established using several criteria. Phase microscopic observations indicate that our preparations are not contaminated with cytoplasmic structures, and electron microscopy shows that the nuclear fine structure is well preserved. Chemical composition data indicate that the RNA/DNA, protein/DNA, acid-soluble protein/DNA, and phospholipid/DNA ratios are very similar to those previously reported for nuclei obtained by other isolation methods (24). The major peptides of the nuclei isolated by these methods are consistent from preparation to preparation, indicating a very low degree of proteolysis. The partition of nuclear and cytoplasmic enzymes indicates nuclear integrity, and the activity of the enzymes shows that the buffer system is not detrimental to enzymic activity. The successful isolation of nuclei from Novikoff hepatoma ascites cells and mouse brain tissue emphasizes the applicability of the hexylene glycol procedures to material recalcitrant to classical isolation techniques. We have obtained excellent preparations of nuclei from every tissue attempted. Thus we feel that the hexylene glycol procedure is a general technique for isolating nuclei from all cell types and that it yields preparations which meet established criteria for nuclear purity.

ACKNOWLEDGMENTS

This work was supported by the following grants: Population Center Grant Hd-7495, NIH HD-8188, NIH HD-7857, NIH Ca-18455 (NCI), NIH CA-18744 (NCI), NSF BMS 75-05622, American Cancer Society CH-14, and American Cancer Society VC-163.

REFERENCES

1. Wang, T. Y. in "Methods in Enzymology," Vol. 12: Nucleic Acids (L. Grossman and K. Moldave, eds.), Part A, p. 417. Academic Press, New York, 1967.
2. Busch, H., in "Methods in Enzymology," Vol. 12: Nucleic Acids (L. Grossman and K. Moldave, eds.), Part A., p. 421. Academic Press, New York, 1967.
3. Muramatsu, M., *Methods Cell Physiol.* **4**, 195 (1970).
4. Roodyn, D. B., *in* "Subcellular Components Preparation and Fractionation" (G. D. Birnie, ed.), p. 15. Butterworth, London, 1972.
5. Busch, H., and Smetana, K., "The Nucleolus," p. 527. Academic Press, New York.
6. Kane, R. E., *J. Cell. Biol.* **25**, 136 (1965).
7. Sisken, J. E., Wilkes, E., Donnelly, G. M., and Kakefuda, T., *J. Cell Biol.* **32**, 212 (1967).
8. Wray, W., and Stubblefield, E., *Exp. Cell Res.* **59**, 469 (1970).
9. Burton, K., *J. Biochem. (Tokyo)* **62**, 315 (1956).
10. Fleck, A., and Munro, H. N., *Methods Biochem. Anal.* **14**, 113 (1966).
11. Lowry, O. H., Rosebrough, N. J., Farr, A. L., and Randall, R., *J. Biol. Chem.* **193**, 265 (1951).
12. Groves, W. E., Davis, F. C., and Sells, B. H., *Anal. Biochem.* **22**, 195 (1968).
13. Laemmli, U.K., *Nature (London)* **227**, 680 (1970).
14. Conn, P. M., and O'Malley, B. W., *Biochem. Biophys. Res. Commun.* **64**, 740 (1975).
14a. Conn, P. M., Parker, M. G., Sheehan, D. M., Wray, W., and O'Malley, B. W., manuscript in preparation.
15. Folch, J., Lees, M., and Sloane, S., *J. Biol. Chem.* **226**, 497 (1956).
16. Fiske, C. H., and SubbaRow, Y., *J. Biol. Chem.* **66**, 375 (1925).
17. Shaw, C. E., and Koen, A. L., *in* "Chromatography and Electrophoresis" (I. Smith, ed.), 2nd ed., p. 325. Wiley (Interscience), New York, 1968.
18. Schwartz, R. J., Tsai, M-J., Tsai, S. Y., and O'Malley, B. W. *J. Biol. Chem.* **250**, 5175 (1975).
19. Sober, H. A., "CRC Handbook of Biochemistry." Chemical Rubber Publ. Co., Cleveland, Ohio, 1968.
20. Kashing, D. E., and Kasper, C. B., *J. Biol. Chem.* **244**, 3786 (1969).
21. Berezney, R., Macauley, L. K., and Crane, F. L., *J. Biol. Chem.* **247**, 5549 (1972).
22. Good, N. E., Winget, G. D., Winter, W., Connolly, T. N., Izawa, S., and Singh, R. M. M., *Biochemistry* **5**, 467 (1966).
23. Wray, W., *Methods Cell Biol.* **6**, 283 (1973).
24. Tata, J. R., in "Methods in Enzymology," Vol. 31: Biomembranes, Part A (S. Fleischer and L. Packer, eds.), p. 253. Academic Press, New York, 1974.

Chapter 4

Isolation of Nuclei and Preparation of Chromatin from Plant Tissues

JOHN T. STOUT AND CAROLYN KATOVICH HURLEY

Laboratory of Genetics,
University of Wisconsin,
Madison, Wisconsin

I. Introduction

The preparation of the chromosomal components from plant tissues requires the isolation of relatively undamaged nuclei free of other subcellular organelles and debris followed by the preparation of chromatin from those nuclei. Selection of a protocol to accomplish these objectives depends to a large degree on the chemical substance being sought as well as on the plant species that is the starting material. The procedures reported here were selected as suitable for histone isolation from maize (*Zea mays*) tissues.

A. The Isolation of Plant Nuclei

Methods for obtaining nuclear preparations were judged by three criteria: purity, condition, and yield.

1. PURITY

In addition to being free of cytoplasmic proteins, ribosomes, and membranes, nuclei must be separated from particles of similar densities such as chloroplasts, mitochondria, starch grains, and cell wall debris. The protein/DNA and RNA/DNA ratios of nuclear preparations from plant tissues should ideally fall within the ranges 3–4 and 0.2–0.3, respectively (*1–4*).

2. CONDITION

The nuclei should appear uniformly round, nearly equal in size, undistorted, and intact under the light microscope. The ultrastructural appearance of isolated nuclei should be generally similar to nuclei in the source tissue; they should have an intact nuclear envelope and the nucleoli should appear normal (*5*). Enzymic activity associated with only the nucleus should be re-

87

tained (3–6). Chromosomal proteins and nucleic acids should give little or no indication of degradation.

3.   YIELD

In plants, recovery of isolated nuclei has ranged from less than 10% to about 40% of the nuclei in the starting material (4, 7). The yield of the component being sought from isolated nuclei is also less than theoretical; for example, 80–90% of the DNA in the nuclei can be recovered as chromatin (1,8). Special precautions are sometimes necessary for the recovery of enzymes since they may leach out of the nuclei or become inactivated. For example, Matthysse and Phillips (6) found that tobacco and soybean nuclei retained the factor (s) required for auxin control of RNA synthesis only if the nuclei were isolated in the presence of the auxin. Retention of the factor (s) in pea bud nuclei did not require auxin.

Based on the above considerations, we found that techniques using buffers containing $CaCl_2$ and sucrose (1,9–11) are unsuitable for the isolation of maize nuclei. Yields are low and the nuclei suffer from obvious distortion and fragmentation.

A method currently in use in our laboratory which gives nuclei in better condition and at higher yields is a modification of the process developed by Kuehl (7; Method "K"). Judged by Feulgen staining, "crude" nuclear pellets are a mixture of nuclei, chloroplasts, a few cell wall pieces, and starch grains. The nuclei and chloroplasts are present in roughly equal numbers while the relative concentration of starch grains varies according to the source tissue. Yields of nuclei from maize tissues range from 0.5×10^7 nuclei/gm (wet weight) for green leaves to 5×10^7 nuclei/gm for immature tassels. The protein/DNA ratio of such crude nuclear preparations from inbred W22 seedlings is about 4–5; for nuclear preparations derived from tassel tissue it is about 3–4. The RNA/DNA ratio is 1.0 for seedling nuclei and about 0.6 for nuclei from tassels. Twenty percent of the DNA in the starting material is usually recovered in the pellet. Ultracentrifugation of the crude nuclei through 2.3 M sucrose solutions brings the protein/DNA and RNA/DNA ratios to 3 and 0.6, respectively, for nuclei from W22 seedlings. Thirty to forty percent of the crude nuclei are recovered, a yield also observed by Kuehl (7) and Dick (2).

Rubenstein (unpublished; Method "R") has further modified the nuclear isolation procedure to provide maize nuclei from which DNA molecules having molecular weights of $3–10 \times 10^7$ may be isolated. The yield of DNA is about 1 mg DNA/10 gm (wet weight) of leaves.

We occasionally employ the method of Hamilton et al. (4; Method "H"). The protein/DNA and RNA/DNA ratios of the nuclei are similar to those for nuclei prepared by Method "K." The advantage of this method is the rapidity with which nuclei can be prepared.

B. Isolation of Chromatin

We have found that direct isolation of chromatin from maize tissues as described by Bonner *et al.* (*12*) for peas results in low yields and severely degraded histones. Although extraction of chromatin from maize nuclei can be accomplished with 2.0 *M* NaCl (*1*), a widely used method for animal nuclei, we have found the CaCl$_2$–precipitation method of Towill and Nooden (*1*) more efficient in our hands. Such chromatin has protein/DNA and RNA/DNA ratios of about 2.0 and 0.2, respectively. According to Bonner *et al.* (*12*), pure chromatin should have a protein/DNA ratio of less than 2.5 and an RNA/DNA ratio of 0.005 to 0.2. We commonly recover 70–80% of the DNA in the nuclear pellet as chromatin–DNA.

One limitation of the Towill and Nooden method we use is that it involves the mechanical shearing of chromatin to assure its complete solubilization. Noll *et al.* (*13*) have shown that such treatment destroys the native arrangement of histone aggregates on the DNA. This shearing, however, does not affect the isolation of undegraded histones. Investigators requiring chromatin in the native state should avoid shearing procedures.

C. Applications

Various formulations based on Kuehl's homogenizing buffer have been used to isolate nuclei from tissues of tobacco, peas, onions, spinach (*7*), and broad beans (*2*). Using Method "K" (described below), we have produced nuclei, chromatin, and histones from chicken erythrocytes and the leaves of maize, barley, oats, wheat, amaryllis, bluegrass, tulips, teosinte, and *Tripsacum dactyloides* (the latter two are maize relatives). We have, however, failed to isolate histones from chromatin produced from the isolated nuclei of leaves of a number of trees (maple, birch, elm, fir, and spruce). The reason for this is not known.

We have isolated nuclei and chromatin from samples of maize roots, coleoptiles, immature tassels, and immature endosperm and from masses of less than 5 gm to kilogram lots from thousands of plants. For experiments involving analysis of the histones and chromosome constitutions of individual plants, green leaves but not the culms (stalks) can be used, reserving for karyotyping the immature tassels, the tissue of choice for cytology in maize.

II. Isolation of Nuclei and Chromatin from Maize Tissues

All operations are carried out at 0–4 °C unless otherwise indicated. Plastic gloves should be worn as a precaution against introducing nucleases and proteases into the preparations.

A. Small Scale Preparations

1. Nuclear Isolation Method "K"

Fresh plant tissue (100 gm) is chopped into 5-mm pieces with a sharp knife or scissors. The tissue pieces are immediately immersed in 600 ml of Buffer "K" (see Section IV). To facilitate penetration of the buffer into the tissues, a vacuum (60 mm Hg) is drawn on the mixture for 5 minutes, released, and then redrawn for 10 minutes. The purged, immersed tissue should appear water-soaked.

After storing overnight to allow the octanol in the buffer to stabilize the nuclei (7), the mixture is homogenized in a 1-gallon Waring Commercial Blendor (Waring Products Co., Winsted, Connecticut) at full speed (17,000 rpm) for 30 seconds. Small samples (less than 20 gm) should be homogenized in a VirTis "45" Homogenizer (The Virtis Co., Gardiner, New York) at full speed for 30 seconds. We have confirmed Kuehl's (7) observation that high blending speeds favor high nuclear yields. The homogenate is filtered through a Tyler 270-mesh (53-μm openings) stainless-steel sieve (Combustion Engineering, Inc., Mentor Ohio.). The residue on the screen is resuspended in 100 ml of buffer "K" and homogenized again. The second homogenate is filtered and the two filtrates are then combined.

The filtrate is centrifuged at 1000 g for 10 minutes in 50-ml conical centrifuge tubes or for 30 minutes in 1-liter bottles. The supernatant is gently poured off and replaced with 80 ml of buffer "K" containing 0.15% Triton X-100. The pellet is resuspended either by swirling or homogenization with a glass–Teflon tissue disrupter until all the clumps have disappeared. The resuspension is centrifuged as before. The resulting pellet is washed several times in Triton-containing buffer until the supernatant is clear and colorless. The use of Triton X-100 selectively solubilizes chloroplasts, prevents nuclear aggregation, removes membrane fragments and proteins adhering to the nuclei, and helps to reduce degradation by cytoplasmic enzymes (14). The resulting crude nuclear pellet is a satisfactory source for isolating chromatin. We have found that storing this crude nuclear pellet at $-20°C$ for several months does not produce significant degradation of histone.

When required, the crude nuclei may be purified by pelleting them through 2.3 M sucrose according to Towill and Nooden (1). To do this, the crude nuclei are resuspended in buffer "K" with 0.15% Triton X-100 (5×10^8 nuclei/ml). This suspension is layered over an equal volume of 2.3 M sucrose containing 3 mM CaCl$_2$–10 mM Tris-HCl (pH 7.2)–0.15% Triton X-100. The interface is mixed and the tubes are centrifuged for 1 hour in a Beckman SW 27 rotor (Beckman Inst. Inc., Palo Alto, California) at 17,000 rpm. The pellet is composed of nuclei of uniform size, starch grains, and a few cell wall fragments. Chloroplasts, membranes, and fragments of nuclei remain in the gradient. This step is necessary when an accurate estimate of DNA, RNA, and protein is to be made using the usual colorimetric assays.

2. NUCLEAR ISOLATION METHOD "H"

Nuclei suitable for the extraction of histone have also been isolated using a method in which the long preincubation in homogenizing buffer as used above is eliminated (4).

Fresh leaves (100 gm) are immersed in 1 liter of cold diethyl ether for 1 minute. The ether is decanted and the leaves washed in cold buffer "H" (see Section IV) to rinse away the ether. The tissue is suspended in more buffer "H" and homogenized at full speed in a Waring Commercial Blendor for 30 seconds. The homogenate is filtered through a 53-μm sieve and centrifuged at 1000 g for 10 minutes. The nuclear pellet is washed two or three times with buffer "H" containing 0.15% Triton X-100 until the supernatant is clear and colorless. This crude nuclear pellet may then be pelleted through 2.3 M sucrose as above to improve purity.

3. NUCLEAR ISOLATION METHOD "R"

Dr. I. Rubenstein has adapted Method "K" for the isolation of high molecular weight maize DNA (unpublished preliminary method). Initial infiltration of the tissue is made using the original formulation of Kuehl's homogenizing buffer (buffer "K" with 0.1 M in place of 0.4 M sucrose) with 400 μg/ml ethidium bromide (caution—may be carcinogenic!) and 2 mg/ml polyvinylpyrollidone (PVP) added.

Healthy green plants at the microsporocyte stage should be used. The above buffer is vacuum infused into the sliced tissue for about 30 minutes. After the nuclei have bound the ethidium bromide, they should be protected from light as a precaution against nicking the DNA. All subsequent work should be carried out in a dimly illuminated area of the laboratory that is protected from outside or artificial lighting. The infusion buffer is poured off and replaced with homogenizing buffer without ethidium bromide and PVP. The suspended tissue is immediately homogenized in a VirTis blender at full speed for 1.5 minutes in a cooled blender jar, and the homogenate is filtered through a 60-mesh sieve followed by a 25-μm Nitex screen (Tetko, Inc., Elmsford, New York). The residue left on the filters is reextracted and the combined filtrates are centrifuged (10 minutes at 1000 g). The nuclear pellet is washed with homogenizing buffer containing Triton X-100 until the supernatant is clear. The pellet of crude nuclei stained with ethidium bromide is resuspended in 0.2 M sucrose–2 mM $CaCl_2$–10 mM Tris-HCl (pH 7.4) at a concentration of 5 × 10⁶ nuclei/ml and frozen.

DNA, when desired, may be isolated from thawed nuclei by adding 2 volumes of the nuclear suspension (about 10⁷ nuclei/ml) to 2 volumes of digestion buffer (see Section IV) and 1 volume of 1 mg/ml Proteinase K (E. M. Laboratories, Inc., Elmsford, New York) per milliliter of digestion buffer. This mixture is dialyzed against the dialysis buffer (see Section IV) at room temperature for 16–24 hours. The DNA digest is then centrifuged at 6000 g for 20 minutes and the supernatant is extracted three times with an equal volume of redistilled (water-saturated at pH 8.5) phenol and once with an equal volume of ether. Residual phenol and ether are removed by dialyzing the aqueous DNA solution at 4°C against fresh dialysis buffer or 0.1 M NaCl–0.1 M EDTA–20 mM Tris-HCl (pH 7.4).

The DNA solution at this point has a concentration of 15–25 μg/ml and contains little RNA. The DNA can be purified according to density by centrifugation in a CsCl gradient, and the high molecular weight DNA can be separated from smaller fragments by centrifugation in a sucrose gradient.

4. Preparation of Chromatin from Isolated Nuclei

Chromatin is prepared according to Towill and Nooden (1). The nuclear pellet is resuspended in lysing buffer (see Section IV) (10^9 nuclei/30 ml of buffer), incubated for 30 minutes, and homogenized in a VirTis "45" Homogenizer at low speed (30 V) for 30 seconds. The solution of sheared chromatin is centrifuged at 1000 g for 10 minutes. The pellet can be reextracted. The combined supernatants of soluble chromatin are made 10 mM in $CaCl_2$, mixed, and stored overnight. The chromatin precipitate is collected by centrifugation at 1000 g for 10 minutes.

B.   Large-Scale Preparations

To isolate gram quantities of histone from isolated nuclei, a scaled-up version of Nuclear Isolation Method "K" has been used. Histone yield is maximized and processing time and buffer components are conserved if maize nuclei are obtained from immature tassels (at the microsporogenesis stage) and from the pale sheath leaves surrounding them. This tissue yields about 5×10^7 nuclei/gm and has little chlorophyll and starch. Planting should be done according to a schedule that considers the number of plants to be harvested each week and the length of time required for the various inbreds used to reach the proper stage of development. We have found that three technicians, working 35 hours per week, can process 3000–4000 plants each week.

Plants are judged to be at the early microsporocyte stage when one can feel the tip of the tassel inside the central sheath of leaves. The immature tassel and its sheath leaves are harvested by grasping the central sheath of four or five leaves and pulling straight up. The leaf segments distal to the tip of the tassel are removed, and the tassels are kept on ice until processed. Tassels are sliced into 1-cm-wide segments, a dozen or so at a time, using the slicing attachment of an industrial food mixer such as a Triumph mixer (Triumph Manufacturing Co., Cincinnati, Ohio). Three hundred sliced tassels are immersed in 28 liters of cold buffer "K" in a 10-gallon pressure cooker. The vessel is closed and a 60-mm vacuum is drawn on it for 20 minutes. The entire contents of the pressure cooker are transferred to another container and are stored overnight.

Three Waring Commercial Blendors are sufficient to homogenize the tissues from 1500–2000 plants infused the previous day. Half-gallon lots of infused tassels, leaves, and homogenizing medium are blended for 30 seconds and filtered, first through a cheesecloth-covered screen, and then through a series of sieves (Tyler 20-, 60-, and 270-mesh) assembled on a shaker. The residues from all filters are combined, resuspended, homogenized again, and filtered.

The filtrate is centrifuged at 15,000 g at a flow rate of 4 liters/minute in a floor-model continuous-flow Sharples Super-Centrifuge (Sharples Equipment Div., Pennsalt Chem. Corp., Warminster, Pennsylvania). The pellet is resuspended in buffer "K" containing 0.15% Triton X-100, 1 volume of paste to 5 volumes of buffer. A Waring blender run at low speed is used to aid resuspension. The nuclear suspension is centrifuged in 1-liter bottles at 1000 g for 30 minutes and the pellet washed until the supernatant is clear. Two liters of buffer are required per wash for 100 tassels.

Chromatin is liberated from the nuclei by suspending them in lysing buffer (see Section IV). After incubating for 1 hour, the solution is blended at low speed in the 1-gallon blender for about 15 seconds and then clarified by centrifugation in an air-driven laboratory model Sharples centrifuge operating at a pressure of 21 lb/in². The pellet is reextracted with one-tenth the initial volume of lysing buffer. The chromatin solution is made 10 mM in $CaCl_2$ and allowed to stand overnight in the cold. Precipitated chromatin is collected by pelleting in the laboratory Sharples centrifuge as described above.

III. Concluding Remarks

The methods selected for the isolation of plant nuclei and chromatin must be suited to the plant tissue used. We have found the three nuclear isolation methods outlined above very suitable for the isolation of nuclei from maize. These methods appear to work well in some other plants and probably could be applied with success to still other untested plant species. The method we employ for chromatin extraction from maize efficiently produces a very satisfactory product which can be used for the isolation of histones; however, it may not be suited for all plant nuclei (2) or for those who desire unsheared chromatin.

IV. List of Buffers

A. Buffer "K"

This is a modified form of Kuehl's (7) buffer. It consists of 2.3% acacia–0.4 M sucrose–2 mM CaCl$_2$–4 mM octanol-1–20 mM Tris brought to pH 7.6 with glacial acetic acid. The insoluble tannins contaminating the acacia are removed by clarifying the buffer in a floor model Sharples centrifuge at 15,000 g with a flow rate of 4 liters/minute. For smaller amounts, clarifying can be done by centrifugation at the same force (15,000 g) for 10 minutes using the GSA rotor in a Sorvall RC-2B centrifuge (DuPont Co., Newtown, Connecticut). This buffer is clearer than that prepared by purifying the acacia before addition to the buffer. Table sugar can be substituted for reagent-grade sucrose.

If the buffer is stored at 4 °C it can become contaminated with microorganisms rather quickly. Addition of 1 mM NaN$_3$ appears to retard this contamination adequately. If desired, protease inhibitors such as NaHSO$_3$ (15) or phenylmethylsulfonyl fluoride (16) may be used. We routinely add 50 mM NaHSO$_3$ to buffer "K" just prior to use. The pH of the buffer drops to about 5 but this does not appear to affect the isolation (7).

B. Buffer "H" [from Hamilton et al. (4)]

1.14 M sucrose	5 mM 2-mercaptoethanol
5 mM MgCl$_2$	10 mM Tris-HCl (pH 7.6)

C. Digestion Buffer

10 mM NaCl	0.5% SDS
10 mM EDTA	10 mM Tris-HCl (pH 8.5)

D. Dialysis Buffer

20 mM NaCl	10 mM Tris-HCl (pH 8.6)
1 mM EDTA	

E. Lysing Buffer

3 mM EDTA
10 mM Tris-HCl (pH 8.0)
Made 50 mM in NaHSO$_3$ just before use.

NOTE ADDED IN PROOF

We have produced moderately high molecular weight maize DNA having little degradation by a simple method derived from that used by Blin and Stafford (17). The DNA produced in this way serves as a suitable substrate for EcoRI digests. Liquid-nitrogen-frozen immature tassels (100 gm) are pulverized by grinding them through a series of stainless-steel sieves (Tyler 20-, 60-, and 270-mesh) kept at less than $-20°$C in Dry Ice. Dry Ice is used as a pestle. The cold powder is mixed with 200 ml (60°C) digestion buffer (0.125 M EDTA, 2% SDS, pH 8.0, 2 mg/ml PVP) containing 500 μg/ml Pronase (Calbiochem, San Diego, California) heat-treated to inactivate DNases. The viscous solution is incubated overnight at 60°C, and the nucleic acids are isolated from the lysate by chloroform–isoamyl alcohol extraction (18).

ACKNOWLEDGEMENTS

This work has been supported by a U.S. Public Health Service Training Grant GMO1156 to J.T.S., National Science Foundation research grant GB43398 to J. L. Kermicle and J.T.S., grants from the American Cancer Society and National Institutes of Health (training grant GM20069) to Dr. Oliver Smithies, and NIH training grant GM00398 to C.H.

We are also indebted to the many scientists who contributed ideas and technical assistance, especially Dr. Sharon Desborough who initially suggested the use of Kuehl's method; Dr. Leigh Towill, who made available his procedures for the preparation of nuclei and chromatin prior to publication; Dr. Oliver Smithies and Dr. J. L. Kermicle for their continued excellent advice; Dr. Irwin Rubenstein for his pre publication contribution on DNA isolation; and to Dr. John Garver and Art Olson of the Pilot Plant, Department of Biochemistry, University of Wisconsin-Madison, for providing guidance in the use of equipment essential for the processing of large quantities of plant tissue.

This is publication number 2009 from the Laboratory of Genetics.

REFERENCES

1. Towill, L. E., and Nooden, L. D., *Plant Cell Physiol.* **14**, 851 (1973).
2. Dick, C., *Arch. Biochem. Biophys.* **124**, 431 (1968).
3. Tautvydas, K. J., *Plant Physiol.* **47**, 499 (1971).
4. Hamilton, R. H., Kunsch, U., and Temperli, A., *Anal. Biochem.* **49**, 48 (1972).
5. Busch, H., *in* "Methods in Enzymology," Vol. 12: Nucleic Acids (L. Grossman and K. Moldave, eds.), Part B, p. 65. Academic Press, New York, 1968.
6. Matthysse, A. G., and Phillips, C., *Proc. Natl. Acad. Sci. U.S.A.* **63**, 897 (1969).
7. Kuehl, L., *Z. Naturforsch.* **19b**, 525 (1964).
8. Frenster, J. H., Allfrey, V. G., and Mirsky, A. E., *Proc. Natl. Acad. Sci. U.S.A.* **50**, 1026 (1963).

9. Busch, H., *in* "Methods in Enzymology," Vol. 12: Nucleic Acids (L. Grossman and K. Moldave, eds.). Part A. p. 421. Academic Press, New York, 1967.

10. Spelsberg, T. C., and Sarkissian, I. V., *Phytochemistry* **9**, 1385 (1970).

11. Stern, H., *in* "Methods in Enzymology," Vol. 12: Nucleic Acids (L. Grossman and K. Moldave, eds.). Part B, p. 100, Academic Press, New York, 1968.

12. Bonner, J., Chalkley, G. R., Dahmus, M., Fambrough, D., Fujimura, J., Huang, R. C., Huberman, J., Jensen, R., Marushige, K., Ohlenbusch, H., Olivera, B., and Widholm, J., *in* "Methods in Enzymology," Vol. 12: Nucleic Acids (L. Grossman and K. Moldave, eds.), Part B, p. 3. Academic Press, New York, 1968.

13. Noll, M., Thomas, J. O., and Kornberg, R. D., *Science* **187**, 1203 (1975).

14. Tata, J. R., Hamilton, M. J., and Cole, R. D. *J. Mol. Biol.* **67**, 231 (1972).

15. Panyim, S., Jensen, R. H., and Chalkley, G., *Biochim. Biophys. Acta* **160**, 252 (1968).

16. Nooden, L. D., van den Broek, H. W. J., and Sevall, J. S., *FEBS Letts.* **29**, 326 (1973).

17. Blin, N., and Stafford, D. W., *Nucleic Acids Res.* **3**, 2303 (1976).

18. Phillips, R. L., Kleese, R. A., and Wang, S. S., *Chromosoma* **36**, 79 (1971).

Chapter 5

Methods for the Isolation of Nuclei from Ciliated Protozoans

DONALD J. CUMMINGS

Department of Microbiology,
University of Colorado Medical Center,
Denver, Colorado

I. Introduction

In recent years, the molecular biology and biochemical properties of ciliated protozoans have been studied extensively in many laboratories. A major reason for this is that most of these organisms have two functionally distinct types of nuclei: macro and micro. The macronucleus controls the vegetative processes of cell growth and in general develops from the micronucleus following a sexual process (1–3). Although the macronuclei from different ciliates are morphologically highly diverse, the macronucleus is commonly oval, polyploid, and contains up to 13,000 times the haploid amount of DNA (3–5). Structurally, macro- and micronuclei contain the same substructures as nuclei from other organisms: i.e., nuclear membranes, nucleoli, chromatin elements, etc. In some ciliates, giant polytene chromosomes develop in the anlagen (6); however, most macronuclei are organized into large (0.5–1 μm) and small (0.1–0.2 μm) bodies, which are considered to be nucleolarlike and chromatin bodies, respectively (7). The micronucleus usually has a more centrally located chromatin body, governs the germ line, divides mitotically, and is generally diploid. Electron micrographs of a macronucleus and micronucleus isolated from *Paramecium aurelia* (8) are presented in Fig. 1. Because of these and other properties, ciliated protozoans can serve as model systems for the study of gene amplification, gene diminution, histone and nonhistone protein function, morphological development of chromosomes, and even as models of cellular aging (1,9–11).

This interest in the use of ciliate protozoans as model systems in many

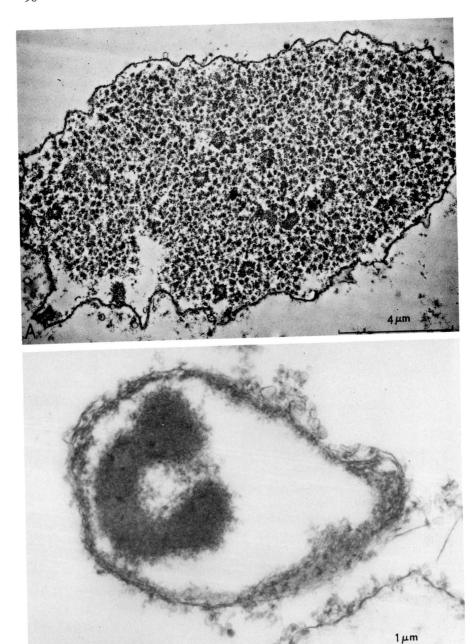

FIG. 1.   Electron micrograph of isolated nuclei from *P. aurelia*, species 1. (A) Macro-nucleus, showing the large and small bodies; (B) micronucleus, showing the more central chromatin body.

areas of cell and developmental biology has resulted in several recent reviews. Cummings and Tait (*12*) discussed in detail several aspects of the isolation of nuclei from *P. aurelia*; Gorovsky *et al.* (*13*) reviewed their methods for the isolation of nuclei from *Tetrahymena*; Nozawa (*14*) summarized procedures from the isolation of several subcellular components from *Tetrahymena*, including nuclei; and most recently, Buetow (*15*) described several procedures used in the isolation of nuclei from a variety of protozoa and algae. Consequently, in this chapter I will not discuss general problems in isolation but will rather present specific methods utilized with *P. aurelia*. Wherever pertinent, I will compare these procedures and results with those obtained with other organisms and ciliates. The primary emphasis will be on the rationale necessary for the isolation of intact nuclei and the specific requirements regarding their use after isolation.

II. Isolation Procedures

A. Rationale

The demand for purified nuclei is not unique to protozoans, and isolation procedures have been in use for some time (*16*). What is perhaps unique to the ciliated protozoans is the size of the macronucleus. In general, there is a direct proportionality between the size of the cell and the size of the macronucleus. From *Tetrahymena*, the macronucleus has dimensions of 6–10 μm (*17*). *P. aurelia* macronuclei are about 15 to 35 μm in size (*8,18*), while those of *Spirostomum ambigium* are greater than 2 mm in length (*3*). Moreover, the protozoans can have a cell wall that is quite refractory to rupture by either mechanical shear or hypoosmotic swelling. In our previous review (*12*) we listed four criteria for the isolation of macronuclei: (1) efficient cell rupture with minimal lysis of the nuclei; (2) isolation of at least 50% of the nuclei with minimal contamination from other subcellular structures; (3) protection of the nuclei so that compositional and functional studies were representative of *in situ* nuclei; and (4) reproducibility and applicability to many species and other ciliates.

B. Procedure for Macronuclei

1. Disruption

We approached the problem of cell disruption by testing several detergents in hypotonic buffer (*8*). Triton X-100, Tween-80, sarkosyl, nonidet NP-40, and sodium deoxycholate all led to cell disruption but also caused lysis of the macronucleus. Lysis of the macronucleus could be avoided using CaCl$_2$, spermidine, and sucrose to protect the nuclei. However, when these were included the detergents caused swelling and increased fragility but did not lead to cell lysis, and some method utilizing mechanical shear had to be employed. We noted that one or two rapid rotations on a Vortex mixer were sufficient to rupture these cells, but for uniformity in method we routinely homogenized the suspension by a few strokes of a loosely fitting Teflon pestle attached to a Tri-R homogenizer at setting −3. Better preservation of nuclei was obtained when a mixture of an ionic detergent (sodium deoxycholate) and a nonionic detergent (nonidet NP-40) was used. All operations were performed at 0–4 °C.

2. ISOLATION

Simple layering of the homogenate onto sucrose gradients did not result in very high yields of macronuclei. Blobel and Potter (*19*) observed that rat liver nuclei were trapped at the sucrose–homogenate interface by the accumulation of endoplasmic reticulum and mitochondria. They avoided this problem by adding sucrose to their homogenate in order to float these components. We therefore mixed the *P. aurelia* cell homogenate with 2.4 *M* sucrose prior to layering onto 2.1 *M* sucrose solution. This resulted in a yield of macronuclei of about 60%. This one-step sucrose gradient was centrifuged for a brief period, and macronuclei were collected as a pellet, free of most cytoplasmic contamination. This method and a modification of it were discussed at length previously (*8,12*). A flow chart illustrating the steps is presented in Fig. 2. This modification resulted from our (*20*) work on DNA-dependent DNA polymerases. Chang and Bollum (*21*) noted that certain DNA polymerase activity was lost when nuclei were isolated in detergents

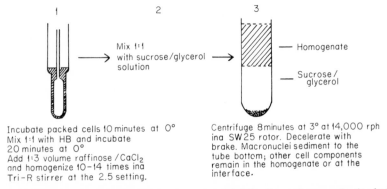

Incubate packed cells 10 minutes at 0°
Mix 1:1 with HB and incubate
20 minutes at 0°
Add 1:3 volume raffinose /CaCl$_2$
and homogenize 10–14 times ina
Tri–R stirrer at the 2.5 setting.

Centrifuge 8 minutes at 3° at 14,000 rph
ina SW 25 rotor. Decelerate with
brake. Macronuclei sediment to the
tube bottom; other cell components
remain in the homogenate or at the
interface.

FIG. 2. Isolation scheme for obtaining macronuclei. HB refers to homogenization buffer containing 0.02% nonidet NP-40–0.02% sodium deoxycholate–3 m*M* CaCl$_2$–1 mg/ml spermine (tetrahydrochloride). This same procedure was described previously (*12*).

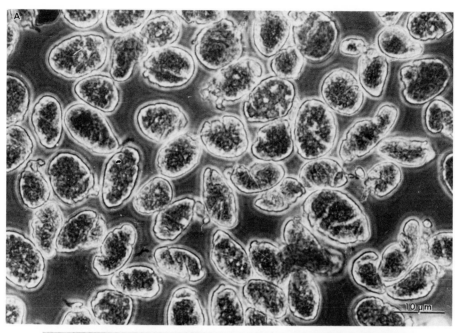

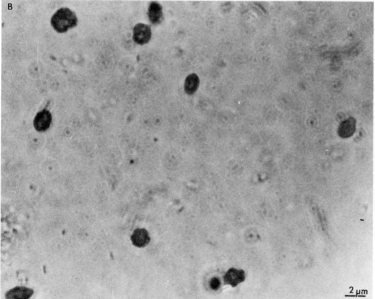

FIG. 3. Light micrograph of isolated nuclei from *P. aurelia*, species 1. (A) Macronuclei; note the general uniformity of size and shape. (B) Micronuclei. Lower yields were obtained than with macronuclei, and contaminating particles can be seen in the background. These nuclei were stained with acetoorcein prior to microscopy.

at concentrations greater than 0.5%. Our earlier method (8) used 0.55% detergents. In attempting to lower this concentration without reducing the yield of macronuclei, we noted (12) that preincubation of the cells in homogenization buffer prior to homogenization was necessary. Thus, we were able to reduce the total concentration of detergents (nonidet NP-40 and sodium deoxycholate) to 0.04%. A mixture of raffinose and additional $CaCl_2$ aided in the preservation of the macronuclei. Figure 3A is a light micrograph of the macronuclei obtained as a pellet in step 3 (Fig. 2). Note that most of these nuclei are uniform in shape and size, with very little contamination.

3. ASSESSMENT OF METHOD

The general method depicted in Fig. 2 has been utilized with seven species of *P. aurelia*, one stock of *P. caudatum*, and one stock of *Euplotes*. In all cases, high yields of macronuclei, structurally similar to *in situ* macronuclei, were obtained. Thus, this procedure has been shown to be broadly applicable. In addition, we (20) have shown that macronuclei could be obtained with the same efficiency from cultures in logarithmic or stationary phases of growth and that high yields of DNA polymerase activity were demonstrated. We will discuss later, a gum arabic–octanol method used by Gorovsky *et al.* (13) in which little success was achieved from stationary cultures of *Tetrahymena*. In this method it is extremely important to remove all traces of the gum arabic and octanol from the isolated nuclei before reliable compositional studies or *in vitro* enzyme analyses can be performed (13). This method, however, is the best reported for the preservation of macronuclear membranes.

Aside from reproducibility and functional criteria, an essential requisite is that this method leads to reliable compositional studies. Both of our published methods (8, 12) have resulted in the same DNA content of macronuclei, ranging from 18 to 26%, from three different species. An early report on *P. aurelia* macronuclei suggested a DNA content of only 4.4% (22) compared with a total cell content of 1.6%. When one considers that greater than 90% of the cell DNA is contained in the macronucleus (4), the value of 4.4% *a priori* appears to be low. The range of values obtained with our method agrees well with that reported for *Tetrahymena* (23), as well as for that obtained with *P. aurelia*, species 8, macronuclei isolated using the gum arabic–octanol method (18). Skoczylas and Soldo also measured the RNA content of macronuclei and obtained a value of about 25%, much higher than the 10–17% obtained by our method (8,12). As indicated earlier, the gum arabic–octanol method leads to better preservation of the nuclear membrane, and this may result in contamination of the membrane with cytoplasmic ribosomes (24).

Earlier, by using the results of Isaacks *et al.* (25) as an example, we (12) pointed out the importance of compositional studies in assessing a method for nuclei isolation. To rupture *P. aurelia*, species 4, Isaacks *et al.* used 1% Triton X-100–12mM $CaCl_2$–0.5 M sucrose–4% polyvinyl pyrrolidone followed by homogenization in a Waring blender for 2 minutes. This procedure resulted in much fragmentation of the macronuclei, which led to poor recoveries and low amounts of DNA. Only 21% of the macronuclei were accounted for, and the macronuclear DNA content was determined to be less than 10%. In fact, these authors calculated that the macronuclei accounted for only 30% of the total cellular DNA, compared with the greater than 90% obtained by others (*4*). In agreement with our results, Isaacks *et al.* (*25*) reported that the RNA content of macronuclei was about 11%. The homogenization in a Waring blender probably led to the high degree of macronuclear fragmentation and loss of DNA. Gorovsky *et al.* (*13*) and Skoczylas and Soldo (*18*) used a Waring blender to successfully obtain macronuclei from *Tetrahymena* and *P. aurelia*, respectively, but they used much shorter blending times in the presence of gum arabic and in the absence of detergent. We have found that some methods for cell disruption such as a Parr bomb or a homogenizer (*12*) are harmful for the isolation of intact macronuclei, and it may be that extensive blending in the presence of 1% Triton X-100 should be added to this list.

C. Procedure for Micronuclei

Several laboratories have devised the same general method for the isolation of micronuclei from *Oxytricha*, *Stylonychia*, *Tetrahymena*, and *Paramecium* (*8,9,12,13,26*). This method involves cellular homogenization, removal of the macronuclei by centrifugation or filtration, and purification of the macronuclear-free supernatant. We have discussed these methods in detail previously (*12*) and for now present only a flow chart (Fig. 4) and a light micrograph of micronuclei isolated from *P. aurelia* (Fig. 3B) by this method. Little information on the other ciliates is available, but for both *Tetrahymena* (*13*) and *Paramecium* (*12*) the yields of micronuclei are quite variable, ranging from 30 to 100%. The reasons for this are not clear. We have tried several methods for cell disruption, including the Parr bomb, milk homogenizer, and that illustrated in Fig. 4. All resulted in the same variability in yield; low yields tended to give the greatest contamination with macronuclear fragments, trichocysts, etc. These problems must be solved before we can have confidence in comparisons between micro- and macronuclei.

1. BIOCHEMICAL COMPARISONS OF MICRO- AND MACRONUCLEI

In spite of the difficulties with isolation procedures for micronuclei, several valuable studies have been performed. Lauth *et al.* (*9*) have shown that the kinetic complexity of micronuclear DNA from *Oxytricha* is 2 to

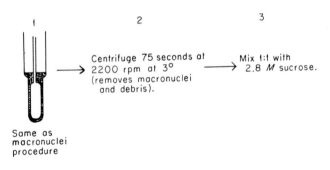

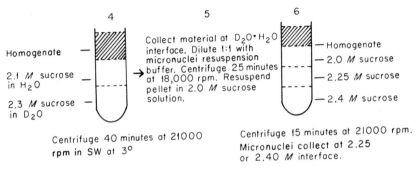

FIG. 4. Isolation scheme for obtaining micronuclei. The field in Fig. 3B was of the band collected from the 2.25–2.40 M sucrose interface in step 6. Most of the micronuclei collected at this interface, but some were retained at the 2.0–2.25 M interface. Macronuclei collected as a pellet and could be readily avoided. Other methods for separating micronuclei from macronuclei were discussed previously (12).

15×10^{11} daltons whereas macronuclear DNA had a value of only 3.6×10^{10} daltons. This value for macronuclear DNA agreed quite well with that obtained for *Paramecium* macronuclear DNA (27). Gorovsky has compared *Tetrahymena* micronuclei with macronuclei in several respects and has found differences in histone content (13), number of ribosomal RNA genes (28), and the amount of methylated bases in nuclear DNA (29). We were unable to demonstrate any differences in DNA methylation between *Paramecium* micro- and macronuclei (30). It is interesting to point out that for both *Tetrahymena* and *Paramecium*, the only methylated base found in the nuclear DNA was [6N]methyladenine. For all other eukaryotic nuclear DNA, 5-methylcytosine is the only base methylated, and this single finding illustrates the evolutionary importance of studying protozoan DNA.

III. Other Criteria for Selection of Method

A. Detergent Versus Nondetergent Methods

In Section II,B,3 I mentioned the use of gum arabic–octanol for the isolation of macronuclei from *Tetrahymena* and *Paramecium* (*13, 18*). Nozawa (*14*) divided methods for obtaining macronuclei from *Tetrahymena* into two categories: nondetergent (gum arabic–octanol) and detergent. He emphasized that detergent methods failed to reproducibly preserve nuclear membranes; moreover, he showed that nuclear membranes could be isolated from macronuclei obtained using gum arabic–octanol. In my earlier method using 0.55% detergents (Nonidet NP-40 plus sodium deoxycholate), I reported that *Paramecium* macronuclei retained their membrane but no quantitative studies were performed.

For this review I compared a variety of methods for the isolation of macronuclei from *P. aurelia*. I was interested in several factors: (1) which polyamines best stabilized the nuclei; (2) the effect of detergents and what agents could be used to preserve nuclear membranes; and (3) the use of gum-arabic with and without detergents. Table I summarizes the results using

TABLE I

ANALYSIS OF MACRONUCLEAR MEMBRANES
FROM *P. aurelia*, SPECIES 4

Method[a]	Percentage cell disruption	Membrane present		Membrane absent	Percentage with membrane
		Complete	Partial[b]		
I	97	1	6	467	1.5
II	97	0	5	342	1.5
III	97	9	33	318	12
IV	> 95	163	202	19	95

[a] The following methods were used for isolation of macronuclei:

I Packed cells resuspended in 2 volumes of 0.06% detergent–750 μg/ml spermine (tetrahydrochloride)–0.10 M raffinose–3 mM CaCl$_2$; homogenized in a Tri-R stirrer.

II Same as I except spermidine(trihydrochloride) used in place of spermine.

III Same as II except 0.10 M sucrose also present during homogenization.

IV Packed cells resuspended in 2 volumes of 6% gum arabic–1.5 mM MgCl$_2$–0.15 M sucrose–1 mg/ml spermidine. Just prior to homogenization, 0.065 ml/10 ml octanol added (*13*).

In each case after homogenization 1.5 volumes of 2.5 M sucrose–150 μg/ml spermidine–3 mM CaCl$_2$ (pH 7.0) were added, and the homogenate was layered onto 2.1 M sucrose [containing spermidine and CaCl$_2$ (pH 7.0)] and centrifuged for 6 minutes at 14,000 rpm in a Beckman SW27 rotor, decelerated with brake. Macronuclear pellets were fixed overnight in 6% glutaraldehyde–0.15 M sodium phosphate (pH 6.8) at 4°C.

[b] Partial membrane indicates that the membrane was not visible completely about the macronucleus. This was due primarily to the cutting angle in preparing sections.

P. aurelia, species 4, in which the method used in Fig. 2 was compared with the effect of additional sucrose and the gum arabic–octanol procedure. It is clear that our method I does not adequately preserve nuclear membranes and that protection by either spermine or spermidine exhibited little difference. Addition of sucrose led to an 8-fold increase in the number of nuclei with membrane attached, thus verifying my previous results (8). Utilization of the gum arabic–octanol procedure indicated that over 90% of the macronuclei had intact membranes, in agreement with Gorovsky *et al* (13), Skoczylas and Soldo (18) and Nozawa (14). These results are illustrated in Fig. 5.

Other parameters were examined using *P. aurelia*, species 1. Again, no difference was noted when either spermine or spermidine was used to stabilize the nuclei, and the same relative percentages of nuclei with intact membranes were obtained as with species 4. With method IV, it can be observed (Table II) that the presence of detergents per se had no effect on the preservation of nuclear membranes by gum arabic. Initial attempts to study these methods in the absence of either detergents or octanol indicated that while membranes were preserved (V and VI, Table II), cell disruption was not satisfactory. Cell disruption was improved by passage of the cell suspension through a milk homogenizer (12) rather than using the Tri-R stirrer, again with preservation of nuclear membranes (VII and VIII, Table II). However, passage through the milk homogenizer generated more macronuclear fragments than did homogenization in the Tri-R stirrer. No significant differences were noted in the yield of macronuclei when methods I through IV in Tables I or II were used. From these results, it can be concluded that gum arabic protects the nuclear membrane against the action of detergents more efficiently than does raffinose and/or sucrose. Gum arabic is a mixed polymer of arabinose, galactose, rhamnose, and glycuronic acid, and its specific action is not clear. Gorovsky *et al.* (13) mention that dextran sulfate can substitute for gum arabic, and this suggests that the preservation of nuclear membrane may be due to a preferential interaction with polymeric molecules rather than low molecular weight substitutes. This cannot be the entire story, however, since we have observed membranes on macronuclei isolated using only sucrose with octanol (D. J. Cummings, unpublished observations).

B. Effect of Cell Age

For other studies in my laboratory, the cells used in the methods summarized in Table II were recently isolated, exautogamous clones (1, 11). Aside from the presence or absence of nuclear membranes, the most striking observation noted was that there were two different types of macronuclei isolated from these young cultures (Fig. 6). About 19% of the macronuclei

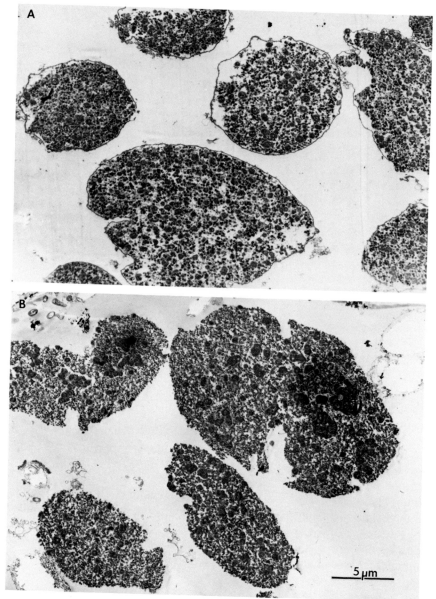

FIG. 5. Electron micrographs of macronuclei isolated from *P. aurelia*, species 4. (A) Method IV, Table 1, where gum arabic–octanol was used. Note the presence of the nuclear membrane in each nucleus. (B) Method I or II, Table 1, where detergents were employed to assist in cell lysis. Note the apparent absence of nuclear membranes. For these nuclei (and those in Fig. 1A and Fig. 6B) the large and small bodies were distributed throughout the nucleus. This contrasts with *Tetrahymena* macronuclei where the large bodies are often observed to be arrayed along the periphery of the nucleus (*17*).

DONALD J. CUMMINGS

TABLE II

ANALYSIS OF MACRONUCLEAR MEMBRANES
FROM *P. aurelia*, SPECIES 1

Method[a]	Percentage cell disruption	Membrane present		Membrane absent	Percentage with membrane
		Complete	Partial		
I	>90	15	12	184	12
II	>90	15	8	188	11
III	>95	238	26	32	89
IV	>95	208	9	14	94
V	<25	34	9	3	93
VI	<50	56	13	5	93
VII	>90	50	17	3	96
VIII	70–80	43	14	2	97

[a] I Packed cells resuspended in 2 volumes of 0.06% detergents–750 μg/ml spermine(tetra-hydrochloride)–0.25 M sucrose–3 mM CaCl$_2$. Homogenized in Tri-R stirrer without preincubation period.

II Same as I but 750 μg/ml spermidine(trihydrochloride) used.

III Packed cells resuspended in 2 volumes of 6% gum arabic–1.5 mM MgCl$_2$–1.5 mg/ml spermidine–0.15 M sucrose. Just before homogenization in the Tri-R stirrer, 0.065 ml/10 ml octanol were added.

IV Same as III except that detergents were added to 0.06% just prior to homogenization.

V Same as III but without octanol.

VI Same as II but without detergents.

VII Same as V but homogenized by one passage through a milk homogenizer.

VIII Packed cells resuspended in 2 volumes of 0.15 M raffinose–0.44 M mannitol–0.25% bovine serum albumin–1 mg/ml spermidine and homogenized by one passage through a milk homogenizer.

obtained from cells of fission age 28, and 8% (Table III) from cells of fission age 36, were distinguished by their lack of organization into large and small subnuclear particles. Instead, the chromatin material was distributed into an amorphous, fibril mass. Because of their occurrence in cultures of young fission age, these macronuclei were tentatively termed "immature." Most important, regardless of the method of isolation, all these immature macronuclei had well-preserved nuclear membranes (Fig. 6, Table III). The nature of these fundamental differences between mature and so-called immature macronuclei will demand considerable study before we can assess their role in aging.

FIG. 6. Electron micrograph of nuclei isolated from *P. aurelia*, species 1, at fission age 28. (A) A so-called "immature" nucleus. Note the absence of large and small bodies. This nucleus was isolated following detergent treatment, but nuclei isolated using gum arabic–octanol had an identical appearance. Two hundred and fifty such nuclei were recorded, and all had complete nuclear membranes regardless of the method of isolation. (B) A normal "mature" nucleus, isolated using the gum arabic–octanol procedure.

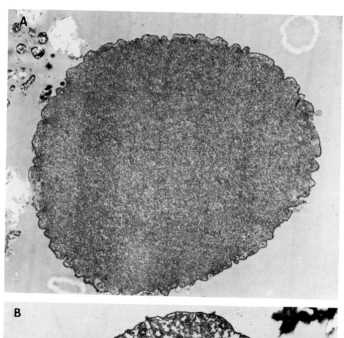

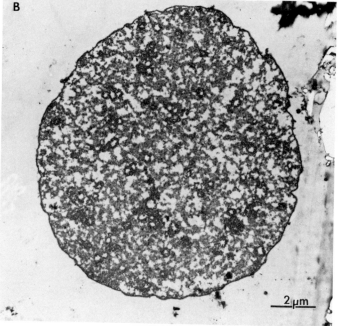

TABLE III

TYPES OF MACRONUCLEI PRESENT
IN *P. aurelia*, SPECIES 1

	Normal			"Immature"[c]		
		Percentage with membrane			Percentage with membrane	
Fission age	Number[a]	Detergent	Gum arabic	Number[a]	Detergent	Gum arabic
27–29	1175	2–12	92	227	100	100
35–37	249	2–12	94	23	—	100
>45	Many	2–12	—	N.O.[b]	—	—

[a] Several micrographs were taken at random, and the corresponding number of normal and "immature" macronuclei were tabulated. Sufficient micrographs were taken to obtain reliable percentages of macronuclei with and without a nuclear membrane.

[b] We have examined many thousands of macronuclei obtained from *P. aurelia*, species 1, of fission age 45 or greater and have never observed (N.O.) an "immature" nucleus.

[c] These "immature" nuclei were not included in the percentages calculated in Table II.

IV. Conclusions and Perspectives

I have compared a variety of methods for the isolation of nuclei from ciliated protozoans. Certain factors have become obvious. First the method for cell disruption can grossly affect the yield of nuclei as well as their composition. Preservation of nuclear membrane can vary considerably depending on the method utilized. However, this cannot be considered a critical factor in all cases. Witness the presence of nuclear membrane on immature macronuclei from *P. aurelia*, species 1, of young fission age regardless of the method of preparation (Table III). It may be that not all macronuclei are affected by detergents in the same way as those from *Tetrahymena* and *P. aurelia*. No such information was given on the isolation of macronuclei from *Oxytricha* (9) or *Stylonychia* (26) using detergents, but nuclear membranes may well have been preserved. The reaction of nuclei to different detergents has been shown to vary in other systems. For examples, Holtzman et al. (31) reported that a mixture of Tween-40 and sodium deoxycholate removed the membrane from HeLa cell nuclei, whereas Blobel and Potter (19) found that these detergents lysed rat liver nuclei. Penman (24) showed that HeLa cell nuclei lacking the outer nuclear membrane had little or no contamination with cytoplasmic ribosomes.

It is important to emphasize that the inability to demonstrate a nuclear membrane does not mean that the complete membrane is absent. In their studies with HeLa cell nuclei, Holtzman et al. (31) compared the effect of detergents on the morphology of isolated nuclei. They observed the presence of the typical bilayered membrane on nuclei isolated in the absence of detergent. This membrane was not present after detergent treatment, but a less distinct inner membrane was detected on some nuclei. We did not observe such a membrane on macronuclei isolated by use of detergents. However, these macronuclei retained their shape and integrity (see Fig. 3A and Fig. 5) even after extensive resuspension manipulations. Moreover, we were able to recover high molecular weight DNA (27) and DNA polymerase activities (20), and our compositional and kinetic complexity studies agreed with careful studies performed with other ciliates (9,13,18,26).

In conclusion, nuclei from ciliated protozoans can be isolated by a variety of methods. In some respects it is important to compare detergent versus non-detergent methods. Both have their advantages with regard to cytoplasmic contamination, preservation of membrane, in vitro enzyme activity, etc. The required use of these isolated nuclei dictates the choice of method.

ACKNOWLEDGMENTS

I want to thank V. A. Chapman, S. S. DeLong, and V. Sundararaman for their assistance with the electron and light microscopy. The original research reported here was supported by National Science Foundation Grant GB 42042.

REFERENCES

1. Sonneborn, T. M., in "Handbook of Genetics" (R. C. King, ed.), Vol. 2, p. 469. Plenum, New York, 1974.
2. Sonneborn, T. M., in "Handbook of Genetics" (R. C. King, ed.), vol. 2, p. 433. Plenum, New York, 1974.
3. Raikov, I. B., in "Research in Protozoology" (T-T. Chen, ed.), p. 1. Pergamon, Oxford, 1969.
4. Allen, S., and Gibson, I., Biochem. Genet. 6, 293 (1972).
5. Allen, S., and Gibson, I., in "Biology of Tetrahymena" (A. M. Elliot, ed.), p. 307. Dowden, Stroudsburg, Pennsylvania (distr. by Wiley, New York), 1973.
6. Ammermann, D., Chromosoma 33, 209 (1971).
7. Jurand, A. and Selman, G. G., "The Anatomy of Paramecium aurelia." Macmillan, New York and St. Martin's, New York, 1969.
8. Cummings, D. J., J. Cell. Biol. 53, 105 (1972).
9. Lauth, M. R., Spear, B. B., Heumann, J., and Prescott, D. M., Cell 7, 67 (1976).
10. Gorovsky, M. A., J. Protozool. 20(1), 19 (1973).
11. Sonneborn, T. M., Methods Cell Physiol. 4, 241 (1970).
12. Cummings, D. J., and Tait, A., Methods Cell Biol. 9, 281 (1975).
13. Gorovsky, M. A., Yao, M-C., Keevert, J. B., and Pleger, G. L., Methods Cell Biol. 9, 311 (1975).
14. Nozawa, Y., Methods Cell Biol. 10, 105 (1975).
15. Buetow, D. E., Methods Cell Biol. 13, 283 (1976).
16. Siebert, G., Methods Cancer Res. 2. 287 (1967).
17. Gorovsky, M. A., J. Cell. Biol. 47, 619 (1970).
18. Skoczylas, B., and Soldo, A. T., Exp. Cell Res. 90, 143 (1975).
19. Blobel, G., and Potter, V. R., Science 154, 1662 (1966).

20. Tait, A., and Cummings, D. J., *Biochim. Biophys. Acta* **378**, 282 (1975).
21. Chang, L. M. S., and Bollum, F. J., *J. Biol. Chem.* **246**, 5835 (1971).
22. Stevenson, I., *J. Protozool.* **14**, 412 (1967).
23. Lee, Y. C., and Scherbaum, O. H., *Biochemistry* **5**, 2067 (1966).
24. Penman, S., *J. Mol. Biol.* **17**, 117 (1966).
25. Isaacks, R. E., Santos, B. G., and Musil, G., *J. Protozool.* **20**(3), 477 (1973).
26. Ammermann, D., Steinbrück, G., von Berger, L., and Hennig, W., *Chromosoma* **45**, 401 (1974).
27. Cummings, D. J., *Chromosoma* **53**, 191 (1975).
28. Yao, M-C., Kimmel, A. R., and Gorovsky, M. A., *Proc. Natl. Acad. Sci. U.S.A.* **71**, 3082 (1974).
29. Gorovsky, M. A., Hattman, S., and Pleger, G. L., *J. Cell Biol.* **56**, 697 (1973).
30. Cummings, D. J., Tait, A., and Goddard, J. M., *Biochim. Biophys. Acta* **374**, 1 (1974).
31. Holtzman, E., Smith, I., and Penman, S., *J. Mol. Biol.* **17**, 131 (1966).

Chapter 6

The Isolation of Nuclei from Fungi

MYRTLE HSIANG AND R. DAVID COLE

Department of Biochemistry,
University of California, Berkeley,
Berkeley, California

I. Introduction

It is well known that histones are found in the chromosomes of higher eukaryotes but not in bacterial chromosomes. It is not at all clear, however, whether or not all primitive eukaryotes contain histones. Moreover the question of their presence in primitive eukaryotes has taken on added significance as our ideas of the structural roles of histones have begun to crystallize. Current views of chromatin (1–4) hold that it consists of a string of subunits each of which is made of about 200 base pairs of DNA wrapped around a core of histones. A common notion (3) is that the core contains two copies of each of four major kinds of histone: H2A, H2B, H3, and H4. These histones are nearly invariant in covalent structure from plants to animals, and the subunits of chromatin to which they give rise are also frequently thought to be uniform in structure, composition, and distribution. The implication of these views is that the functions of the subunit structure are common to all eukaryotes—functions such as mitosis, and especially the packing of large amounts of DNA into the restricted space of the nucleus. Obvious questions arise then if primitive eukaryotes, with condensed nuclei and more or less normal mitosis, lack histones as has been reported in some cases (5). Structural studies on chromatin from primitive eukaryotes, such as fungi, would seem to be a prime hunting ground for structure-function relationships of chromosomal proteins. Aberrations in mitosis (6), differentiation, etc., might be correlated with structures, occurrence, and distribution of proteins in a way analogous to the use of bacterial mutations in the identification of protein functions.

The preparation of nuclei from which to extract chromosomal proteins is essential for the establishment of their nuclear origin, and especially to avoid confusion between chromosomal and ribosomal proteins. To compare different stages in the cell cycle, the nuclear isolation needs to preserve as much as possible the native structural characteristics of the nuclei. The technical problems in procedures for nuclear isolation are by no means trivial, however. There are reports that some primitive eukaryotes do not contain histones, other reports that some contain incomplete complements of histone, and still other reports that some contain the normal complement (7). The disparate reports may reflect true differences between the occurrence of histones in the variety of organisms studied, or they may indicate that the success of isolation methods varies among organisms because of differences in proteolysis, solubility, etc. The problem of isolating fungal nuclei and chromatin might have yet another face; current methods might isolate proteins that appear to be histones when they are not. Leighton et al. (8) make a case that all of the reports of histones isolated from fungi (other than slime molds) actually describe ribosomal proteins. Because of such uncertainty we must not conclude at the present time that techniques are available for nuclear isolation that are applicable generally to all systems, and we cannot even be sure that any of them have been fully successful. We will consider here some selected techniques that have been used with some measure of apparent success for particular systems. "The Isolation of Yeast Nuclei and Methods to Study Their Properties" by John H. Duffus in Volume 12 of this series (9) reviewed techniques for yeast nuclei. The present chapter therefore will concentrate on other fungi, the true slime mold *Physarum polycephalum*, and the bread mold *Neurospora crassa*.

II.   Isolation of Nuclei from *Physarum polycephalum*

In the last few years, a steadily increasing number of scientists have concentrated their research on true slime molds, especially on *Physarum polycephalum*. The reason for this growing interest is based in part on its complicated and intriguing life cycle (10).

During the vegetative stage of *P. polycephalum*, many cells coalesce into a single giant cell containing up to several million nuclei. In this cell, called a *plasmodium*, all the nuclei divide synchronously (11). Upon starvation, the plasmodium undergoes differentiation into spherules (the diploid dormant stage) or into sporangia (the haploid spores). In the proper environment, the spores germinate to a haploid ameba (myxameba), or develop into a haploid dormant stage (the thick-walled cystamebaes). The fusion of two ameba, of different mating types, results in a zygote with a diploid nucleus, which will divide without concomitant cell division and give rise once more to the multinucleated plasmodium.

A. Isolation of Nuclei from Growing Plasmodia (3 to 6 Hours after Mitosis)

1. PROCEDURE A OF MOHBERG AND RUSCH (12)

1. About 5 gm of cell mass were obtained from one rocker-grown plasmodium or from 10 stationary Petri dish cultures (13) and homogenized in 200 ml of "control" medium (0.25 M sucrose–10 mM CaCl$_2$–10 mM Tris buffer (pH 7.2)–0.1% (w/v) Triton X-100) in a 1-litre Waring blender for 30 seconds at 70 V (high speed) in the cold room. The homogenate was kept on ice for 10–15 minutes for the foam to settle and then centrifuged 5 minutes at 50 g to remove the unbroken portions of plasmodium.

2. The supernatant was filtered by gravity through two thicknesses of milk filter (Rapid Flo, single gauze-faced, Johnson and Johnson, Chicago, Illinois) clamped in a Buchner funnel.

3. Then 20 ml of the filtrate were placed in a 50-ml conical polypropylene or polycarbonate centrifuge tube and underlaid with 10 ml of a solution of 1 M sucrose–10 mM CaCl$_2$–10 mM Tris (pH 7.5)–0.1% Triton, delivered slowly through a syringe needle. The interface between the two sucrose solutions was diffused slightly by gentle agitation with the syringe needle, and the tubes were centrifuged for 10 minutes at 50 g. After the withdrawal of 2 ml of solution from the bottom of the centrifuge tube with a wide-tipped pipette (to remove sediment and to prevent nuclei from forming a collar just above the bottom of the tube), nuclei were pelleted from the supernatant by centrifugation 10 minutes at 2200 rpm or 1,000 g.

4. The nuclei were resuspended in 80 ml of control medium by slow stirring (55 V) in a 250-ml blender cup. After underlaying 10 ml of 1M sucrose again, the resuspended nuclei were centrifuged again—first at 500 rpm for 10 minutes and then at 2200 rpm for 10 minutes over the underlaid 1 M sucrose solution.

2. PROCEDURE B OF MOHBERG AND RUSCH (12)

The procedure is identical to procedure II,A.1 at steps 1 and 2, but these steps are followed by different third and fourth steps.

3. The filtered homogenate was centrifuged at 2200 rpm for 10 minutes to pellet the nuclei.

4. The nuclear pellet was washed twice with 80 ml of control medium.

3. ISOLATION OF NUCLEI FROM GROWING PLASMODIA BY THE METHOD OF JOCKUSCH AND WALKER (14)

1. The harvested plasmodia from 4×20 ml of liquid culture were washed by gentle agitation in 200 ml of 0.25 M sucrose–10 mM EDTA (pH 7.2) for 5 minutes and collected by centrifugation.

2. The washed cells were resuspended in 0.25 M sucrose–10 mM Tris (pH 8.0)–10 mM MgCl$_2$–0.1% Triton X-100 and lysed in a blender (2 minutes, setting 0.5, 90–120V). The homogenate was centrifuged at 50 g for 5 minutes and filtered through two thicknesses of cheesecloth.

3. The nuclei were then sedimented through a cushion of 1 M sucrose –10 mM Tris (pH 8.0)–10 mM MgCl$_2$–0.1% Triton X-100 by centrifugation at 1000 g for 15 minutes.

B.   Isolation of Nuclei from Growing Plasmodia at Mitosis (12)

Two to four Petri dish plasmodia in prophase through telophase (13) were blended in 200 ml of control medium for 2 minutes at 45–50 V (low speed) or for 45 seconds at 60 V. The homogenate was then centrifuged and filtered according to either procedure II,A,1 or II,A,2.

C.   Isolation of Nuclei from Growing Myxamebas

1. Myxamebas which were grown on killed *E. coli* on Millipore membranes were harvested at 48 hours (logarithmic phase) or 72 hours (stationary phase). At least 20 million ameba cells were washed by suspension in a Vortex mixer in 200 ml of ice water and centrifugation at 2000 rpm for 10 minutes until the wash water was free of bacteria.

2. The amebas were resuspended in 200 ml of homogenization medium containing 0.4% Triton and blended at 70 V for 1 minute.

3. Nuclei were isolated from the homogenate by procedure II,A,1 for plasmodial nuclei, except that nuclei were pelleted by centrifugation at 2200 rpm for 20 minutes instead of 10 minutes.

D.   Isolation of Nuclei from Encysted Amebas (12)

1. Encysted amebas were broken by use of a French pressure cell as described below in Section II,F.

2. Nuclei were isolated from homogenate as in Section II,C.

E. Isolation of Nuclei from Microplasmodia and from Starving and Sporulating Plasmodia (12)

1. Microplasmodia from three 20-ml shaker cultures, or 15 Petri dish cultures of "starved" or "sporulating" plasmodia, were scraped into 200 ml of 0.25 M sucrose–10 mM EDTA (pH 7.5)–0.1% Triton and stirred for 1 minutes in a 1-liter blender cup at the slowest speed possible (homogenization at this step drastically reduced the yield of nuclei). Plasmodia were then sedimented at 1500 rpm for 5 minutes. This EDTA–Triton wash was necessary to remove part of the slime before the cell homogenization.

2. Microplasmodia were homogenized in the same way as growing plasmodia, but starving plasmodia were blended for 2 minutes at 70 V in homogenization medium with 0.4% Triton.

3. Nuclei were isolated by procedure II,A,1.

F. Isolation of Nuclei from Spherules (12)

The diploid dormant stage of spherules was produced by allowing microplasmodia to exhaust the nutrient medium. Microsclerotium, which is a mass of spherules, formed on the walls of the flasks.

1. The culture broth was aspirated from two 10-day-old shaken cultures, leaving the collar of microsclerotia on the walls of the flasks (the medium was too slimy to harvest by centrifugation). Microsclerotia were transferred to 200 ml of EDTA–sucrose–Triton wash solution, blended at 70 V for 1 minute to disperse slime and to destroy any remaining microplasmodia, centrifuged at 1000 rpm for 5 minutes, washed with 200 ml of cold distilled water, suspended in 80 ml of control medium, and given 10 passes in Potter–Elvehjem homogenizer to break microsclerotia and release spherules.

2. The suspension was passed twice through a French pressure cell at 8000 lb/in², the first time to break spherules and the second to strip cytoplasm from nuclei.

3. Then 120 ml of control medium and 6 ml of 10% Triton X-100 were added to the broken spherules suspension and the mixture was blended for 1 minute at 70 V.

4. Nuclei were isolated from the homogenate by procedure II,A,1 for plasmodial nuclei.

G. Comparison of the Methods

Interphase plasmodial nuclei prepared by the two procedures (II,A,1 and II,A,2) of Mohberg and Rusch have comparable DNA, RNA, protein, polysaccharide, and polyphosphate contents ($1:0.8:4.5:1.3:0.2$ weight ratios). Under the phase-contrast microscope there were occassional granules, probably due to polyphosphate. In electron micrographs, there were some membranous contaminants, along with threads of slime or fragments of broken nuclei, but there were very few if any mitochondria or free nucleoli. Therefore procedure II, A, 2, which omitted two of the two-step discontinuous sucrose density centrifugations, seems to be satisfactory and simpler. The RNA contents of chromatin preparations from *Physarum* are higher than those from mammalian preparations; the RNA/DNA ratio is 0.8 rather than about 0.05 (*15*). The high ratio does not seem to be due to ribosomal contamination since polyacrylamide gel electrophoresis of proteins extracted by acid or by $CaCl_2$ ($1 M$) (*16, 17*) did not reveal significant levels of ribosomal proteins. More likely the higher RNA/DNA ratio reflects the higher RNA content of *Physarum* nucleoli [RNA/DNA:6/1 (*12*)] as compared to a mammalian nucleus such as rat liver [RNA/DNA: 1.4/1 (*18*)].

A pleasing indication of the quality of the nuclei obtained by the methods quoted derives from an analysis of their histones. The histone complement is similar to that of higher eukaryotes, qualitatively and quantitatively, and there are no proteolytic degradation products detected in electrophoretic analysis. The high protein/DNA ratio is apparently due to nonhistone chromosomal proteins, two of which are actin and myosin (*19*).

Bradbury *et al.* (*20*) studied the effect of various divalent metal ions on the fragility of the nuclei. Concentrations of Ca^{2+}, Mg^{2+} and Zn^{2+} from 0 to 10 mM were used in procedure B of Mohberg and Rusch (described in Section II, A, 2). The stabilization of nuclei by Ca^{2+} and Mg^{2+} increased with ion concentration up to 10 mM. Ca^{2+} stabilized nuclei more than Mg^{2+} but Zn^{2+} (up to 10 mM) could not protect nuclei from disruption in even the most gentle procedure. They found that nuclei prepared in 1 mM Ca^{2+}, or in 1 mM Ca^{2+} plus 1 mM Mg^{2+}, or in 2 mM Mg^{2+} were all suitable for isolation of nucleoli, metaphase plates, and interphase chromatin because their membranes could be lysed easily by osmotic shock in 10 mM Tris (pH 7.2) with two strokes of a Teflon pestle. In contrast to the observation of Mohberg and Rusch, Bradbury *et al.* found that the histones isolated from 10 mM Mg^{2+} nuclei were qualitatively and quantitatively identical to those obtained from 10 mM Ca^{2+} nuclei. However, nuclei prepared from 2 mM Mg^{2+} lack the usual heterochromatic regions when viewed under the light microscope.

Jockusch and Walker (*14*) modified the Mohberg and Rusch procedure by substituting magnesium ion for calcium ion and raising the pH of the homogenizing buffer from 7.2 to 8.0. They also prewashed the plasmodia in EDTA–sucrose solution and sedimented the nuclei only once through 1 M sucrose solution. These modifications resulted in a considerable increase in the yield (based on DNA, the recovery of nuclei had increased from

40–50% to 60–70%). The nuclear membrane so prepared is much more fragile, and the nuclei can be gently lysed by suspending in 10 mM EDTA; this would be advantageous for those studies of chromatin structure where harsh chemical disruption should be avoided. The nuclear membranes prepared by the method of Mohberg and Rusch were highly resistant to lysis. Mohberg and Rusch (12) observed that histones prepared from nuclei that had been isolated in solutions containing magnesium are partially degraded and are obtained in low yield (50% or less). Hence, they recommended that Mg^{2+}-containing medium not be used in isolation of nuclei from *Physarum* for histone studies. However, the histone content of nuclei isolated by Jockusch and Walker (14) in Mg^{2+} medium is qualitatively and quantitatively identical to that of nuclei isolated by Mohberg and Rusch. It seems unlikely then, that Mg^{2+} in itself is deleterious. There are no ribosomal protein contaminants visible in the electrophoretic gel patterns of the histone extracts; therefore, the chance of ribosomal contamination probably is negligible (no RNA analysis was given). The prewash of the harvested plasmodia by EDTA–sucrose was reported to be essential for the modified procedure (to reduce polyphosphate, a potential contaminant of the chromatin and nuclei). Although Mohberg and Rusch observed occasional nuclear blebbing of growing plasmodia by such washes, essentially no loss of histones was observed by Jockusch and Walker from the chromatin so prepared. Of course, losses of nuclear proteins during such isolation procedures are difficult to exclude absolutely. Quantitation of nuclear proteins which are not chromatin-bound nor as tightly bound as histones would be more difficult to establish, and it is critical in considerations of how complete the patterns of nuclear proteins are. Quantitation of these proteins might be better with nuclei isolated from nonaqueous media (21), but the isolation of *Physarum* nuclei in nonaqueous media has not yet been accomplished.

III. Isolation of Nuclei from *Neurospora crassa*

A. Isolation of Nuclei from the "Osmotic" Mutant, Strain E 11200 (Also for Wild Type), by the Method of Reich and Tsuda (22)

1. Fifty-six grams of mycelium of *Neurospora crassa* were harvested by straining through two layers of cheesecloth; the harvest was washed extensively with distilled water, squeezed dry, washed with 0.35 M mannitol and blotted dry with paper.

2. The small chopped fragments were cooled to $4°$ and ground gently in a mortar with three times their weight (150 gm) of acid-washed sand until a smooth paste was obtained. Then 150 ml of $0.35 M$ mannitol was added and the paste was stirred into a thick suspension which was filtered through three layers of cheesecloth to yield 90 ml of an opaque orange fluid.

3. The filtrate was centrifuged at 500 g for 5 minutes in a refrigerated, 9-in Lourdes rotor. The central portion of sediment which was opaque and gelatinous was resuspended in its own supernatant; an outer white ring was discarded.

4. The suspension was centrifuged at 2000 g for 7 minutes to give a supernatant, a superficial sediment, and denser sediments. The supernatant was recentrifuged at 5000 g for 30 minutes to give a pellet that was suspended in $0.35 M$ mannitol along with the superficial and central sediments from the first centrifugation.

5. The combined sediments from above (crude nuclei) were pelleted by recentrifugation of the suspension at 3200 g for 7 minutes (resuspension resulted in 10% lysis of the nuclei). The yield is 10%.

B. Isolation of Nuclei from Wild-Type Strain 74 A by the Method of Dwivedi, Dutta, and Bloch (23)

1. The harvested and washed mycelium was washed with chilled solutions of sucrose–EDTA $[0.5 M$ sucrose–1 mM EDTA–10 mM Tris (pH 6.5)$]$ and squeezed dry.

2. The small chopped pieces were ground gently with two to three times their weight of fine, acid-washed sand. The paste was mixed with 5 volumes of sucrose–EDTA solution (pH 6.2) and strained, first through eight layers of cheesecloth and then through four layers of silk.

3. The filtrate was centrifuged at 2000 g for 30 minutes and the crude nuclei were pelleted. The yield based on DNA was not measured, but it was estimated to be less than 1%.

4. Nuclei were purified by suspension in 10 volumes of $0.5 M$ sucrose–5 mM $CaCl_2$–2 mM EDTA (pH 6.2) and centrifugation at 5000 g for 2 minutes. The suspension of the pellet was further purified by centrifuging through $1.65 M$ sucrose solution at 2200 g for 1 hour.

C. Isolation of Nuclei from Slime Mutant, No. 1118, by the Method of Hsiang and Cole (24)

1. The spheroplasts of the slime mutant were suspended in 7 volumes of $0.3 M$ sucrose–65 mM Na β-glycerophosphate (pH 7.0)–1 mM $CaCl_2$ and homogenized for 4 minutes in an Omni-Mixer at variac setting of 55 V. The

supernatant resulting from centrifugation at 700 g for 10 minutes was further centrifuged at 3000 g for 30 minutes to pellet the nuclei. The yield of the crude nuclei was 46% based on DNA.

2. The nuclei were resuspended by a Teflon pestle homogenizer in four times the original volume of the same buffer, then homogenized by six strokes in the homogenizer in the same buffer, 0.5% Triton X-100. The yield was 25% based on DNA.

D. Discussion

Neither of the two methods published for the isolation of nuclei from wild-type *Neurospora* is entirely satisfactory from the standpoint of yield; one gave less than 10% and the other gave 1%. The use of snail gut enzyme to obtain protoplasts as intermediates would be one approach to overcome the problem of breaking the *Neurospora* cell wall, but the preparation of protoplasts involves the incubation of the cells for some time under conditions in which *in vivo* protein degradation might take place. Nevertheless the method might have some use, especially in combination with the use of selective protease inhibitors.

The method of nuclear preparation from slime mutant as described by Hsiang and Cole (*24*) is simple, and it gives a reasonable yield (50% from crude nuclei, and 25% after washing the nuclei with Triton X-100). The RNA/DNA ratio (0.09) observed in this preparation seems favorable, approaching the ratios (0.03–0.09) generally found in higher eukaryotic systems (*15*). Neither the characterization of isolated nuclei nor optimization of concentrations of sucrose, divalent metal ions, chelators, and detergents in the preparation are completely worked out. The action of nuclease and protease inhibitors on the nuclear fine structure might also be profitably studied. With such refinements the yields of nuclei and the quality of the final chromatin might be still further improved.

The quality of the chromatin produced by these procedures is open to question. Although they might be straightforward representations of the endogenous composition, reports of absence of histone (*23*) or incomplete histone complements (*24*) in chromatins obtained by these methods can be questioned on the grounds that the chromatin might have been partially degraded. Yeast chromatin has been shown (*25, 26*) to contain H2A, H2B, and H4 histones, but neither H1 nor H3 was found; somewhat similarly, H2B and a variant of H2A were identified in *Neurospora* but the presence of H1, H3, and H4 was not demonstrated (*24*).

Whether or not the four histones H2A, H2B, H3, and H4 are present is of critical concern to current postulates of chromatin structure. While there are some data to the contrary (*27, 28*), a very common idea of the subunit

31. Vidali, G., and Neelin, J. M., *Can. J. Biochem.* **46**, 781 (1968).
32. Olins, A. L., Breillatt, J. P., Carlson, R. D., Senior, M. B., Wright, E. B., and Olins, D. E., *in* "The Molecular Biology of the Mammalian Genetic Apparatus—Its Relationship to Cancer, Aging and Medical Genetics" (P. T'so, ed.). Elsevier—North Holland Publ. Co, Amsterdam, 1977.
33. Panyim, S. Jensen, E. H. and Chalkley, R., *Biochim. Biophys. Acta* **160**, 252 (1968).
34. Fahrney, D. E., and Gold, A. M., *J. Am. Chem. Soc.* **85**, 997 (1963).
35. Cohen, J. A., Oosterbaan, R. A., and Berends, F., in "Methods in Enzymology," Vol. 11: Enzyme Structure (C. H. Werner Hirs, ed.) p. 686. Academic Press, New York, 1967.
36. Chevaillier, P., and Philippe, M. *Methods Cell Biol.* **10**, 69 (1975).

Part B.   Nuclear—Cytoplasmic Exchange

Chapter 7

Methods for Nuclear Transplantation in Amphibia

J. B. GURDON

Medical Research Council
Laboratory for Molecular Biology,
Hills Road, Cambridge, England

I.   Introduction

The initial reasons for developing the technique of nuclear transplantation were to test the genetic content of somatic cells and to analyze interactions between nucleus and cytoplasm. The technique is capable of contributing to our understanding of the function of chromosomal proteins partly on account of the gene reprogramming that results from nuclear transplantation, and partly because the same technique permits the injection of purified proteins or DNA into living cells. The potential applications of this technique are discussed in more detail in the last section of this article.

The earliest nuclear transplantation experiments were carried out on protozoa (*1*). The nucleocytoplasmic exchange of macromolecules has been extensively studied by this means in recent years by Goldstein (*2*) with *Amoeba*. The technique for nuclear transplantation in *Amoeba* has been described by Goldstein (*3*). After several previous attempts, the first substantial success in transplanting nuclei in multicellular organisms was achieved by Briggs and King (*4*), working with the amphibian *Rana pipiens*. Since then some success has been obtained with *Drosophila*; enucleated eggs injected with blastoderm nuclei have yielded normal adults, but normal development has not yet been promoted by the nuclei of postblastoderm determined cells. Details of methods for transplanting nuclei in insects have been described by Zalokar (*5*), Illmensee (*6*), and Okada, Kleinman, and Schneiderman (*7*).

Less success has been obtained with mammal eggs. It has not yet been possible to achieve normal development from a nucleus and enucleated egg. References to methods and results may be found in Graham (8) and Bromhall (9).

Among multicellular organisms, the large size and ready availability of the eggs and oocytes of amphibia give them great advantages for nuclear transplantation over those of other animals. The following account is therefore limited to a description of nuclear transplantation methods in amphibia. The techniques described here are applicable in detail to eggs and oocytes of *Xenopus laevis*. Minor differences in technique apply to eggs of *Rana pipiens* [see reviews by King (10, 11)].

The technical details given below should be sufficient for workers to carry out nuclear transplantation and the injection of macromolecules, but it must be emphasized that much practice is needed to obtain the best results. Manual skill is especially important for obtaining normal development from single nuclei in enucleated eggs and for injections into the germinal vesicle (nucleus) of oocytes. The injection of multiple nuclei into eggs (which are cultured for only a few hours) or into oocyte cytoplasm does not demand special dexterity.

Note that substantial differences exist in the methods required to transplant single as opposed to multiple nuclei and in the methods required to inject eggs as opposed to oocytes. The term *egg* refers to a single cell surrounded by jelly. The term *oocyte* refers to a growing (or fully grown) cell which is located in the ovary of a female where it is surrounded by several thousand follicle cells. Hormone injection [see Gurdon (12) for details] causes fully grown oocytes to be released from the ovary, passed down the oviduct where they are surrounded by jelly, and released into the surrounding water. Eggs, unlike oocytes, can be fertilized and can commence development.

II. The Preparation of Nuclei for Injection

A. Single Nuclear Transfers to Eggs

The standard procedure is as follows. A piece of tissue is isolated, and its cells are dissociated in a saline medium[1] lacking Ca^{2+} and Mg^{2+} but supplemented with $5 \times 10^{-4} M$ EDTA. Adult tissue cells or cultured cells may need to be dissociated in trypsin by standard procedures. Dissociated donor

---

[1] The exact composition of the most suitable saline medium differs according to species. Modified Barth Saline (MBS) is suitable for *Xenopus*; its composition is given in Table I.

TABLE I

COMPOSITION OF MODIFIED BARTH SALINE (MBS) SOLUTION

Component	gm/liter	Final concentration (mM)
NaCl	5.13	88.0
KCl	0.075	1.0
NaHCO$_3$	0.20	2.4
MgSO$_4$ · 7 H$_2$O	0.20	0.82
Ca(NO$_3$)$_2$ · 4 H$_2$O	0.08	0.33
CaCl$_2$ · 6 H$_2$O	0.09	0.41
HEPES buffer (+ NaOH)	2.38	10
Streptomycin sulfate	0.01	—
Benzyl penicillin[a]	0.01	

[a] Not stable in solution for more than a few days.

cells may be kept for over half an hour in MBS (Table I) lacking Ca^{2+} and Mg^{2+} but containing 0.1% bovine serum albumin (BSA); an agar-covered or siliconized glass slide is used.

B. The Transfer of Multiple Nuclei to Oocytes or Eggs

For experiments involving the biochemical analysis of the activities of transplanted nuclei, it is very desirable to have many nuclei in one oocyte or egg. Eggs injected with multiple nuclei do not develop normally, because cleavage is necessarily irregular as a result of the multiple spindles that are formed around many nuclei. However, the initial responses of nuclei to egg cytoplasm are normal, even when several hundred nuclei are injected. This has proved particularly valuable in studies on the initiation of DNA synthesis. To study the transcriptional activity of transplanted nuclei, it is useful to transplant many nuclei into an oocyte. Oocytes are very stable cells capable of remaining morphologically and synthetically active for several weeks in culture. Nuclei injected into oocytes do not divide, and there is therefore no dilution of the chromosomal proteins and other molecules present in injected nuclei, a situation which does not apply to nuclei injected into eggs.

Multiple nuclei can be injected by sucking many cells into a pipette as for single nuclear transfers, but this is an inefficient method. Recently a detailed study has been made of procedures for making nuclear suspensions from which several hundred nuclei may be conveniently injected into a single oocyte or egg. Nuclei were prepared by different methods and injected into oocytes which were incubated for several days. They were then labeled with [^{3}H]uridine, fixed, sectioned, and autoradiographed. The different methods of preparing nuclei were then related to the morphological appearance, and RNA-synthesizing activity, of the injected nuclei. The conclusion from this study (*13*) was that the following lysolecithin method is generally applicable to all cells that can be prepared as a single cell suspension and that it gives better results than other methods, as judged by the criteria mentioned above. The use of detergents and homogenization procedures gives particularly bad results.

C. The Lysolecithin Procedure for Preparing Nuclei to Inject into Oocytes or Eggs [Details from Gurdon (*13*)]

1. One million cells are collected from a suspension culture or trypsinized from a bottle, excess serum is added to inactivate the trypsin, and the cells are pelleted.

2. The cells are resuspended in 0.5 ml of 0.25 M sucrose–75 mM NaCl–0.5 mM spermidine trihydrochloride–0.15 mM spermine tetrahydrochloride at 18°–24°C.

3. Ten micrograms of lysolecithin are added (10 μl of a 1 mg/ml solution).

4. Then the cell suspension is swirled by hand every 15 seconds for a total of 90 seconds at 18°–24°C. Cells are lysed at this step and are referred to as nuclei.

5. One milliliter of the solution specified at step 2 above but supplemented with 3% BSA, at 0°C, is added and mixed into the nuclear suspension.

6. The nuclei are pelleted and resuspended in 20 μl of the solution specified at step 5.

7. About 2 μl of the nuclear suspension are added to 10 μl of 0.2% trypan blue in phosphate-buffered saline, and the percentage of lysed cells (seen as blue-stained nuclei) is scored. This should be between 95 and 99%.

8. The nuclear suspension is transferred as a drop onto a siliconized glass slide which is kept on ice, where the nuclei may serve as a source for injection for at least half an hour.

III. Enucleation

In many experiments the nucleus of the oocyte or egg must be removed before injecting other nuclei. This is obviously true, for example, for any developmental test of a somatic nucleus transplanted to an egg. It is also useful to be able to find out whether the activity of nuclei or macromolecules injected into oocytes is dependent on, or independent of, the presence of the oocyte nucleus (germinal vesicle).

A. Unfertilized Eggs

The nucleus of an unfertilized egg is present as metaphase chromosomes on a spindle (in second meiotic metaphase). The spindle and chromosomes are located at the animal pole of the egg, and are usually visible as a minute dark spot within a light-colored area in the center of the black animal hemisphere. A needle can be used to pull out the region of cytoplasm in this area, including the spindle and chromosomes. However, ultraviolet (UV) irradiation provides a much more convenient method of enucleation than this manual technique, and it has been applied to the eggs of *Xenopus* (14), *Pleurodeles* (15), and *Amblystoma* (16).

For enucleating *Xenopus* eggs, the following procedure is recommended. Unfertilized eggs are collected within 10 minutes of oviposition in tap water. Four eggs are placed, without water, at the corners of a 10 mm^2 piece of glass slide. Each egg is carefully oriented so that its animal pole is facing directly upwards. It remains in this position sticking to the glass by its jelly. All free water must be removed from around each egg so that it cannot move during irradiation. The piece of glass carrying the eggs is transferred to a standard-sized microscope slide and placed under a downward-shining beam of UV light. After the desired exposure, the piece of glass carrying the eggs is moved to the microscope stage for transplantation. During irradiation the jelly is partly dissolved by the UV light and water is released. Therefore, after irradiation any egg which has rotated or moved so that its animal pole is no longer facing directly into the UV beam is rejected.

For many years this laboratory has obtained reliable enucleation with the UVS-100 medium-pressure mercury arc lamp supplied by Hanovia (Slough, England). The light is set up as shown in Fig. 1A. It is essential to note that all filters, diaphragms, and lenses that may be supplied are removed, and that any finger prints etc. should be completely cleaned off the quartz arc tube. The arrangement shown in Fig. 1A provides a fairly evenly illuminated area of about 3 cm in diameter in which the eggs must be placed for irradiation. If the amount of light in the illuminated area is measured with a black-ray UV meter (Ultra-violet Products, Inc., San Gabriel, California, model J-225), a reading of 12,000 ergs/sec/cm^2 is obtained. This is probably an underestimate of the dose received by eggs because the window of the monitor is larger than the evenly lit area and is quite unequally sensitive to UV light over the window area. The dose of Hanovia UV light (arranged as described) needed for enucleation in usually a little more than that required to make unfertilized eggs readily penetrable to a micropipette (see below). This dose will vary considerably from one batch of eggs to another according to the amount and composition of their jelly, but it is usually in the range of 20–40 seconds. (See Note Added in Proof, p. 139.)

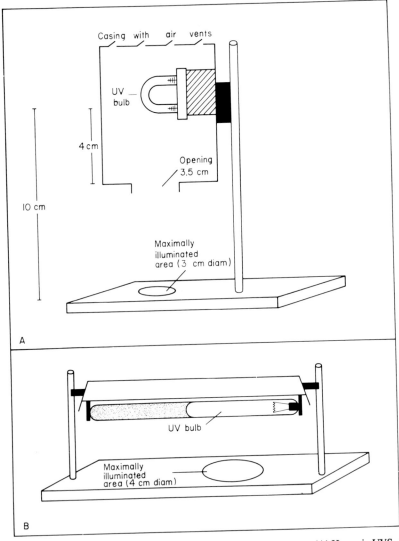

FIG. 1.   Diagrams of UV light apparatus suitable for *Xenopus* eggs. (A) Hanovia UVS–100; this light causes penetrability of the egg jelly and membranes, and destruction of the egg pronucleus. (B) Mineralite UVSL–15; this light causes destruction of the egg pronucleus, but has no penetrability effect.

The UV light is actually performing two independent functions. First, it dissolves part of the jelly and makes the jelly and membranes surrounding an egg easily penetrable by a 25-μ micropipette. Nonirradiated unfertilized eggs are badly damaged by the penetration of a pipette of this size. Second, the UV light destroys the egg chromosomes. The wavelength of irradiation required for these two functions appears to be different, since other UV sources, such as the mineralite (model UVSL-15, from Ultraviolet Products, Inc., Andover Road, Winchester, England), have the second but not first of these effects. It is likely, but not proved, that enucleation is most effective at a wavelength of 2540 Å and that penetrability (including jelly dissolution) is promoted most efficiently by light of a different (possibly lower) wavelength. It is important to bear in mind that the efficiency of a UV bulb may decrease with age and that the enucleation and penetrability activities may not decrease coordinately. Once an egg has received UV for penetrability, it must be injected and immersed in medium within 2 minutes or its surface will become dry and it will die.

Occasionally eggs laid by some females are not efficiently enucleated by the above procedure, probably because the composition of their jelly (or extraneous, attached materials) shields the UV light from the egg nucleus. If this happens, a supplementary dose of enucleating UV may be provided by exposure to light from the UVSL-15 mineralite source just referred to. All covering is removed from one side of the tubular bulb, which is arranged as shown in Fig. 1B. This set-up provides an area (about 4 cm in diameter) of maximum illumination from which a reading of 16,000 ergs/sec/cm² is recorded. Since this light has no dejellying (penetrability) activity, eggs may be exposed to it for a much longer time than is possible with the Hanovia light. We find that eggs tolerate up to 2 minutes exposure to this lamp arranged as in Fig. 1B. For nuclear transplantation, eggs must *first* be exposed to the mineralite and *then* to the Hanovia sources to make them penetrable. An effective procedure is to expose eggs to the mineralite for 30–60 seconds and then to give them the minimum penetrability dose (5–10 seconds) with the Hanovia light.

The effectiveness of UV sources is most conveniently tested by irradiating eggs 5 minutes after fertilization. Successful enucleation gives haploid embryos. Artificial fertilization has been described by Wolf and Hedrick (17) and by Gurdon and Woodland (18).

B. Oocytes

Methods of enucleating oocytes have been described by various authors, for example, Dettlaff et al. (19) and Smith and Ecker (20), etc. However, the methods so far described do not appear to give a permanent and complete healing of the injection wound. Recently Ford and Gurdon (21) have described an improved method which gives sufficiently good healing for enucleated oocytes to survive for several days, and even for nuclei to be injected after enucleation, with subsequent survival for 4 days or more.

C. Enucleation of Oocytes [See (*21*) for details]

1. Oocytes in MBS are removed from an ovary and separated into single large oocytes with associated follicular material. The follicular material is carefully peeled off each oocyte by hand using finely pointed forceps.

2. Defolliculated oocytes are transferred to one-half strength MBS (1:1) with H_2O). A syringe needle (26G$\frac{1}{2}$, 13/45) is inserted by hand into the animal pole of the oocyte at an angle of 45° and withdrawn leaving a slit of about one-fifth of the diameter of the oocyte.

3. The oocyte is squeezed equatorially with forceps until the translucent germinal vesicle partly emerges from the slit. It is then allowed to squeeze itself out, a process which is complete in about 20 minutes.

4. As soon as the germinal vesicle has separated from the oocyte, the oocyte is transferred to a healing medium (10 mM NaCl–1 mM MgSO$_4$– 90 mM K$_2$HPO$_4$ and KH$_2$PO$_4$ to give a pH of 7.2) for 1$\frac{1}{2}$ hours at 19 °C.

5. Oocytes (whose enucleation slits should now be completely healed) are then transferred to MBS and cultured further, injected with nuclei, etc.

IV. The Transplantation Procedure

A. Equipment

After the preparation of donor cells or nuclei, and recipient eggs or oocytes, the third step is the transplantation of the former into the latter. In view of the large size of eggs and mature oocytes (1 mm diameter), very simple equipment can be used (Fig. 2). The procedure is carried out under a stereomicroscope at a magnification of about ×25. The microscope should have a strong spotlight and variable-intensity understage lighting. The micropipette for transplantation is held in a simple manipulator. The most convenient type is one which gives a 5-fold reduction in the movement of the hand, in *all* directions—e.g., the microdissector of Microinstruments Limited (7 Little Clarendon Street, Oxford, England; cost $250). With such an instrument, the micropipette can be inserted into an egg or an oocyte at a 45° angle. The pipette is connected, by a push fit, to 1 mm plastic tubing, the other end of which is attached to a screw-controlled syringe. A suitable syringe is the "Agla" (micrometer syringe outfit, Wellcome Reagents, Beckenham, Kent, England; cost $50) to which a spring has been fitted around the piston. The whole syringe, plastic tubing, and most of the pipette are filled with light paraffin oil (0.850 gm/ml), as shown in Fig. 2.

The design of the micropipette is of great importance (Fig. 3). For oocytes,

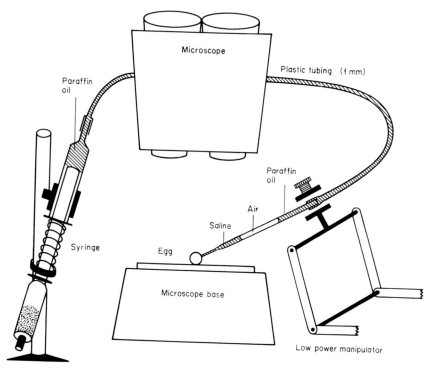

FIG. 2.   Diagram of a nuclear transplantation outfit. A syringe is connected by oil-filled tubing with a micropipette, which is held in a manipulator and viewed through a stereomicroscope, as explained in the text.

the thinnest portion of the pipette must not be long, since it is otherwise too flexible to penetrate the follicle cell layers. On the other hand, for injecting single nuclei to eggs, the thinnest portion of the pipette should be long and without any taper. For single nuclear transfers the shape of the pipette orifice is critical. It must be smooth in contour so as not to tear the donor cell, but must also be sharp so as not to damage the recipient egg. The optimal shape and dimensions of micropipettes are shown in Fig. 3. We have assembled from commercially available components a microforge by which pipettes of this type can be easily made. Details of the microforge assembly have been published (*21a*) as has the technique by which micropipettes are made from inexpensive 1 mm melting-point glass tubing (*22*).

B.   Procedure

The transplantation of single nuclei to eggs is carried out as follows. A whole single cell is sucked into a pipette whose internal diameter is just

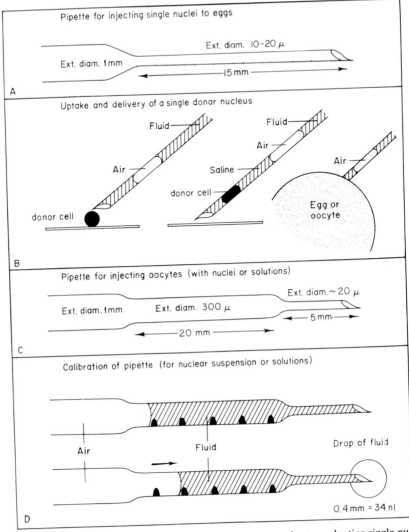

FIG. 3. Construction of glass micropipettes. (A) A pipette for transplanting single nuclei to unfertilized eggs. (B) Operation of the single-nucleus transfer pipette and its air-bubble marker. (C) A pipette for injecting oocytes with solutions or with suspensions of nuclei. (D) Calibration of the oocyte pipette shown in C.

smaller than that of the donor cell, so that the cell is ruptured but its nucleus remains intact and surrounded by cytoplasm. The whole broken cell is injected into the recipient egg. Enucleated eggs will cleave only if a nucleus is injected; donor cell cytoplasm on its own has no effect. In most experiments the volume of cytoplasm injected is less than 10^{-5} of the volume of the egg cytoplasm. The injection of whole cells has no effect because the donor nucleus is effectively isolated from the recipient egg cytoplasm by its cell wall.

The transplantation of single nuclei to eggs requires some skill and practice. The difficulty lies in judging how *little* donor cell distortion has a good probability of just breaking the cell but not of removing too much cytoplasm from around the nucleus. The optimal amount of distortion is indicated in Fig. 3B. Once the donor cell is in the end of the pipette, the pipette is inserted into the egg so that its tip will be located slightly above (nearer the animal pole than) the center of the egg. Since yolk and pigment render eggs (and oocytes) opaque, the position of the tip of the pipette has to be guessed. For the same reason the donor nucleus cannot be seen, and an air bubble is taken into the shaft of the pipette to control the injection of the donor nucleus by the syringe (Fig. 3B). It is most convenient if the donor cells are kept on an agar-covered slide on a microscope stage throughout the procedure and if the UV-irradiated eggs on another slide are moved onto and off the stage as required.

For experiments with oocytes, the appropriately shaped pipette (Fig. 3C) is inserted while the oocyte is steadied with forceps. The recipient oocyte (or group of oocytes) must be placed as dry as possible on a slide, since any fluid will cause them to move out of the way of the pipette, even when held by forceps. When injecting suspensions of nuclei, it is necessary to calibrate the injection pipette (Fig. 3D). This may be done by making small marks with a glass-writing pen on the wider section of the pipette shaft, a procedure which is very rapid and is accurate enough for most purposes. Fluid is squeezed out from the pipette so that the meniscus moves from one mark to the next; the volume of the space between the two marks is determined from the diameter of the fluid drop which hangs from the end of the pipette [see Gurdon (22) for details]. A suitable volume for injection is 40–60 nl, in which up to 1000 nuclei may be contained. When a pipette is filled to the top of the marked gradations, it can be used to inject 4–6 oocytes before refilling.

When injecting oocytes, it is possible, with practice, to inject *into* the germinal vesicle (nucleus), as described by Gurdon (13). To do this the pipette is inserted at a right angle to the surface of the oocyte exactly over the center of the animal hemisphere. The pipette is inserted to about half the depth of the oocyte. At the same time the oocyte is squeezed gently with forceps. By this procedure nuclei can be deposited in the germinal vesicle (which subsequently disperses its contents) in over 70% of all attempts. It is very easy to inject into the *cytoplasm* only of an oocyte by entering equatorially and aiming well below the germinal vesicle which lies near the animal pole.

Injection of eggs or oocytes is greatly facilitated if the person doing injections is assisted by someone else who puts eggs or oocytes onto a slide and UV-irradiates the eggs. About 100 cells/hour can be injected under these conditions.

V. Culture

Eggs are relatively unaffected by the composition of their medium. The one specified in Table I (MBS) is suitable. It is important to note that cleaving embryos derived from UV-dejellied eggs must be transferred to hypotonic solution (e.g., 1/10 strength MBS), since gastrulation is abnormal in full-strength solution.

Oocytes are more sensitive to their culture medium. In spite of numerous attempts [Colman (23)], no culture medium has been found which offers any improvement over MBS for short-or long-term culture. For example, the culture medium best suited to cultured line amphibian cells (e.g., 7/10 Dulbecco's modified Eagle's medium) results in the deterioration of oocytes within a few days, whereas they can survive and remain metabolically active for over 1 month in MBS.

VI. Labeling

Eggs are largely impermeable to labeled molecules which are incorporated into nucleic acids or proteins. However, labeled molecules can be conveniently injected at any stage because the UV treatment given for nuclear transplantation renders the jelly and vitelline membrane penetrable by a pipette. Labeled precursors should be dried down and taken up in 90 mM NaCl–15 mM Tris-HCl (pH 7.5). If 50 nl of label are injected into embryos at 5 mCi/ml, several thousand cpm per embryo can be incorporated into RNA, and very much more into protein in a few hours.

Oocytes can be conveniently labeled by incubation in MBS containing amino acids or nucleosides, but nucleotides must be injected. If an oocyte is incubated in 5 μl of [^{35}S]methionine at 1 mCi/ml, the oocyte will incorporate 10^6 cpm into protein within a few hours. Incubation in [^{3}H]uridine or [^{3}H]guanosine labels RNA well, but 90–95% of the labeled RNA is in the follicle cell layers. About 2–3 $\times$ 10^3 cpm of RNA can be labeled in a single *fully defolliculated* oocyte in 24 hours.

VII. Applications of Nuclear Transplantation Techniques in Amphibia to the Study of Chromosomal Proteins

There are three applications of nuclear transplantation methods to the study of chromosomal proteins. The first is related to the major changes in chromosomal function that take place when nuclei are transplanted to eggs or oocytes. For example, the nucleus of a specialized adult skin cell is reprogrammed after transfer to eggs in the sense that its mitotic products (daughter nuclei) participate in muscle, nerve, lens cell, etc., differentiation, and therefore express genes that are not expressed at all in skin cells (24). It has been shown that immediate changes in chromosomal function take place soon after nuclear transplantation; this applies, for example, to the induction of DNA synthesis (25) and the repression of ribosomal RNA synthesis (26). These changes are accompanied by the loss of proteins from transplanted nuclei (27,28) and by the accumulation of proteins from the egg cytoplasm in transplanted nuclei (29). These changes in nuclear activity and other associated events including nucleocytoplasmic protein exchange have been reviewed (22). There are few, if any, other experimental conditions under which such a major reprogramming of nuclear activity takes place. The hope is therefore eventually to describe in detail which chromosomal proteins are lost and which cytoplasmic proteins are taken up at the time of nuclear reprogramming, and hence to gain some idea of the function of the proteins involved. It is not at all easy, however, to follow the movement of *individual* proteins under these conditions.

The second application of amphibian nuclear transfer techniques aims to get round this difficulty by using the activity of transplanted nuclei as a test system in which to assay, under living conditions, the activity of purified chromosomal proteins. For this purpose oocytes are more suitable than eggs, whose nuclei are not transcriptionally active until mid–late cleavage stages. In principle, the endogenous activity of oocyte chromosomal genes could be used to assay the function of injected proteins. But this is not realistic in practice, because individual gene expression must be recognized by newly synthesized proteins, and any change in gene activity at the transcriptional level will not be easily detected over the large background of proteins synthesized from the accumulated stock of stable messages in oocytes. Therefore, a much more sensitive assay is provided by injecting multiple nuclei into oocytes in such a way that they remain transcriptionally active. The amount of mRNA carried over with the injected nuclei is small in relation to the amount synthesized and translated into protein as a result of new gene transcription. In this way it would be possible to inject purified proteins and to detect their effect on known gene transcription. An experimental system in which these requirements are fulfilled has been described by Gurdon, De Robertis, and Partington (30). It was shown that histones and cytoplasmic nonhistone proteins rapidly enter and accumulate in the injected test nuclei. Probably all chromosomal proteins prepared in a native condition would thus gain access to the test genes whose activity they may regulate.

This third application of nuclear transplantation methods to the study of chromosomal proteins involves the substitution of pure DNA for whole nuclei. If pure DNA could be injected into eggs or oocytes in such a way that replication and transcription take place, this would open up two direct means of investigating the function of chromosomal proteins. On the one hand, purified proteins could be associated *in vitro* with DNA or preferably with defined segments of DNA (for example, cloned genes), and the effect of the associated protein could be assayed under natural *in vivo* conditions. Alternatively, purified DNA could be injected into eggs or oocytes, which would be incubated for the amount of time needed for the DNA to express an activity. The injected DNA would then be reextracted with proteins and other macromolecules associated with it. If injected DNA performs different activities according to whether it is in oocytes or eggs, this might make it possible to recover DNA-associated proteins responsible for different functions. Initial experiments with DNA injected into eggs have succeeded in providing strong evidence for replication (31,32). More recently, DNA of particular kinds (ribosomal DNA) has been injected into fertilized eggs, and transcripts complementary to the injected DNA have been recovered from the resulting blastulae (33). SV40 transcripts have been found in oocytes in which SV40 DNA was injected into the nucleus (germinal vesicle), but little, if any, effect is observed if DNA is injected into the oocyte cytoplasm (34). The technique for injecting DNA into eggs and oocytes is as described for multiple nuclear transfers. However, to achieve a substantial success in injecting DNA into the germinal vesicle, a lot of pratice is required.

In conclusion, the most immediately useful applications of nuclear transplantation methods for investigating the function of chromosomal proteins will probably involve multiple nuclear transfers and the injection of purified genes or chromosomal proteins.

NOTE ADDED IN PROOF

Since writing this article, Dr. H. Woodland informs me that the original specification of the Hanovia UVS-100 bulb has changed, and does not have the same biological effects as the type of bulb originally tested (14). Hanovia is currently able to supply bulbs of the original specification (described as UVS–100, *ozone producing*, Part number 60261) at about $60 each. The modified bulbs which do not produce ozone give a much weaker emission at low wavelengths.

References

1. Comandon, J., and De Fonbrune, P., *C. R. Seances Soc. Biol. Ses Fil.* **130**, 740 (1938).
2. Goldstein, L., *Exp. Cell Res.* **85**, 159 (1974).
3. Goldstein, L., *Methods Cell Physiol.* **1**, 97 (1964).
4. Briggs, R., and King, T. J., *Proc. Natl. Acad. Sci. U.S.A.* **38**, 455 (1952).
5. Zalokar, M., *Proc. Natl. Acad. Sci. U.S.A.* **68**, 1539 (1971).
6. Illmensee, K., *Wilhelm Roux' Arch. Entwicklungsmech. Org.* **170**, 267 (1972).
7. Okada, M., Kleinman, I. A., and Schneiderman, H. A., *Dev. Biol.* **37**, 43 (1974).
8. Graham, C. F., *Adv. Biosci.* **8**, 263 (1971).
9. Bromhall, D., *Nature (London)* **258**, 719 (1975).
10. King, T. J., *Methods Cell Physiol.* **2**, 1 (1966).
11. King, T. J., *in* "Methods in Developmental Biology" (F. H. Wilt and N. K. Wessells, eds.), p. 737. Crowell, New York, 1967.
12. Gurdon, J. B., *in* "Methods in Developmental Biology" (F. H. Wilt and N. K. Wessells, eds.), p. 75 Crowell, New York, 1967.
13. Gurdon J. B., *J. Embryol. Exp. Morphol.*, **36**, 523–540 (1976).
14. Gurdon, J. B., *Q. J. Microsc. Sci.* **101**, 299 (1960).
15. Signoret, J., and Picheral, B., *C. R. Hebd. Seances Acad. Sci.* **254**, 1150 (1962).
16. Signoret, J., Briggs, R., and Humphrey, R. A., *Dev. Biol.* **4**, 134 (1962).
17. Wolf, D. P., and Hedrick, J. L., *Dev. Biol.* **25**, 348 (1971).
18. Gurdon, J. B., and Woodland, H. R., *in* "Handbook of Genetics" (R. C. King, ed.), Vol. 4, pp. 35–50. Plenum, New York, 1975.
19. Dettlaff, T. A., Nikitina, L. A., and Stroeva, O. G., *J. Embryol. Exp. Morphol.* **12**, 851 (1976).
20. Smith L. D., and Ecker, R. E., *Dev. Biol.* **19**, 281 (1964).
21. Ford, C. C., and Gurdon, J. B., *J. Embryol. Exp. Morphol.*, **37** (in press).
21a Gurdon, J. B., *Lab. Practice*, 63–64 (1974).
22. Gurdon, J. B., "The Control of Gene Expression in Animal Development." Oxford Univ. Press, London and New York, and Harvard Univ. Press, Cambridge, Massachusetts, 1974.
23. Colman, A., *J. Embryol. Exp. Morphol.* **32**, 515 (1974).
24. Gurdon, J. B., Laskey, R. A., and Reeves, O. R., *J. Embryol. Exp. Morphol.* **34**, 93 (1975).
25. Graham, C. F., Arms, K., and Gurdon, J. B., *Dev. Biol.* **14**, 349 (1966).
26. Gurdon, J. B., and Brown, D. D., *J. Mol. Biol.* **12**, 27 (1965).
27. Gurdon, J. B., *Proc. R. Soc. London Ser. B.* **176**, 303 (1970).
28. Di Berardino, M. A., and Hoffner, N. J., *Exp. Cell Res.* **94**, 235 (1975).
29. Merriam, R. W., *J. Cell Sci.* **5**, 333 (1969).
30. Gurdon, J. B., De Robertis, E. M., and Partington, G. A., *Nature (London)* **260**, 116 (1976).
31. Gurdon, J. B., Birnstiel, M. L., and Speight, V. A., *Biochim. Biophys. Acta* **174**, 614 (1969).
32. Laskey, R. A., and Gurdon, J. B., *Eur. J. Biochem.* **37**, 467 (1973).
33. Gurdon, J. B., and Brown, D. D., *Symp. Mol. Biol. Genet. Apparatus* **2**, 111–123.
34. Mertz, J., and Gurdon, J. B., *Proc. Natl. Acad. Sci., U.S.A.* (in press).

Chapter 8

Methods for Studying Nucleocytoplasmic Exchange of Nonhistone Proteins in Embryos

MARIE A. DIBERARDINO, NANCY J. HOFFNER, AND
MICHAEL B. MATILSKY

Department of Anatomy,
The Medical College of Pennsylvania,
Philadelphia, Pennsylvania

I. Introduction

One of the central problems that has long concerned developmental biologists has been the manner in which nucleocytoplasmic interactions determine cell differentiation during embryogenesis. One view currently held proposes that cytoplasmic determinants enter embryonic nuclei and activate particular genes. The most likely candidates that could control both the activation and specificity of genomic transcription during embryogenesis are the nonhistone proteins, whose function in this respect has been demonstrated in a number of eukaryotic systems employing adult cells (1–4). During embryogenesis such proteins could originate from those inherited from oogenesis (5), those synthesized in the embryo on maternal templates, and those translated on new templates made by the embryonic genome (6). As embryogenesis proceeds, cell differentiation would be characterized by the possession of specific structural as well as specific regulatory proteins. The latter could lead to the establishment of cell-specific types by controlling genomic transcription and also could confer stability on the genome in the differentiated state.

With essentially this view as a possible model for the initiation and control of cell differentiation, we recently initiated studies directed toward elucidating the role of nonhistones in nuclear differentiation during early embryogenesis (7). The conceptual framework under which these experiments were conducted was as follows. If particular nonhistones contribute to the process of nuclear differentiation, their accumulation in the nucleus should be restricted to particular times in the developmental process. If one of these temporal events could be determined, then the importance of these non-histones to the process of nuclear differentiation could be tested by transplanting these nuclei into the cytoplasm of enucleated eggs. Such proteins, if important to nuclear differentiation, would be expected to migrate from the transplanted nucleus into the cytoplasm of the egg during the first cell cycle, when transplanted nuclei are induced to undergo DNA replication while displaying verying degrees of reversibility (*8–11*). This hypothesis was tested by transplanting into enucleated *Rana* eggs late gastrula endodermal nuclei whose nonhistones were labeled *in vivo* with [³H]tryptophan. Autoradiograms demonstrated that most nonhistones leave the nucleus during its reprogramming in the egg cytoplasm prior to the first cleavage of the egg; however, other types of proteins labeled with [³H]lysine remain for the most part in the nucleus (Fig. 1). Cytochemical studies indicated that some of the nonhistones that leave transplanted nuclei are acidic proteins, while some of the proteins that remain in the nucleus are histones. The retention of lysine-containing proteins by transplanted nuclei is consistent with two other reports. *Rana* nuclei labeled with [³H]leucine (*12*) and *Xenopus* nuclei labeled with [³H]arginine and [³H]alanine (*13*) retain most of their label after transplantation into eggs. The latter two amino acids appeared from gel electrophoretic studies to be incorporated mostly into histones.

The reprogramming of a transplanted nucleus involves complex nucleo-cytoplasmic exchanges. When frog nuclei are transplanted into eggs whose cytoplasmic proteins are previously labeled with leucine, labeled cyto-proteins accumulate in the nuclei (*14,15*) and this event occurs during nuclear enlargement, the period during which the induction of DNA synthesis occurs (*10*). The type of leucine-labeled cytoproteins which migrate into transplanted nuclei was not characterized at that time. How-ever, it is now known that some of these migrating proteins are nonhistones (N. J. Hoffner and M. A. DiBerardino, unpublished). Thus, at the present time, the reactivation of transplanted nuclei appears to involve complex nucleocytoplasmic exchanges of proteins in which most nuclear nonhistones of determined cells egress while most histones remain. The migration of cytoproteins into these nuclei involves, in part, nonhistones. Presumably, these proteins play an important role in the control of gene activity. How-ever, the form in which these proteins traverse the cellular compartments, the identification of specific proteins, and how they control gene expression remain to be elucidated.

The amphibian embryo has long been an extremely feasible form for studying nucleocytoplasmic interactions. Fertilization and development

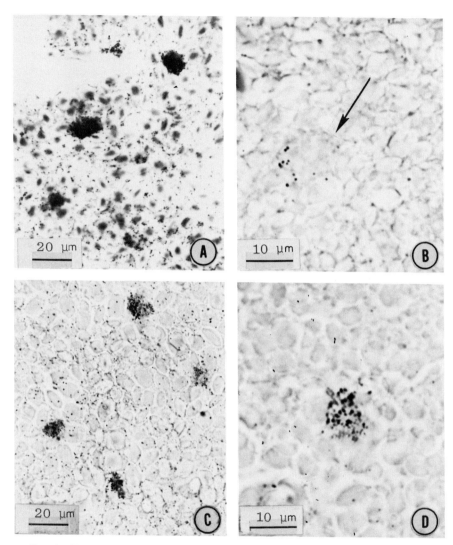

FIG. 1.   Late gastrula endodermal donors and nuclear transplants. (A, C) Cells previously labeled with [³H]tryptophan and [³H]lysine, respectively. Note the heavy density of grains in the nuclei. (B, D) Nuclear transplants fixed 120 minutes after activation of the recipient eggs. Note that few [³H] tryptophan grains are present in the transplanted nucleus (arrow) in (B), but the [³H]lysine transplanted nucleus is heavily labeled (D).

occur outside of the female. Their eggs and embryos are large and can be conveniently manipulated with microsurgical techniques under a stereomicroscope. Radioactive substances can be injected easily. In addition to the mature egg and embryonic phases, these forms possess large immature ovarian oocytes that contain very large nuclei (germinal vesicles) into which the accumulation of injected substances can be monitored. Numerous studies have been conducted in these oocytes to determine the selective properties of the nuclear membrane. One recent and significant report demonstrated that labeled nuclear and cytoplasmic proteins extracted from oocytes and injected into comparable oocytes accumulate and remain in their respective compartments of origin. Furthermore, many of the proteins that enter the germinal vesicle and accumulate there are acidic proteins (*16*).

Since embryonic and adult nuclei can be transplanted into eggs or oocytes (*8,17–19*). An experimental test of nucleocytoplasmic exchange can be conducted in a living system during development. With regard to nonhistone proteins, future studies should eventually lead to the identification of individual nonhistone proteins and explain their function in controlling genomic activity.

II. Materials

A. Microscopic Assembly

The injection of radioactive substances into amphibian embryos and oocytes and the transplantation of nuclei require a research-quality stereomicroscope to which is connected a micromanipulator, a microinjection apparatus, two AO Spencer microscope lamps, and a foot control for focusing the microscope.

Several models of stereoscopic microscopes may be used for microinjections, e.g., Zeiss, Leitz, and Wild. In all cases, the microscope should give good resolution and a wide, flat field with considerable depth of focus and provide magnifications usually from $\sim \times 6$ to $\times 100$. When transplantations of nuclei from advanced embryonic cells and adult cells are made, magnifications up to $\times 200$ are convenient for handling these small cell types.

A variety of micromanipulators are commercially available; however, a sturdy one which allows for precision control at high magnifications is required. We have found that the Leitz micromanipulator (E. Leitz, New York, New York) is well adapted for amphibian microinjections. A Leitz microinjection assembly equipped with a 2-ml syringe provides precision control for aspiration and injection of solutions and nuclei.

Since both hands of the operator are occupied simultaneously with both the injection apparatus and the manipulator, a foot control of the microscopic focus is needed. We have used the Gomco Electra Focus (Buffalo, New York) which operates the focus of the microscope electrically. A slight depression of the foot plate controls the focusing arm of the microscope with high precision even at high magnifications.

B. Microneedles and Micropipettes

Microneedles are required for activating eggs by pricking, for removal of the egg nucleus, and for severing exovates formed after injection. Microneedles are drawn from solid soft-glass rods with an outside diameter (OD) of ~1 mm. The glass rod and capillary tubing are obtainable from Drummund Scientific Co., Broomall, Pennsylvania. Capillary tubing, of Pyrex glass (OD 1 mm, ID 0.8 mm ±0.01 mm), is the source from which micropipettes are constructed. These dimensions result in thin-walled capillary tubing, a prerequisite for fashioning proper pipettes.

Approximately 4-in. lengths of either glass rod or tubing are placed in the center of the 1-mm slot of a heating element in a Livingston needle puller and drawn out to desired needle points or pipette orifices. The needle puller we employ was designed by Dr. L. G. Livingston and built by Otto Habel (Biology Dept., Swarthmore College, Swarthmore, Pennsylvania). The final tip of a glass micropipette is fashioned with a microforge assembled with a stereomicroscope (Sensaur Microforge de Fonbrune, Aloe Scientific, St. Louis, Missouri). Additional details concerning the microscopic assembly and construction of micropipettes have been published previously (9,20).

C. Calibration of Volume in Micropipettes

The following equipment is needed for calibrating volumes in glass micropipettes.

1. A 3 × 1 in. glass microscope slide on one end of which a small flat piece of hard plastic (25 × 25 × 1 mm) is glued. A slit is fashioned lengthwise in the middle of the plastic to accommodate the micropipette.

2. Suction-ejection apparatus (Fig. 2). This consists of a 3¼-in. rubber bulb to which is attached a 6-in. stainless-steel tube of proper diameter to fit the bulb. At the distal end of the tube, a #18 gauge stainless-steel needle is soldered. A small (2-cm long) piece of rubber tubing (0.10-in. ID)

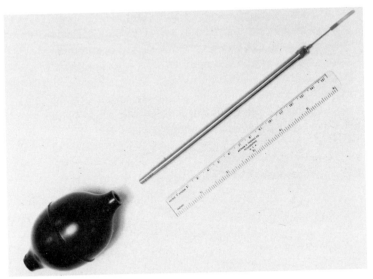

Fig. 2.   The suction-ejection apparatus consists of a 3½-in. rubber bulb to which is attached a 6-in. stainless-steel tube. At the distal end of the tube, a #18 gauge stainless-steel needle is soldered. A small (2-cm long) piece of rubber tubing (0.10 in. diameter) is fitted onto the needle and a glass micropipette is then inserted into the free end of the rubber tubing.

is fitted onto the needle. The diameter of the rubber tubing is small enough to fit tightly around an inserted micropipette.

3.   A 1.0-μl Hamilton glass syringe pipette in 0.01-μl gradations.

4.   Tech-Pen ink, Mark-Tex Corp., Englewood, New Jersey.

The procedure is as follows. A small drop of distilled water is delivered onto a piece of parafilm which is placed on one end of the microscope slide. The slide is positioned on the stage of a stereomicroscope and the micropipette is touched to the drop of water which moves into the pipette by capillary action. The pipette is positioned in the groove and the meniscus is marked by a thin line of ink drawn around the outside of the pipette with a glsss needle.

The marked micropipette is now inserted into the rubber tubing of the suction-ejection apparatus. Next, the 1-μl syringe pipette is used to deliver the desired injection volume (0.2 μl in this case) onto the parafilm. Under the scope, the water is drawn up into the micropipette. The micropipette is returned to the microscope slide and the new meniscus is demarcated with an ink line. The calibrated volume resides between the two ink marks. During the injection procedure, the syringe is depressed until the fluid passes from the second to the first ink mark, thus delivering the calibrated volume into the embryo or oocyte.

D.   Vessel for Holding Radioactive Amino Acids

The volume of diluted radioactive amino acid stock which is required for a series of injections in one experiment is extremely small, approximately $6 \mu l$ or less. In order to handle such small quantities and prevent evaporation during the course of an experiment, the following vessel for holding radioactive liquids has been devised from polyethylene capsules (5-mm diameter) which are routinely used for embedding material for electron microscopy (EMS polyethylene embedding capsules, size 100, Electron Microscopy Sciences, Fort Washington, Pennsylvania). The capsules and slides are first washed with detergent and rinsed thoroughly with distilled water. Next, for our purposes, the capsule is cut to about 6 mm in height yielding a capsule which holds about 50 μl of fluid. The capsule is then anchored in a piece of plasticene which adheres firmly on a glass microscope slide. The slide is placed on the stage of the microscope while the operator aspirates the desired volume of radioactive fluid into a calibrated micropipette. After the pipette is loaded, the capsule is closed and the slide is shifted onto the laboratory table and replaced with the operating dish containing the embryo or oocyte to be injected.

E.   Media

The composition of amphibian Ringer's medium is given in Table I, and that of Steinberg's medium and modified Steinberg's medium is given in Table II.

III.   Procedures for Obtaining Embryos and Oocytes

The procedures in this and subsequent sections will be those conducted on the leopard frog, *Rana pipiens*. When appropriate, reference to other amphibian species will be made. A more comprehensive coverage of this section has been published previously [*Rana*, (22), *Urodeles*, (23), *Xenopus*, (24)].

A.   Breeders

Northern *R. pipiens* are purchased from dealers before the frogs hibernate (autumn) and again just as the frogs emerge from hibernation (spring) and before they mate. Sexually mature adults obtained twice annually at these periods contain abundant mature ova and spermatozoa; when stored under cold laboratory conditions they can provide functional gametes for

TABLE I

AMPHIBIAN RINGER'S
MEDIUM[a]

NaCl	6.50 gm
KCl	0.14 gm
CaCl$_2$	0.12 gm
NaHCO$_3$	0.10 gm
Water	1000 ml[b]

[a] Sterilize with Millipore filtration.

[b] Glass-distilled or deionized.

TABLE II

STEINBERG (*21*) AND MODIFIED STEINBERG MEDIA[a]

	Steinberg	Modified Steinberg
NaCl (17.0%)	20 ml	20 ml
KCl (0.5%)	10 ml	10 ml
Ca(NO$_3$)$_2$·4 H$_2$O (0.8%)	10 ml	—
MgSO$_4$·7 H$_2$O (2.05%)	10 ml	—
HCl (1.00 *N*)	4 ml	4 ml
Water	946 ml	966 ml
Tris(hydroxymethyl) aminomethane	560 mg	691 mg
Streptomycin sulfate	50 mg	—
Penicillin-G, sodium	50,000 units	—

[a] The salts are dissolved in either glass-distilled water or deionized water and heat-sterilized. The inclusion of antibiotics is optional. If they are included, dissolve them separately in a portion of the water, sterilize by Millipore filtration, and combine them with the salt solution after cooling.

experiments throughout the year (10–12 months). We usually stock frogs from two sources (J. M. Hazen, Co., Alburg, Vermont, and Nasco, Fort Atkinson, Wisconsin) as a precaution, in case local health conditions of the frogs are suboptimal.

When the frogs arrive in the laboratory, they are sorted and washed with running cold tap water. Storage in small groups is advisable to prevent the spread of disease if it should occur; e.g., five females or six males can be stored in individual 2-quart aquarium bowls. In recent years the so-called "red leg" malady has significantly limited the population size of wild and laboratory frogs. We have found that the inclusion of salts and neomycin sulfate in tap distilled water reduces the malady in a large proportion of the stock. The medium is composed of 25% of full-strength Ringer's solution including 1 ml (200 mg/ml) of neomycin sulfate per 1000 ml of water. Neomycin sulfate is sold under the trade name Biosol liquid (Upjohn) as an oral veterinary drug.

Sufficient medium is added to almost cover the frogs. The frogs are stored at 4°C and the medium is changed three times per week with previously chilled (4°C) medium. No feeding is required when the frogs are maintained in cold storage.

B. Embryos

Embryos can be obtained under laboratory conditions by induced ovulation of oocytes in sexually mature females and artificial insemination of these oocytes by sperm from mature males. Rugh (25) has devised the standard injection technique for induced ovulation and the procedure for artificial insemination. The following is an abbreviated account based mainly on his procedure. A more extensive description of these procedures is covered elsewhere (22,26).

1. INDUCED OVULATION

The anterior lobe of the pituitary gland contains the ovulation-inducing hormones. The pituitaries of females are about two times as large and two times as potent as those of males, and the dosage of pituitaries needed to induce ovulation varies with the season. In addition, the size and potency of different anterior lobes of the pituitaries vary. The injected females likewise vary in response as well as in the quality of the eggs which they produce. Because of these facts it is advisable to inject at least two females for each experiment.

The ovulatory effect of pituitary factors can be augmented by injecting a female with a combination of progesterone and anterior lobes of the glands (27). The doses which we have found to be effective when injected R. pipiens are kept at 18°C are listed in Table III.

Sacrifice the frog by pithing the spinal cord from the hindbrain and then decapitate the frog. Expose the ventral side of the brain and remove the pituitary gland located in the center of the posterior region of the brain and directly underneath the cross formed by the parasphenoidal and transverse bones [see DiBerardino (22), pp. 59–61 for a detailed description of dissection of pituitary glands]. Place the gland into 2 ml of amphibian Ringer's solution and remove the adhering endolymphatic tissue and the pars intermedia and pars nervosa. The anterior lobe can now be injected

TABLE III

RECOMMENDED INJECTION DOSES FOR *R. pipiens*

Months	Number of female pituitaries	Progesterone
September through November	2–3	2 mg
December through February	2	2 mg
March and April	1	2 mg
May through August	1	—

into a recipient female or stored in amphibian Ringer's solution in the frozen state for at least 6 months.

Draw the gland(s) with about $\frac{1}{2}$–1 ml of amphibian Ringer's solution into the barrel of a 2-ml glass syringe and then apply a hypodermic needle (#18–#23) to the syringe. Inject the gland(s) intraabdominally in the posterior region.

Keep the injected frog(s) in a cool place, preferably at 18–20 °C for 38–48 hours. At this time most of the eggs should be in the uterus and fertilizable. If eggs are required sooner, the females can be kept at about 25 °C; however, the chance of obtaining overripe eggs is greatly increased at warm temperatures.

The eggs can now be forced from the uterus through the cloaca by "stripping." This is achieved by applying gentle and steady pressure on the abdomen in the direction of the cloaca. In a few seconds the female will discharge her eggs which are encased in jelly. The eggs of an ovulated female can be used adequately for approximately 5–7 days if she is stored at 4 °C and is not exposed to laboratory temperature for long intervals. A period of 45–60 minutes at 18 °C is sufficient time for temperature equilibration following 4 °C storage.

2. ARTIFICIAL INSEMINATION

Select a mature male and pith. Make an incision in the middle third of the abdominal cavity. Move the intestines aside and the pair of testes will be evident. They are yellow, oval bodies, located near the anterior borders of the kidneys. Dissect the pair of testes and place them into 10% amphibian Ringer's solution. A 10% solution of either Holtfreter [see Rugh (25), p. 15], Niu-Twitty (28), or Steinberg (21) medium can be substituted for 10% amphibian Ringer's.

Macerate the testes thoroughly in about 5 ml of 10% Ringer's solution with fine forceps and scissors until a milky suspension is obtained. Allow the suspension to stand for about 15 minutes during which time additional spermatozoa are released and become active. Sperm suspensions prepared in the above manner and stored at 4 °C can be successfully used for 48 hours, provided they are not kept at laboratory temperature for long intervals. The suspension should be warmed at laboratory temperature for 15 minutes before use.

Artificial insemination is easily accomplished. With a narrow-mouth pipette agitate the sperm suspension and pipette a thin film of this suspension into a large watch glass or Syracuse dish. Strip about 100 eggs and discard. Next, strip about 100–150 eggs directly into the sperm suspension. If the eggs ooze out in thick clusters, pipette additional sperm suspension on top of the clustered eggs.

After 15 minutes flood the eggs with 10% amphibian Ringer's solution. Approximately 1 hour after insemination, decant the solution and rinse twice with spring water or dechlorinated tap water. Dechlorination of the tap water is accomplished by adding 0.7–0.9 gm of sodium thiosulfate to 30 gallons of tap water; allow the water to stand a minimum of 24 hours before use. Loosen the eggs from the bottom of the dish. Transfer the eggs with a wide-mouth pipette into a glass vessel containing spring water or de-chlorinated tap water and cover the bowl with a glass lid to reduce evaporation and to exclude dust. Place no more than 30–40 eggs per 200 ml of water. The embryos can be reared in this vessel until the feeding stage, and the water need not be changed during this interval.

In *R. pipiens*, during a developmental period of 12 days at 18 °C, 25 stages are clearly distinguished and tabulated on the basis of morphology, average body length, and age. This period starts with fertilization and includes the main embryonic events of cleavage, gastrulation, and neurulation and concludes with the onset of feeding, when the operculum fold has just grown over the gills and all the main embryonic organ systems are completely developed. The stage seriation for *R pipiens* at 18 °C was originally prepared by Shumway (29) and has been reproduced in a number of publications [e.g., Hamburger (26) and Rugh (25)] and in part in various textbooks dealing with comparative embryology. Similar series of developmental stages have been tabulated for *R. sylvatica* (30) *X. laevis* (31), *Pleurodeles waltlii* (32), and *Ambystoma mexicanum* (33).

Since amphibians are poikilothermic animals, the developmental rate of their progeny can be controlled within limits by environmental temperature. The range of temperature tolerance varies among species; *R. pipiens* can tolerate temperatures of 6°–28 °C [see DiBerardino (22) for original references to Atlas and Moore concerning studies of temperature tolerance in amphibian embryos].

C. Oocytes

Sexually mature *R. pipiens* females are removed from refrigeration, pithed, and decapitated. When the peritoneal cavity is exposed, the large ovaries containing large numbers of full-grown, immature oocytes will be immediately visible. The whole ovaries or parts of the ovaries are cut out and placed into finger bowls or large petri dishes containing amphibian Ringer's salt solution. The oocytes will remain viable in this solution for about 4 days if kept at 18°C (*34*).

The oocytes within the ovary are surrounded by a layer of follicle cells and a thecal layer containing blood vessels. The presence of these layers external to the oocyte does not interfere with some types of experiments. However, removal of these layers does facilitate the injection of individual oocytes with radioactive material. When the presence of these layers interferes with the interpretation of results of experiments, e.g., homogenizing oocytes for biochemical studies, then the cells should be removed. Absence of these layers does not affect progesterone-induced maturation of these oocytes (*34–36*).

Removal of the follicular and thecal layers can be accomplished (1) manually or (2) chemically:

1. With two pairs of watchmaker's forceps, separate small sections of ovarian tissue and place in Syracuse dishes containing Ringer's. This operation and defolliculation of the oocytes are carried out under a stereomicroscope at ×10. Full-grown (1.6–1.8 mm), immature (prior to meiotic maturation) oocytes are released one at a time by gently pulling away, with two pairs of forceps, the two layers surrounding each oocyte. This takes practice, since the oocytes are fragile and easily punctured or torn apart by the forceps. The released oocytes are then pipetted into a dish containing fresh Ringer's medium. Almost all of the follicle cells can be removed manually, although a few cells can still be seen attached to the oocyte when the oocyte is examined in sectioned and stained preparations by light microscopy at ×100.

2. To defolliculate large numbers of oocytes simultaneously certain enzymes have been utilized successfully in certain experiments. Oocytes placed in Pronase (50 μg/ml of Steinberg) for 3–4 minutes (*37*) or in 0.2% bacterial collagenase dissolved in Ringer's for 2 hours (*35*) are adequately defolliculated. However, the use of collagenase lowers the percentage of development beyond the initial cleavage (*38*). Since these chemicals introduce still another variable into an already complex experimental system, manual defolliculation is preferred in our studies.

After defolliculation, oocytes can be stored for many months prior to biochemical analysis by freezing and storing them at −70°C in small (0.4 ml) plastic tubes containing small amounts of Ringer's medium.

IV. Radioactive Labeling of Embryos and Oocytes

A. Embryos

Amphibian eggs and embryos are impermeable to nucleosides and amino acids; therefore, these substances must be injected. Although this account is concerned with the injection of [³H]tryptophan, the procedures described are applicable to the injection of a wide spectrum of substances and isotopic forms. Also, an embryo can be injected at any stage; however, the introduction of radioactive substances into the uncleaved egg and into the blastocoele and archenteron cavities of blastulas and gastrulas, respectively, allows for widespread diffusion of the injected material.

1. INJECTION INTO FERTILIZED EGGS

Eggs are stripped from an ovulating female and inseminated. Fifteen minutes after insemination flood the eggs with 10% amphibian Ringer's solution and remove most of the jelly surrounding the eggs with watchmaker's forceps and fine scissors.

Transfer dejellied eggs to a syracuse dish (operating dish) containing a layer of 2% agar dissolved in Steinberg's salt solution (Ringer's salt solution for oocytes). Under the microscope bore a few small holes in the agar to hold single eggs snugly. Remove the operating dish from the microscope stage and replace it with the vessel containing the radioactive solution. With the aid of the micromanipulator insert the micropipette (10–20 μlD) into the vessel and aspirate the solution to the top calibrated mark of the pipette. We routinely inject a final pH 7.0 solution of 0.083 μCi of [³H]tryptophan diluted with sterile Steinberg salt solution to a final volume of 0.2 μl into each egg or embryo. This volume and amount of radioactivity do not interfere with subsequent cleavage.

After the micropipette is loaded, remove the vessel from the microscope stage and replace it with the operating dish containing the eggs. Position the micropipette over one egg and insert the pipette at the equator of the egg with a slight angle downward. This approach to injection is required to avoid injury to the pronuclei located in the animal hemisphere. Depress the barrel of the syringe and inject the calibrated volume of solution. Withdraw the pipette and to prevent leakage sever the exovate formed at the site of penetration of the pipette. The exovate may be severed either by cutting it with a pair of glass needles or by turning the egg upside down. When the polarity of the fertilized egg is reversed to the extent that the vegetal hemisphere is uppermost, the egg will rotate within its vitelline membrane and the exovate will sever from the surface coat.

Transfer the injected eggs into a separate dish containing dechlorinated tap water. Some batches of eggs tend to be soft and healing of the injection site is slow. In these cases, the operated eggs should be transferred to Steinberg salt solution and, if the experimental time permits, transferred to dechlorinated tap water after about 2 hours.

2. Injection into Embryos at the Blastula and Gastrula Stages

A large portion of the animal hemisphere of mid-blastulas is occupied by a blastocoele cavity. Small glass micropipettes can be easily inserted into this cavity through the animal hemisphere which consists of a thin roof containing about three layers of cells. Subsequent to removal of the jelly surrounding the embryo, insert a micropipette approximately at a 45° angle away from the center of the animal pole and deliver 0.2 μl (pH 7.0) of 0.083 μCi of [^{3}H]tryptophan in Steinberg solution into the blastocoele. Withdraw the pipette and transfer the embryo to a separate dish containing dechlorinated tap water. Usually it is not necessary to sever the exovate.

When injecting late gastrulas, place the embryo with the yolk plug down and perpendicular to the agar. Then insert the pipette approximately 500 μm inward into either the future dorsal or ventral side in the center region and inject the solution. The injected medium will be deposited in the archenteron cavity. Withdraw the pipette and transfer the operated embryo to a separate dish containing dechlorinated tap water. Severing of the exovate may be done if necessary to prevent leakage.

The length of incubation will vary with the type of experiment being conducted. We have found that a 3-hour incubation with [^{3}H]tryptophan or [^{3}H]lysine results in dense labeling of most endodermal nuclei of late gastrulas as determined with autoradiographic techniques.

When blastulas or gastrulas are injected with 0.2 μl of 0.083 μCi of either [^{3}H]lysine or [^{3}H]tryptophan, normal development is not significantly affected. In control series 88 out of 89 injected embryos attained stages of organogenesis, and 89% ($\overline{X}$) of the total injected embryos developed into feeding larvas.

B. Oocytes

Oocyte proteins can be labeled with radioactive amino acids in two ways: (1) incubation of oocytes in radioactive medium, or (2) microinjection of radioactive material. An oocyte, unlike a mature egg, has the capacity to transport amino acids across its cellular membrane. However, use of the incubation method of labeling does not permit one to control the amount of label each oocyte receives. Microinjection is a relatively accurate method for repeatedly delivering the same quantity of radioactivity into each oocyte.

The major advantage of the incubation method is the elimination of the need for the sophisticated apparatus that is required for microinjection. Microinjection, on the other hand, provides a better-controlled experimental system.

1. INJECTION INTO OOCYTES

With a wide-mouth pipette place a defolliculated oocyte into a hemispherical well of an operating dish containing a layer of 2% agar and amphibian Ringer's medium. Remove the operating dish from the stage of the stereomicroscope and replace it with a capsule containing the radioactive liquid for injection into the oocytes. When labeling nonhistones of oocytes inject [³H]tryptophan. Dilute [³H]tryptophan stock with amphibian Ringer's so that there is 0.083 μCi in each 0.2 μl of injection medium (pH 6.8–7.0). The amount of injected radioactivity and/or the radioactive atom can be varied for different experimental requirements. The above conditions label oocytes satisfactorily for light microscopic autoradiography and permit normal maturation events *in vitro* (see controls below). In addition, the same conditions, utilizing other amino acids, allow satisfactory resolution for scintillation analysis of radioactive proteins separated by acrylamide gel electrophoresis (R. Greene, personal communication). Furthermore, the introduction of 0.2 μl of additional volume to the oocyte causes little leakage. Approximately 14–16% of radioactivity is lost from injected oocytes during the incubation period.

With the aid of the micromanipulator and microinjection apparatus, aspirate 0.2 μl of [³H]tryptophan from the capsule into a calibrated micropipette (10–20 μm ID) at approximately ×6 magnification. Next, replace the radioactive vessel with the operating dish and inject the oocyte in the vegetal hemisphere at approximately ×10 magnification. The oocyte is injected in the vegetal hemisphere to prevent injury to the germinal vesicle which is the large nucleus located within the animal hemisphere. After removal of the pipette from the injected oocyte, the oocyte is transferred through two rinses of amphibian Ringer's and then to a small Stender dish containing amphibian Ringer's where it is incubated for the desired period. Depending on the character of the experiment, oocytes can be incubated at about 18°C for a few minutes to about 1–2 weeks. Regarding the micropipette, its remaining contents are expelled into a waste dish to prevent oocyte cytoplasm from clogging the opening and then it is reloaded for the next injection.

At the termination of an experiment, the injected oocytes can be fixed and processed for autoradiographic analysis (see Section VI) or frozen and stored at −70 °C for biochemical analysis.

2. CONTROLS

Two sets of controls are run simultaneously with each experiment. Five oocytes are injected with amphibian Ringer's and another group of five oocytes is injected with [³H]tryptophan. Immediately after injection, the oocytes are exposed to a 10 μg/ml progesterone solution (Wyeth, aqueous solution, 25 mg/ml) for 1 hour. If the injection procedure has not injured the oocyte, then these oocytes should undergo meiotic maturation within 36 hours. Externally visible maturation events include: (1) the germinal vesicle rising to the surface of the animal hemisphere producing a large white spot at 10–12 hours post progesterone (p.p.); (2) germinal vesicle breakdown, 12–16 hours p.p.; (3) first polar body extruded 20–24 hours p.p. and first "black dot" visible; and (4) second polar body extruded 36 hours p.p. (*34–36*).

V.   Nucleocytoplasmic Exchange of Nonhistones

Over 95% of late gastrula endodermal nuclei accumulate dense amounts of [³H]tryptophan which is incorporated into nuclear nonhistone proteins (*7*). To test whether such endodermal nuclear proteins would egress from the nucleus when residing in foreign cytoplasm, these labeled nuclei were transplanted singly into the cytoplasm of mature enucleated eggs. Most of the nonhistone proteins egressed from the transplanted nucleus and entered the egg cytoplasm. In contrast, transplanted endodermal nuclei labeled with [³H]lysine retained most of the lysine. These experiments showed that during the first cell cycle of the egg, when transplanted nuclei are induced to undergo DNA synthesis, some of the reprogramming process involves the loss of endodermal nuclear nonhistones.

The above test of nucleocytoplasmic exchange of nonhistones involves the techniques of amphibian nuclear transplantation (*8,9,20*) and autoradiography. The former procedure will be described below and the latter one in Section VI.

A.   Preparation of Enucleated Host Eggs

The preparation of host eggs that will receive transplanted nuclei requires two important steps. First, the egg must be activated, and second, the egg nucleus must be either removed or destroyed *in situ.* Activation of the recipient egg is of prime importance, because this event causes profound changes in the condition of the cytoplasm and these changes are essential for successful nuclear transplantation. In *R. pipiens,* successful activation is achieved by simply pricking the egg with a clean glass needle (8). This pricking also suffices for most frog species tested. In *Xenopus,* ultraviolet (UV) irradiation is the most reliable way to simultaneously activate the egg and destroy the egg nucleus (39). In *Ambystoma* and *Pleurodeles* the application of mild electric shock is the preferred method for activating eggs of these urodelen forms (40,41). Successful removal of the egg nucleus is achieved most easily in *R. pipiens* by enucleation with a glass needle (42). In addition, successful inactivation of the egg nucleus can also be obtained by microlaser irradiation (43). Finally, in other species, namely, *Xenopus* (39), *Pleurodeles,* and *Ambystoma* (44). UV irradiation is the most reliable way to inactivate the egg nucleus. The following account summarizes the procedures for activation, enucleation, and nuclear transplantation in *R. pipiens.* A more comprehensive description can be found elsewhere (9,20).

1. ACTIVATION AND ENUCLEATION (FIG. 3)

Eggs from an ovulating female are stripped into a Syracuse dish containing 10% amphibian Ringer's solution (10% Niu-Twitty or 10% Steinberg solution can be substituted). The outer jelly layers of the eggs will swell after immersion in the solution and usually anchor the eggs to the bottom of the dish. Under a stereomicroscope with a magnification of approximately ×20, immediately prick each egg with a clean glass needle at some point in the animal hemisphere away from the animal pole region.

Most eggs parthenogenetically stimulated by pricking will be activated, and within 12–20 minutes at 18°C the eggs will rotate within their vitelline membranes. In addition, the cortical granules directly over the egg nucleus in its second metaphase of meiotic maturation migrate away from the center of the animal pole area, permitting the operator to see a pit which appears as a small "black dot." This pit is situated directly over the egg nucleus and is visible for about 15–30 minutes after activation.

Once the small, distinct black dots are visible, the egg nuclei can be removed. A strong light source, such as that provided by two AO Spencer Universal lamps connected to a variable transformer, will permit the operator to see the black dots distinctly. Select eggs that are firmly attached to the dish. Under a magnification of 40–60 diameters insert a clean glass needle with a fine point and firm shaft diagonally through the jelly membranes, vitelline membrane, and surface coat of the egg at a point slightly to one side of and directly beneath the dot. Move the needle straight up through the dot making a small tear in the surface coat. Immediately, an exovate containing

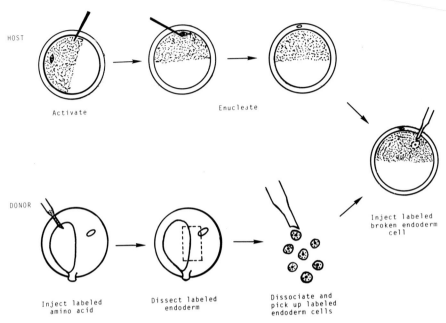

HOST

Activate   Enucleate

Inject labeled
broken endoderm
cell

DONOR

Inject labeled   Dissect labeled   Dissociate and
amino acid   endoderm   pick up labeled
                                            endoderm cells

FIG. 3.   Labeling of donor cells and nuclear transplantation. *Lower*: Radioactive amino acid is injected into the archenteron cavity of a late gastrula embryo. At the termination of the incubation, the labeled endoderm (broken lines) is dissected and the tissue is dissociated into individual cells. *Upper*: Mature egg is activated by pricking with a glass needle. About 20 minutes later, the egg nucleus in its second meiotic metaphase is removed manually with a glass needle, thus producing an enucleated host egg. Finally, a single endodermal cell labeled with a radioactive amino acid is aspirated into a glass micropipette causing the cell membrane to break. The broken cell and its nucleus are injected into the animal hemisphere of the enucleated host egg.

the egg nucleus will form and lodge in the perivitelline space and/or inner jelly layer. This operation when performed by an experienced person results in 99–100% successful enucleations.

Subsequent to the enucleation procedure, place the enucleated eggs in a separate dish and remove the outer layers of jelly with watchmaker's forceps and a pair of fine-pointed scissors. Next, rinse the eggs in Steinberg's medium. They are now ready to receive a transplanted nucleus.

B.   Donor Cells (Fig. 3)

Embryos derived from inseminated eggs are reared to the desired developmental stage. For the purpose of our studies we have found that the optimal incubation time of $[^3H]$tryptophan for late gastrulas is 3 hours. Accordingly, donor embryos are injected with the tritiated amino acid (Section IV,A,2) 3 hours prior to the dissection of donor cells. The donor embryo is placed in Steinberg's medium lacking Ca^{2+} and Mg^{2+} (modified Steinberg's solution MS), and the vitelline membrane is removed with watchmaker's forceps. In the case of a late gastrula embryo, insert two pairs of watchmaker's forceps in the anterior region and rip open the embryo. The endoderm containing large white cells will be clearly visible on one side of the embryo. With a pair of glass needles cut out the desired area of endoderm. Gently suck up the explant with a pipette and rinse it twice in fresh MS and then place it in a final MS solution for 30 minutes during which time the donor area will dissociate into single cells. The dissociated cells can then be sucked up gently in a Pasteur pipette (1-mm orifice) and transferred to the agar floor of an operating dish containing Steinberg's solution.

C. Nuclear Transplantation (Fig. 3)

Glass micropipettes whose internal diameters range from 35–40 μm are optimal for transplanting gastrula endodermal cells. This diameter is approximately one-half to one-third less than the diameter of the cell, and the cell membrane will break when the cell is sucked into the pipette. The breakage of the cell is essential because the nucleus when transplanted into the egg must be free in the egg cytoplasm; otherwise, nucleocytoplasmic interactions will not occur. However, it is important that the donor cytoplasm remain around the nucleus in the pipette, so that the nucleus is not exposed to the possible deleterious effects of the injection medium. The drawing up of the cell into the pipette and the actual nuclear transplantation should be conducted at $\times 100$ magnification.

After the donor cell is carefully broken in the tip of the micropipette, the pipette is moved with the control lever of the micromanipulator toward one of the previously prepared enucleated eggs residing in one of the hemispherical depressions in the agar of the operating dish. The pipette is then inserted into the center of the host egg cytoplasm at a point along the animal-vegetal axis approximately one-third of the distance from the animal pole. Depress the plunger of the syringe slightly and very slowly, while watching the movement of the meniscus of the fluid column within the shaft of the pipette. The broken cell kept near the orifice of the pipette can now be liberated with a minimal amount of injection medium into the egg.

As the pipette is withdrawn, the surface coat usually clings to the outer wall of the pipette and forms a canal adhering to the vitelline membrane. This canal can easily be severed with a pair of glass needles to prevent leakage. The injected egg is then sucked up with a wide-mouth pipette, rinsed in dechlorinated tap water, and placed individually in a small Stender dish containing dechlorinated tap water for the period of the experiment.

The validity of nuclear transplantation studies depends on adequate protection afforded to the nucleus by its own cytoplasm during the transfer procedure (8). Unless this condition is met, the transplanted nucleus will not support normal development of the recipient egg or even abnormal and extensive development. Since the objective of these initial studies was to determine whether nonhistones migrate from the transplanted nucleus to the cytoplasm during its reprogramming in the first cell cycle of the egg under conditions that promote normal and extensive development, the cytoplasm was transplanted along with its nucleus. Thus, the transfer procedure involves the injection of cytoplasmic and nuclear proteins from the donor cell into the enucleated egg. In addition, there exists the possibility that free amino acids may be introduced into the recipient eggs, since a chase of the labeled amino acids is not employed, because the addition to the dissociated cells of unlabeled amino acids sufficient for a chase increases the tonicity of the solution considerably. Thus, the above conditions permit a test of the developmental events of a living nucleus. The injection of the labeled cytoplasmic proteins or possibily unbound labeled amino acids from the donor cell into the host egg does not interfere with the main question posed in these experiments; namely, do nonhistones leave a nucleus after it is transplanted into the young cytoplasm of the egg? Once this type of information is obtained under conditions that permit a test of a living nucleus, modifications of the technique can be applied to related experiments; e.g., donor cells and host eggs can be treated with an inhibitor of protein synthesis and isolated nuclei may be injected.

VI.   Cytological Procedures

The optimal approach to studying nucleocytoplasmic exchange of proteins or other labeled substances is to look for these events during the first cell cycle of the egg while the nucleus still contains a nuclear membrane. One approach to this analysis is to fix the nuclear transplants and subsequently process them for autoradiography. In conducting these experiments in *R. pipiens* the nuclear transplantations are made approximately between 30 and 90 minutes after activation of the host egg. The nuclear transplants are then fixed 120 minutes after egg activation. By 130–145 minutes after activation of the host at 18 °C, nuclear transplants have attained the metaphase stage (11,45). Therefore, fixation at 120 minutes after activation of the egg allows sufficient time for nucleocytoplasmic exchange to occur prior to dissolution of the nuclear membrane.

The following cytological procedures are routinely employed for oocytes, eggs, and embryos. In the case of eggs and embryos, if jelly is still present, it should be removed prior to fixation.

A. Histologic

Fix the material for 2 hours in 10% formaldehyde (37.5% commercially available formaldehyde diluted with distilled water to a concentration of 10%) adjusted to pH 7.0 with 1 N NaOH. The fixative formaldehyde is employed for protein studies because it immobilizes most proteins (46,47).

Subsequent to fixation, change the material in three rinses of distilled water at 1-hour intervals and then maintain in distilled water for 12–24 hours.

Dehydrate in a graded series of ethanol (30 minutes each). Place in amyl acetate, change once after 1 hour, and maintain in amyl acetate overnight or to a maximum of 3 days. Clear in toluene (2 minutes), infiltrate for 1 hour, and embed in paraplast (56–57°C melting point). Serially section the material at 6 μm.

Paraffin Section Adhesive (PSA) has been successfully and routinely used for mounting the ribbons. Other standard adhesives frequently result in loss of sections. Although it was possible to purchase PSA in the past, it is no longer produced commercially, and we are indebeted to Behring Diagnostics, Inc. for the formula which consists of 1000 ml of horse serum, 100 ml of bovine serum albumin (30%), and 1.1 gm of sodium azide (NaN$_3$). The PSA is filtered and diluted (10 drops of PSA in 30 ml of distilled water) before use. The ribbons are spread on a slide flooded with diluted PSA and dried overnight on a slide warming table at 30°C.

Remove paraplast from the slides in xylene and hydrate sections in a graded series of ethanol. Rinse in distilled water and then in cold tryptophan (four times the injected concentration) for 10 minutes followed by another distilled water rinse for 10 minutes.

In *R. pipiens* oocytes and embryos, the dense melanin pigment granules make visualization of grains in the autoradiograms difficult and bleaching is necessary. Expose slides to 30% hydrogen peroxide (Merck, Superoxol) at room temperature (20°–23°C) for approximately 16 hours. Rinse two times in distilled water, dehydrate in a graded ethanol series, and air-dry at least overnight. Control studies employing cytochemical procedures to reveal *in situ* acidic and basic proteins revealed no difference in the staining appearance of these proteins in bleached or unbleached sections.

B. Autoradiography

Dip slides in liquid Kodak NTB$_2$ Nuclear Track Emulsion diluted 1:1 with distilled water and warmed to 44°C. Dry overnight and store in light-tight slide boxes for 1–3 weeks. A 3-week exposure is optimal for most of our studies. At the termination of exposure, develop the slides in full-strength Dektol for 2 minutes, fix for 5 minutes in Kodak Fixer, rinse for 20 minutes in running tap water followed by a distilled water rinse, and air-dry at least overnight. The above procedure is the general one described by Prescott (48) for liquid emulsion.

Place dried slides in aqueous toluidine blue O (Harleco) at a concentration of 0.25% w/v (pH 6.0) for 10–20 minutes, destain for 30 seconds in 95% ethanol, rinse in absolute ethanol, clear in xylene, and mount in Technicon.

C. Controls

To determine whether the ^{3}H atom in the amino acid(s) is incorporated mainly into proteins, the following test should be conducted. After hydration of the sections, expose the slides to 5% trichloroacetic acid (TCA) in a boiling water bath for 15 minutes, a period sufficient to remove nucleic acids as determined by the Feulgen procedure for DNA (49) and Azure B procedure for RNA (50). Sections treated in this manner do not display a reduction of grains compared to untreated autoradiograms, indicating that the amino acid(s) are incorporated mainly into proteins.

D. Analysis

The quantitation of grains in autoradiograms can be determined by visual counting of grains in microscopic areas delimited by an ocular grid or by measuring grain density with a reflectance microscope equipped with a photometer (51). This instrument measures autoradiographic silver grain density with a high accuracy and speeds the collection of autoradiographic data by a factor of 10 or greater compared to visual counting. In either approach, the quantity or density of grains in the nucleus and in representative areas of the cytoplasm are compared after correction for background. Such quantitation indicates whether the nuclear or the cytoplasmic compartment has a heavier concentration of labeled proteins and whether there has been a nucleocytoplasmic exchange of label during the course of an experiment.

VII. Concluding Remarks

The nucleocytoplasmic exchange of proteins, including in some cases nonhistones, has been extensively studied in unicellular forms (52), cell culture (4), and cell hybridization (53) systems. These studies have correlated significant information relating migration of proteins with changes in the cell cycle and changes in gene expression.

The system described in this chapter is a developing metazoan system, and as such, it provides an opportunity ultimately to analyze factors controlling gene expression as they relate directly to the establishment and maintenance of nuclear and cell differentiation in a vertebrate organism. Studies dealing with the nucleocytoplasmic exchange of nonhistone proteins in amphibian embryos have so far been limited and have centered mainly around the reprogramming of nuclei transplanted into eggs. These studies have indicated that when transplanted nuclei are reactivated in egg cytoplasm, there is a bidirectional migration of nonhistone proteins between the nuclear and cytoplasmic compartments. This event results in the egression of most labeled tryptophan from the nuclei of determined cells and the acquisition by these nuclei of labeled tryptophan from the egg cytoplasm. Whether completely functional proteins or subunits of proteins are being exchanged between the two compartments remains to be clarified. The critical problems for the future involve the identification of individual species of nonhistone proteins and a direct assessment of their function in controlling genomic activity and cell differentiation in a metazoan system.

This chapter describes methods for analyzing the nucleocytoplasmic exchange of nonhistone proteins in amphibian embryos. The main procedures outlined include procurement of embryos and oocytes, readioactive labeling of embryos and oocytes by microinjection, nuclear transplantation, and cytological methods for autoradiography. Various biochemical techniques could be applied to the characterization of migrating nonhistones in the present system. In addition, this system with some modifications could also be extended to the study of nucleocytoplasmic interactions of histones, RNAs, and other complex macromolecules.

ACKNOWLEDGMENT

The research summarized was aided by research grants GB 19631 and GB 41838 from the National Science Foundation.

REFERENCES

1. Gilmour, R. S., and Paul, J., *Proc. Natl. Acad. Sci. U.S.A.* **70**, 3440 (1973).
2. Barrett, T., Maryanka, D., Hamlyn, P. H. and Gould, H. J., *Proc. Natl. Acad. Sci. U.S.A.* **71**, 5057 (1974).
3. O'Malley, B. W., and Means, A. R., *Science* **183**, 610 (1974).
4. Stein, G. S., Spelsberg, T. C., and Kleinsmith, L. J., *Science* **183**, 817 (1974).
5. Briggs, R., *in* "Genetic Mechanisms of Development" (F. H. Ruddle, ed.), p. 169. Academic Press, New York, 1973.
6. Malacinski, G. M., *J. Exp. Zool.* **181**, 409 (1972).

7. DiBerardino, M. A., and Hoffner, N. J., *Exp. Cell Res.* **94**, 235 (1975).
8. Briggs, R., and King, T. J., *Proc. Natl. Acad. Sci. U.S.A.* **38**, 455 (1952).
9. King. T. J., *Methods Cell Physiol.* **2**, 1 (1966).
10. Graham, C. F., Arms, K., and Gurdon, J. B., *Dev. Biol.* **14**, 349 (1966).
11. DiBerardino, M. A., and Hoffner, N., *Dev. Biol.* **23**, 185 (1970).
12. Ecker, R. E., and Smith, L. D., *Dev. Biol.* **24**, 559 (1971).
13. Gurdon, J. B., *Proc. R. Soc. London Ser. B* **176**, 303 (1970).
14. Arms, K., *J. Embryol. Exp. Morphol.* **20**, 367 (1968).
15. Merriam, R. W., *J. Cell Sci.* **5**, 333 (1969).
16. Bonner, W. M., *J. Cell Biol.* **64**, 431 (1975).
17. Gurdon, J. B., *Proc. Natl. Acad. Sci. U.S.A.* **58**, 545 (1967).
18. King, T. J., and DiBerardino, M. A., *Ann. N. Y. Acad. Sci.* **126**, 115 (1965).
19. Subtelny, S., and Bradt, C., *Dev. Biol.* **2**, 393 (1960).
20. King, T. J., *in* "Methods in Developmental Biology" (F. H. Wilt and N. K. Wessells, eds.), p. 737. Crowell Co., New York, 1967.
21. Steinberg, M., *Carnegie Inst. Washington Year Bk.* **56**, 347 (1957).
22. DiBerardino, M. A., *in* "Methods in Developmental Biology" (F. H. Wilt and N. K. Wessells, eds.), p. 53. Crowell, New York, 1967.
23. Fankhauser, G., *in* "Methods in Developmental Biology" (F. H. Wilt and N. K. Wessells, eds.), p. 84. Crowell, New York, 1967.
24. Gurdon, J. B., *in* "Methods in Developmental Biology" (F. H. Wilt and N. K. Wessells, eds.), p. 75. Crowell, New York, 1967.
25. Rugh, R., "Experimental Embryology: Techniques and Procedures." Burgess, Minneapolis, Minnesota, 1962.
26. Hamburger, V., "A Manual of Experimental Embryology." Univ. of Chicago Press, Chicago, Illinois, 1960.
27. Wright, P. A., and Flathers, A. R., *Proc. Soc. Exp. Biol. Med.* **106**, 346 (1961).
28. Niu, M. C., and Twitty, V. C., *Proc. Natl. Acad. Sci. U.S.A.* **39**, 985 (1953).
29. Shumway, W., *Anat. Rec.* **78**, 138 (1940).
30. Pollister, A. W., and Moore, J. A., *Anat. Rec.* **68**, 489 (1937).
31. Nieuwkoop, P. D., and Faber, J., "Normal Table of *Xenopus Laevis* (*Daudin*)," 2nd Ed. North-Holland Publ. Amsterdam, 1975.
32. Gallien, K., and Durocher, M., *Bull. Biol. Fr. Belg.* **91**, 97 (1957).
33. Schreckenberg, G. M., and Jacobson, A. G., *Dev. Biol.* **42**, 391 (1975).
34. Smith, L. D., Ecker, R. E., and Subtelny, S., *Dev. Biol.* **17**, 627 (1968).
35. Masui, Y., *J. Exp. Zool.* **166**, 365 (1967).
36. Schuetz, A. W., *J. Exp. Zool.* **166**, 347 (1967).
37. Smith, L. D., and Ecker, R. E., *Dev. Biol.* **19**, 281 (1969).
38. Drury, K. C., and Schorderet-Slatkine, S., *Cell* **4**, 269 (1975).
39. Elsdale, T. R., Gurdon, J. B., and Fischberg, M., *J. Embryol. Exp. Morphol.* **8**, 437 (1960).
40. Signoret, J., and Fagnier, J., *C. R. Hebd. Seances Acad. Sci.* **254**, 4079 (1962).
41. Briggs, R., Signoret, J., and Humphrey, R. R., *Dev. Biol.* **10**, 233 (1964).
42. Porter, K. R., *Biol. Bull.* (*Woods Hole, Mass.*) **77**, 233 (1939).
43. McKinnell, R. G., Mims, M. F., and Reed, L. A., *Z. Zellforsch.* **93**, 30 (1969).
44. Signoret, R., Briggs, R., and Humphrey, R. R., *Dev. Biol.* **4**, 134 (1962).
45. Subtelny, S., and Bradt, C., *J. Morphol.* **112**, 45 (1963).
46. Alfert, M., and Geschwind, I. I., *Proc. Natl. Acad. Sci. U.S.A.* **39**, 991 (1953).
47. Pipkin, J. L., Jr., *Methods Cell Physiol.* **3**, 307(1968).
48. Prescott, D. M., *Methods Cell Physiol.* **1**, 365 (1964).
49. Feulgen, R., and Rossenbeck, H., *Z. Physiol. Chem.* **135**, 203 (1924).

50. Flax, M. H., and Himes, M. H., *Physiol. Zool.* **25**, 297 (1952).
51. Entingh, D., *J. Microscopy* **101**, 9 (1974).
52. Goldstein, L., *in* "The Cell Nucleus" (H. Busch, ed.), Vol. I, p. 387. Academic Press, New York, 1974.
53. Davidson, R. L., and de la Cruz, F. F. eds., "Somatic Cell Hybridization." Raven, New York, 1974.

Chapter 9

Methods for Studying Nucleocytoplasmic Exchange of Macromolecules

CARL M. FELDHERR

Department of Anatomy,
University of Florida College of Medicine,
Gainesville, Florida

I. Introduction

There is evidence that gene activity in eukaryotic cells is regulated by specific nonhistone proteins which are believed to be synthesized in the cytoplasm. Although the interaction of these proteins with chromatin has been the subject of considerable study, there is relatively little information concerning the process by which endogenous proteins enter the nucleus and whether nucleocytoplasmic exchange might itself serve as a regulatory step. With this in mind, methods have recently been developed to investigate the uptake of endogenous cytoplasmic proteins by the nucleus. Before discussing these methods we will briefly consider (*1*) the morphology of the nuclear envelope, which represents the major physical barrier between the nucleoplasm and cytoplasm, (*2*) the inherent difficulties in studying envelope permeability, and (*3*) the permeability concepts that have developed from studies using exogenous tracers. This should put the problems of nucleocytoplasmic exchange in better perspective.

The nuclear envelope is composed of two membranes separated by a perinuclear space several hundred angstrom units thick. The membranes are interrupted by circular pores which are 600–700 Å in diameter and can occupy approximately 20% of the nuclear surface. The pores contain an electron-opaque annular material composed of protein, which is probably the major structural component; RNA, which might have a structural role, but is more likely material in transit to the cytoplasm; and possibly DNA. Clearly the envelope represents a unique intracellular barrier, and one would expect its permeability characteristics to differ markedly from those of other cellular membranes, especially with regard to macromolecules.

The initial difficulty encountered in studying nuclear permeability involves exposing the nuclei to the test substance. Macromolecules present in the extracellular medium enter cells primarily, if not exclusively, by way of pinocytosis vesicles and are not generally released into the ground cytoplasm. Thus, it is necessary to bypass the plasma membrane by either injecting directly into the cytoplasm or by using isolated nuclei. In most instances mass nuclear isolation procedures result in envelope damage which can be detected with the electron microscope. Furthermore, it has been demonstrated from nuclear transplantation studies on amoebas (1) and oocytes (2) that removal of nuclei from the cytoplasm causes a rapid loss of activity. Certainly, isolated nuclei are the material of choice for a variety of experimental procedures; however, it is not at all clear that their *in vivo* permeability characteristics are retained. Unless this can be demonstrated, and it has not yet been possible to do so in any isolated system, the validity of the results will be open to question.

Direct injection of tracers into the cells avoids the more serious problems associated with isolation procedures, and it is preferable if the cells remain viable and it can be demonstrated that the tracers are not bound or metabolized following injection. The major limitation of the injection approach is that relatively few cell types can be conveniently studied. To date, most studies have been performed on oocytes and amebas; these cells are large enough for this type of experimentation and are not adversely effected by microinjection. However, recent progress in microsurgical procedures (3) as well as the development of an elegant technique for "microinjection" by cell fusion (4,5) should permit the investigation of a wider variety of cell types in the future.

Current views of nuclear envelope permeability to large molecules are based largely on injection studies utilizing electron-opaque particles, which were localized with the electron microscope, and exogenous proteins or synthetic polymers labeled with fluorescein or radioactive atoms. Tracers having known chemical and physical properties were selected. The conclusions drawn from these investigations can be summarized as follows:

1. The nuclear pores are the only sites that have been shown experimentally to function as pathways for macromolecular exchange; however, other possible routes, such as direct transit across the membranes of the envelope, cannot be ruled out (6).

2. Passage through the pores is restricted to a central channel formed by the annular material. The width of these channels (i.e., the functional size of the pores with regard to macromolecular exchange) varies in different cell types. In amphibian oocytes the functional pore size is estimated to be approximately 90 Å in diameter (7), whereas in amebas it might be as large as 145 Å (8).

3. As one might expect, molecular size is a major factor determining whether a given substance will enter the nucleus. This is illustrated by a study on roach oocytes (9) in which fluorescein labeled proteins with molecular weights varying from 12,000 to 68,000 were injected into the cytoplasm, and the subsequent appearance of label in the nucleoplasm was determined using fluorescence microscopy. The smaller polypeptides, with molecular weights of approximately 18,000 and below, rapidly penetrated the nuclear envelope, and in some instances they equilibrated between the nucleus and cytoplasm within minutes. Larger proteins entered the nucleus at much slower rates. The cytoplasmic concentration of ovalbumin (MW 43,000) was twice as great as the nuclear concentration even after 5 hours. Bovine serum albumin (MW 68,000) was also detected in the nucleoplasm after 5 hours, but in even lower concentrations than ovalbumin (the nucleus to cytoplasmic ratio was only 0.08). Results consistent with these have been reported by Paine *et al.* (7) and by Bonner (10). Both of the latter investigations were performed on amphibian oocytes.

4. The permeability characteristics of the nuclear envelope do not appear to be fixed. In amebas, for example, the number and size of gold particles taken up by the nuclei vary during different periods of the cell cycle (11) and also during different physiological states (12). These variations appeared to be caused by changes in the properties of the annular material associated with the pores; however, changes in pore number during the cell cycles have been reported and could also effect permeability (13).

Reviews dealing with the structure, function, and composition of the envelope have been published by Feldherr (6), Kessel (14), and Franke (15).

II.   Methods

A.   The Nuclear Uptake of Endogenous Proteins

In order to study the patterns of uptake of endogenous proteins into the nucleus the following general approach has been adapted. First, *Xenopus laevis* oocytes are pulse-labeled with tritiated amino acids. At various times after labeling the nuclei are isolated and run on sodium dodecyl sulfate (SDS)–polyacrylamide gels. The incorporation of different size polypeptides into the nucleoplasm is determined by measuring radioactivity in gel slices. Since it can be shown that the proteins which enter the nucleus are synthesized in the cytoplasm, this system appears to be a valid one for investigating nucleocytoplasmic exchange. The success of this approach is dependent on certain specific features of amphibian oocytes, especially the size of the germinal vesicles which make it possible to perform experiments on fewer than 10 nuclei. In addition, the germinal vesicles can be isolated by hand in a matter of seconds, thus avoiding the problems normally associated with mass isolation procedures.

B. Biological Material

Xenopus laevis, which can be purchased from Nasco (Fort Atkinson, Wisconsin), are maintained in dechlorinated tap water and fed three to four times a week. Ovaries, or portions of ovaries, are removed from anasthetized animals using the surgical procedures described by Gall (16). The animals are anesthetized by packing them on ice until they fail to respond to mechanical stimulation; this usually requires 30–45 minutes. By employing these procedures, each animal can be used for two to four experiments. If necessary, the ovaries can be stored in dry sealed embryological watch glasses at 4° (16). Since most of our studies were of relatively short duration (30 minutes to 6 hours), we used either Millipore-filtered amphibian Ringer's or OR-2 solution (17) as the extracellular medium. For longer term experiments it would be advisable to use the culture medium developed by Eppig (18). Because of their size, late stage-5 and stage-6 oocytes (19) are the cells of choice, but all of the procedures described here can be performed on cells as small as stage-3 oocytes.

For the following reasons, it is advantageous to defolliculate the oocytes before labeling: (1) microinjection and enucleation procedures can be carried out more easily in the absence of the follicle layer; and (2) incorporation of labeled amino acids by follicle cells could introduce a source of error in instances where total oocyte counts are necessary. When only small numbers of cells are required, it is preferable to defolliculate manually using watchmaker's forceps. Alternatively, the follicle layer can be removed enzymatically using either trypsin (20) or collagenase (18).

C. Labeling Procedures

Amphibian oocyte proteins can be pulse-labeled either by microinjection or incubation. The advantage of microinjection is that cells can be provided with known and reproducible amounts of label. Incubation of the oocytes in precursor is a much simpler procedure and, in many instances, entirely adequate.

Microinjection pipettes with tip diameters of approximately 25 μm (ID) are constructed with a de Fonbrune microforge as described by King (2). To deliver reproducible volumes, each pipette is constricted at a predetermined point along its shaft. The volume contained between the tip and the constriction is measured by using isotope solutions (21) or by filling the pipette with oil and measuring the diameter of the drop formed when the oil is injected into water.

In most of our experiments we injected 30–35 nl of amphibian Ringer's solution containing approximately 470,000 dpm of L-leucine-^{3}H (sp act 44 Ci/mmol). It has been demonstrated that the incorporation of L-leucine ^{3}H into material precipitable by trichloroacetic acid (TCA) is essentially completed within 30 minutes. The time required for incorporation is a function of both the pool size and the specific activity of the precursor and can be expected to vary depending on the nature of the amino acid used. The importance of these factors has been discussed by Ecker and Smith (22). Data relating to the pool size of various amino acids in *Xenopus* oocytes have been published by Eppig and Dumont (23).

Sufficient counts for nuclear permeability studies can also be obtained by incubating the oocytes in Ringer's or OR-2 solution containing labeled precursor. Using L-leucine-^{3}H, for example, a 30-minute incubation in medium containing 125 μCi/ml (sp act 44 Ci/mmol) is adequate. In this instance, there is little additional incorporation into precipitable counts if the cells are rinsed in several changes of fresh medium following incubation. This is not always the case; L-methionine-^{3}H can undergo significant incorporation even after the cells are washed in precursor-free medium. Again, this is dependent on the pool size and specific activity of the amino acids being studied. One can prevent incorporation after the initial labeling period by either chasing with cold amino acids or inhibiting protein synthesis with puromycin. Over 90% inhibition can be achieved by microinjecting 160 ng of puromycin per oocyte, or simply incubating the cells in medium containing 100 μg of puromycin per milliliter. It has been demonstrated that the uptake of proteins by the nucleus is not affected by 3-hour treatments with puromycin; longer-term experiments have not yet been performed.

Double-labeling experiments, using ^{3}H and ^{14}C, are difficult to perform since available ^{14}C-labeled amino acids have relatively low specific activities and would require long-term pulses, making kinetic studies difficult. Fortunately, double-labeling can be accomplished using ^{3}H- and ^{35}S-labeled L-methionine. Oocytes are incubated for 30 minutes in medium containing either precursor at a concentration of 1 mCi/ml and a specific activity of approximately 14 Ci/mmol (the highest specific activity available for L-methionine-^{3}H). The specific activity of L-methionine-^{35}S is considerably higher, and it is necessary to dilute this precursor with cold methionine. To prevent further incorporation beyond the 30-minute labeling period, the oocytes are either chased or treated with puromycin.

D. Measurements of Isotope Incorporation in Whole Cells

The total incorporation of labeled amino acids into precipitable macro-molecules can be determined by extracting oocytes with three or four changes of either 7.5 or 10% TCA over a period of approximately 20 hours. The oocytes need not be homogenized prior to extraction. The procedures for measuring radioactivity in whole oocytes are the same as those used for gel slices and will be described below.

The injection of L-leucine-^{3}H resulted in the incorporation of anywhere from 40,000 to over 120,000 cpm per oocyte. Incubation for 30 minutes in either L-leucine-^{3}H or L-methionine-^{3}H gave comparable results, i.e., variable incorporation with an average of about 60,000 cpm per oocyte. These levels of total cell radioactivity are sufficient for nuclear uptake studies. Significant variations in total counts occur among oocytes from different animals, as well as those obtained from the same ovary. As one would expect, similar differences are observed in nuclear counts, and it is essential that this potential problem be considered when designing experiments.

As already indicated, the conditions for labeling oocytes are such that incorporation of the precursor is completed within 30 minutes. Furthermore, it was found in long-term experiments, utilizing L-leucine-^{3}H, that there are no quantitative changes in total counts for at least 2 or 3 days after labeling. To determine whether there are qualitative changes in the labeled polypeptides, SDS–polyacrylamide electropherograms were obtained for whole oocytes 30 minutes and 3 hours after L-leucine-^{3}H injection. During this interval no obvious differences in the gel patterns could be detected.

E. Isolation of Nuclei

In most cell types, characterization of nuclear proteins involves mass isolation procedures which could easily alter the normal distribution of macromolecules between the nucleus and cytoplasm. To avoid this problem, oocyte nuclei are isolated by hand in a medium containing 102 mM KCl–11.1 mM NaCl–7.2 mM K_2HPO_4–4.8 mM KH_2PO_4 (pH 7.0 ±0.1); "5:1 medium," made by mixing 5 parts 0.1 M KCl and 1 part 0.1 M NaCl, can also be used (16). Nuclei can be isolated in less than 30 seconds, and they are immediately fixed in absolute ethanol. There is evidence that no appreciable exchange of protein across the nuclear envelope occurs during this period (24). Another advantage of alcohol fixation, aside from preventing loss of protein, is that the nuclear envelope separates from the nucleoplasm and can be dissected off with watchmaker's forceps. This assures that all contaminating cytoplasm is removed, although a comparison of nuclear counts before and after removal of the envelope indicates that such contamination is not a serious problem (*10,25*). The procedure we use for isolating amphibian oocyte nuclei will be published in a separate chapter in this series and is similar to that described by Gall (*16*).

It is frequently desirable to obtain total precipitable nuclear counts. For this purpose, alcohol-fixed nuclei (with or without the envelopes removed) are extracted for 90 minutes in 10% TCA, prior to being dissolved in scintillation fluid. For nuclei fixed 30 minutes, 3 hours, and 6 hours after injection, we have obtained average counts 770, 5560, and 10,190, respectively.

F. SDS–Polyacrylamide Gel Electrophoresis and Counting Procedures

At various times after labeling, the uptake of different size polypeptides by the nuclei is analyzed by measuring radioactivity in slices of SDS–polyacrylamide gels. Alcohol-fixed nuclei (not extracted with TCA) are dissolved in sample buffer consisting of 0.0625 *M* Tris-HCl (pH 6.8)–2.5% SDS–5% 2-mercaptoethanol–0.02% mercaptoacetic acid. The samples are then heated to 90°C for 5 minutes. Finally, glycerol and bromphenol blue are added to concentrations of 10% and 0.001%, respectively. To obtain staining patterns of nuclear polypeptides it is necessary to run at least 20 nuclei per gel. If the gels are to be sliced and counted, 5–10 nuclei are sufficient. For tube gels, the nuclei are dissolved in 100 μl of sample buffer; when running slab gels 30 μl of buffer are used. Gel patterns for whole oocytes can be obtained by fixing oocytes in absolute ethanol and dissolving them in 50 μl of sample buffer per oocyte.

All samples are run on discontinuous SDS–polyacrylamide gels (3% stacking and 10% running gels). The method described by Laemmli (*26*) is used, with the exception that 0.02% mercaptoacetic acid is added to the electrode buffer (as well as the sample buffer) to prevent reoxidation of the polypeptides. Tube gels (ID 5 mm) are electrophoresed at 1.6 mA/gel until the dye marker has migrated about 47 mm into the running gel (approximately 4 hours). Slab gels are run in a Bio-Rad model 220 gel electrophoresis cell. The slabs (1.5 mm thick × 160 mm wide × 120 mm high) are electrophoresed at 35 mA for approximately 2 hours and 45 minutes, by which time the dye marker migrates about 74 mm. The following molecular weight standards were used in our studies: phosphorylase a, MW 94,000; bovine serum albumin (BSA), MW 68,000; ovalbumin, MW 43,000; and cytochrome c, MW 11,700.

At the completion of the runs the gels are fixed in 7.5% TCA overnight. Staining, when required, is carried out for 2–3 hours in 0.1% Coomassie blue dissolved in 50% methanol–40% ion-free water–10% glacial acetic acid. When tube gels were used, only the molecular weight standards were stained; however, the gels containing labeled material were treated in the same methanol–acetic acid solution to compensate for any changes in gel length. All of the slab gels were stained since standards and experimental material were run in adjacent slots. Due to the fact that relatively few nuclei are used to determine labeling patterns, the bands in these gels are only lightly stained, and there is no interference with subsequent counting procedures. Destaining is carried out in 7.5% glacial acetic acid.

The labeled gels are frozen on Dry Ice and sliced transversely at 1- or 2-mm intervals. Each slice is then placed in a vial along with 10 ml (for slab gel slices) or 20 ml (for tube gel slices) of scintillation fluid containing 3.5% Protosol, and 0.4% Omnifluor (New England Nuclear) in toluene. The vials are then incubated in a 60°C oven for 2–3 hours, and finally they are placed on a reciprocating shaker, at room temperature, until the slices become transparent. Whole oocytes and nuclei were dissolved in 10 ml of scintillation fluid and treated the same way as gel slices. Radioactivity can then be measured using a liquid scintillation counter.

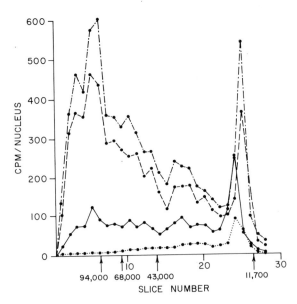

Fig. 1. SDS–gel patterns of L-leucine-³H-labeled polypeptides in oocyte nuclei. The "controls" were fixed 30 minutes (——) and 3 hours (– –) after injection of the label. The effect of simultaneously injecting L-leucine-³H and puromycin is shown in the lower curve (.....). In the experiment shown in the upper curve (– · – · – ·), puromycin was injected 30 minutes after the label. The arrows indicate the migration of standard proteins. (From Feldherr (27).

III. Results

An example of the data obtained using the above procedure is shown in Fig. 1, which illustrates the pattern of nuclear protein uptake, as well as the effects of puromycin [for further details see Feldherr (27)]. The "control" oocytes were enucleated either 30 minutes or 3 hours after being injected with L-leucine-^{3}H. An examination of these two curves shows that (1) even after 30 minutes labeled polypeptides as large as 150,000 MW are present in the nucleoplasm, and (2) with increasing time, the total number of counts per nucleus increases as does the proportion of counts in the higher molecular weight polypeptides. Recent long-term experiments have shown that the nuclear counts continue to increase for about 18 hours after a pulse label. The lower curve in Fig. 1 shows the results obtained 3 hours after L-leucine-^{3}H and puromycin are injected simultaneously. Under these conditions, there is over 90% inhibition of protein synthesis. The upper curve represents an experiment in which L-leucine-^{3}H was injected into the oocytes, followed 30 minutes later (i.e., after incorporation of the label is complete) by an injection of puromycin. The nuclei were isolated $2\frac{1}{2}$ hours after the second injection. These results, which are similar to the 3-hour controls, show that nuclear uptake occurs in the absence of protein synthesis: they also show that the labeled nuclear polypeptides are taken up from a cytoplasmic pool and not synthesized in the nucleoplasm.

The most interesting results obtained so far are related to the uptake rates of large polypeptides by the nucleus. Using data for relative concentrations of different size polypeptides in the nucleus and cytoplasm, it could be calculated that, on a volume basis, polypeptides with molecular weights of 94,000 and above were more concentrated in the nucleus than in the cytoplasm by 3 hours after injection. As indicated in the introduction, such a rapid accumulation of large polypeptides would not be expected from previous permeability studies with exogenous tracers. For example, Bonner (10) has shown that 24 hours after injecting BSA (MW 68,000) into Xenopus oocytes, the nuclear concentration was only about 0.18 that of the cytoplasm.

IV. Conclusions and Applications

A method has been described for studying the uptake of endogenous proteins by oocyte nuclei that avoids many of the technical problems encountered using smaller cells. Initially, the oocytes are pulse labeled either by microinjection or incubation, and a cytoplasmic pool of labeled polypeptides is established. At specific periods after labeling, nuclei are isolated by hand and fixed in ethanol. This procedure requires less than 30 seconds, and the *in vivo* component of the nuclear proteins appears to be retained. Total incorporation of precursor can readily be obtained for whole cells and nuclei using standard liquid scintillation techniques. Qualitative analysis of the labeled polypeptides is performed using SDS–polyacrylamide gels.

So far, studies have been concerned mainly with the uptake rates of proteins by the nucleus. The studies indicate that these substances enter the nucleoplasm at a greater rate than exogenous tracers of comparable size. The rapid accumulation of large endogenous polypeptides could be due to (1) the existence of specific transport processes or (2) simple diffusion. The latter explanation is possible if certain conditions are met regarding the shape and equilibrium distributions of the larger polypeptides. Determining which of these mechanisms is operating is obviously of fundamental importance in understanding nucleocytoplasmic interactions. Using the approach described in this chapter should make it possible to investigate a number of other basic problems relating to protein exchange. For example, the uptake rates of specific polypeptides can be studied utilizing two-dimensional gel procedures. It is feasible, by employing Eppig's technique for culturing oocytes (*18*), to study turnover rates of nuclear polypeptides. Slight modifications of the labeling and nuclear isolation procedures should make it possible to investigate quantitative and qualitative changes in nuclear protein uptake during different stages of oogenesis.

ACKNOWLEDGMENTS

The author would like to thank Mr. Paul Richmond for his assistance in the preparation of this manuscript. The original investigations of the author were supported, in part, by grant GM 21531 from the National Institutes of Health.

REFERENCES

1. Goldstein, L., *Methods Cell Physiol.* **1**, 97 (1964).
2. King, T. J., *Methods Cell Physiol.* **2**, 1 (1966).
3. Diacumakos, E. G., Holland, S., and Pecora, P., *Nature (London)* **232**, 28 (1971).
4. Loyter, A., Zakai, N., and Kulka, R. G., *J. Cell Biol.* **66**, 292 (1975).
5. Schlegel, R. A., and Rechsteiner, M. C., *Cell* **5**, 371 (1975).
6. Feldherr, C. M., *Adv. Cell Mol. Biol.* **2**, 273 (1972).
7. Paine, P. L., Moore, L. C., and Horowitz, S. B., *Nature (London).* **254**, 109 (1975).
8. Feldherr, C. M., *J. Cell Biol.* **25**, 43 (1965).
9. Paine, P. L., and Feldherr, C. M., *Exp. Cell Res.* **74**, 81 (1972).
10. Bonner, W. M., *J. Cell Biol.* **64**, 421 (1975).
11. Feldherr, C. M., *J. Cell Biol.* **39**, 49 (1968).
12. Feldherr, C. M., *Tissue Cell* **3**, 1 (1971).
13. Maul, G. G., Maul, H. M., Scogna, J. E., Lieberman, M. W., Stein, G. S., Hsu, B. Y., and Borun, T. W., *J. Cell Biol.* **55**, 433 (1972).

14. Kessel, R. G., *Progr. Surf. Membr. Sci.* **6**, 243 (1973).
15. Franke, W. W., *Int. Rev. Cytol. Suppl.* **4**, 71 (1974).
16. Gall, J. G., *Methods Cell Physiol.* **2**, 37 (1966).
17. Wallace, R. A., Jared, D. W., Dumont, J. N., and Sega, M. W., *J. Exp. Zool.* **184**, 321 (1973).
18. Eppig, J. J., *In Vitro*, in press.
19. Dumont, J. N., *J. Morphol.* **136**, 153 (1972).
20. Colman, A., *J. Embryol. Exp. Morphol.* **32**, 515 (1974).
21. Smith, L. D., and Ecker, R. E. *Curr. Top. Dev. Biol.* **5**, 1 (1970).
22. Ecker, R. E., and Smith, L. D., *Dev. Biol.* **18**, 232 (1968).
23. Eppig, J. J., and Dumont, J. N., *Dev. Biol.* **28**, 531 (1972).
24. Macgregor, H. C., *Exp. Cell Res.* **26**, 520 (1962).
25. Feldherr, C. M., and Richmond, P. A., *Exp. Cell Res.*, in press.
26. Laemmli, V. K., *Nature (London)* **227**, 680 (1970).
27. Feldherr, C. M., *Exp. Cell Res.* **93**, 411 (1975).

Part C.   Fractionation and Characterization of Histones. I

Chapter 10

Histone Nomenclature

E. M. BRADBURY[1]

Biophysics Laboratories, Portsmouth Polytechnic, Gun House,
Hampshire Terrace, Portsmouth, Hampshire, England

During the past 3 to 4 years strenuous efforts have been made to clarify the confusion of histone nomenclatures by deciding on one nomenclature acceptable to all histone workers.

Two main nomenclatures have been used extensively. In the one developed by Johns, Phillips, and Butler (*1–4*), the very lysine-rich histone was named F1, the slightly lysine-rich histones F2A2 and F2B, and the arginine-rich histones F3 and F2A1. This nomenclature was based on the separation of histone fractions by selective solvent extraction and by elution on carboxymethyl-cellulose columns. The second major nomenclature was developed by Luck and co-workers (*5–7*), who separated histones into three classes, I, II, and III, by column chromatography on Amberlite CG-50. The nomenclature was extended by separating the components of each class of histones (*8,9*) to give the very lysine-rich histone I, the slightly lysine-rich histones IIb1 and IIb2, and the arginine-rich histones III and IV.

An attempt to replace these two nomenclatures by a completely new nomenclature based on the three most abundant amino acids in each histone in order of their abundance (using the one-letter code for amino acids) was short-lived because it was too cumbersome to use and confusion arose when the second and third most abundant amino acids were present in almost equal amounts.

The originators of the above nomenclatures were present at the Ciba

---

[1]*Present address:* Biophysics Laboratories, Portmouth Polytechnic, St. Michael's Building, White Swan Road, Portsmouth, Hampshire, England.

Foundation meeting on "The Structure and Function of Chromatin" (*10*) and the opportunity was taken to hold a detailed discussion on histone nomenclatures. Most participants were in favor of a single nomenclature based in some way on the two original nomenclatures in wide usage. A nomenclature was proposed that combined the elements of both these nomenclatures (Table I). There is a large overlap between the new nomenclature and each of the older nomenclatures which allows for easy acceptance by existing workers. Because of the great similarity of histones from different tissues and organisms the labels in the new nomenclature represent a particular class of histone, e.g., bovine H4 or pea H4. It is unlikely that there will be a large number of additional histone classes, and other unique histone classes can be described by the addition of further arabic numerals, e.g., the lysine-rich histone unique to erythrocytes (*11,12*) becomes H5 and the new histone from trout testis H6 (*13*).

Chemically modified histones can be described by the addition of prefixes, e.g., P.Ac. H4 to show the H4 histone is monoacetylated and monophosphorylated. This can be extended to include the sites of the modifications, e.g., 11, 16 Ac.2 H4.

This new nomenclature was overwhelmingly accepted by the participants of the Ciba Foundation Symposium (*10*) and has been submitted to the IUPAC–IUB Commission on Nomenclature for ratification. As can be seen from current literature, it is being widely used both by the established workers and new young workers. It is to be hoped that existing users of

TABLE I

HISTONE NOMENCLATURE

Nomenclatures	Johns, Phillips, and Butler (*1–4*)	Luck *et al* (*5–7*), Murray *et al.* (*8*), Fambrough *et al.* (*9*)	Ciba Foundation Symposium (*10*)
Lysine-rich histones	F1	Ia	H1
		Ib	
Slightly lysine-rich	F2A2	IIb1	H2A
histones	F2B	IIb2	H2B
Arginine-rich histones	F3	III	H3
	F2A1	IV	H4
Other histones			
Unique lysine-rich histone			
from nucleated erythrocytes	F2C(*14*)	5(*11,12*) or V(*15*)	H5
Lysine-rich histone		T(*13*)	H6

the older nomenclatures will help to complete the transition to a single nomenclature by also adopting the nomenclature H1, H2A, H2B, H3, H4, H5, and H6.

REFERENCES

1. Johns, E. W., *Homeostatic Regul. Ciba Found. Symp. 1969*, p. 128.
2. Johns, E. W. *in* "Histones" (D. M. P. Phillips, ed.). Plenum, New York, 1971.
3. Johns, E. W., and Butler, J. A. V., *Biochem. J.* **82**, 15 (1962).
4. Phillips, D. M. P., and Johns, E. W., *Biochem. J.* **94**, 127 (1965).
5. Luck, J. M., Rasmussen, P. S., Satake, K., and Tsvetikov, A. N., *J. Biol. Chem.* **233**, 1407 (1958).
6. Rasmussen, P. S., Murray, K., and Luck, J. M., *Biochemistry* **1**, 79 (1962).
7. Satake, K., Rasmussen, R. S., and Luck, J. M., *J. Biol. Chem.* **235**, 2801 (1960).
8. Macpherson, A., and Murray, K., *Biochim. Biophys. Acta* **104**, 574 (1965).
9. Fambrough, D. M., Fujimura, F., and Bonner, J., *Biochemistry* **7**, 575 (1968).
10. Fitzsimons, D. W., Wolstenholme, G. E. W., eds, *Ciba Found. Symp. Struct. Function Chromatin* **28**, 1975.
11. Neelin, J. M., *in* "The Nucleohistones" (J. Bonner and P. O. P. Ts'O, eds.), p. 66. Holden-Day, San Francisco, 1964.
12. Neelin, J. M., and Butler, G. C., *Can. J. Biochem.* **39**, 485 (1961).
13. Wigle, D. T., and Dixon, G. H., *J. Biol. Chem.* **246**, 5636 (1971).
14. Hnilica, L. S., *Experientia* **20**, 13 (1964).
15. Vidali, G., and Neelin, J. M., *Eur. J. Biochem.* **5**, 330 (1968).

Chapter 11

The Isolation and Purification of Histones

ERNEST W. JOHNS

Division of Molecular Biology,
Chester Beatty Research Institute,
Institute of Cancer Research–Royal Cancer Hospital,
London, England

I. Introduction

The reasons for fractionating, isolating, and characterizing different histone fractions are numerous, and of course the methods used vary considerably. Many workers may be satisfied with a simple separation of histones in polyacrylamide gel, whereas others may require certain fractions in gram quantities. In short, the aims and resources of an investigation largely dictate the methods to be used, and a chapter such as this, dealing with six different well-characterized proteins, cannot hope to cover in detail all the methods available. However, most investigators in this field appear to require at some time or another a few milligrams of each of the five mammalian histone fractions in a highly purified state to be used as standards. I shall therefore describe briefly the preparative methods which have enabled us to obtain all five histone fractions from calf thymus during one preparation in quantities from a few milligrams to grams in high states of purity. However, when concentrating on individual fractions rather than all five together, the methods can be simplified and modified, sacrificing others fractions for the sake of speed and purity. These modifications are given in Section VIII.

Methods are also given for the preparation of the nucleated erythrocyte specific fractions F2C (Section IX).

All the methods described in this chapter are relatively simple, and no complicated techniques are involved. However, it should be stressed that biological preparations are extremely variable and every effort should be made to standardize procedures. One of the major variables would appear to be the state of the original tissue, and at the risk of appearing too elementary I would emphasise that all tissues should be frozen immediately on removal from the animal to $-80\,°C$ or lower and not thawed again for any reason before use.

Accepting a certain variability then, it is essential to characterize all proteins prepared, and the methods for such characterization are discussed in Section X.

The methods described are those that the author has found most useful, and these, not surprisingly, include mainly the methods developed in his own laboratory. This is not to say that other methods are not equally useful, but the literature on this subject is complex and confused by a multitude of different nomenclatures. For a full discussion of the various methods available, their relative advantages and disadvantages, and for a comparison of the various nomenclatures used, the reader is referred to the following: Phillips (1), Elgin et al. (2), Hnilica (3), Bradbury (4), and the chapter by Bradbury in this volume.

II.   The Isolation of the Five Main Histone Fractions of Calf Thymus during One Preparation

Some years ago we published two methods (5) for the isolation of four histone fractions from calf thymus by the selective extraction of deoxyribonucleoprotein (DNP) using ethanol–HCl mixtures. These four fractions had previously been characterized and designated as follows. F1 and F2B were lysine-rich histones, F3 was an arginine-rich histone, and the fourth fraction which we called F2A was still a mixture of two histones. This was subsequently separated into its two components which we called F2A1 and F2A2. F2A1 was another arginine-rich histone and F2A2 was an intermediate type having a molar ratio of lysine to arginine of about one (6). A selective extraction method was later developed for these two fractions (7).

In our laboratory over the past 8 years these methods have been combined to give a procedure for the isolation of all five fractions during one preparation, but this combined procedure has never been published in detail. Also during this period many modifications to the original procedures have been made which improve the separations obtained. Sanders and McCarty (8) have also adapted these basic extraction methods to develop a procedure for the isolation of all six histones from avian erythrocytes. When all fractions are required from nucleated erythrocytes, this method is recommended since the presence of the nucleated erythrocyte-specific histone F2C, and its partial replacement of histone F1, means that a different procedure is necessary for the separation of this group of histones.

With most selective extraction procedures that produce large quantities of histones, a certain amount of cross-contamination is inevitable, but using the method described here and with care and attention to detail this should never exceed 5% in the worst case. Indeed most samples give single bands on polyacrylamide gel electrophoresis, which with our quantitative methods (9) we estimate as having less than 2% of other protein present. In most cases these figures can be improved by recycling the protein through the relevant precipitation procedure (see Sections III–VI). However, since very large quantities of fractionated material can be obtained by the use of selective extraction techniques, purification by any of a number of procedures is possible.

More recently Oliver et al. (10) have also given their detailed attention to the further purification of histones, based to a large extent on our original method I (5). We now consider method II as the better for the preparation of all five fractions and method I is recommended for the preparation of histone F1 alone (see Section VIII, A). However, some of the purification stages detailed below have been taken in a slightly modified form from their excellent paper on this subject.

Although the method detailed below used 50 gm of calf thymus, producing approximately 200 mg of each of the histone fractions, it can be scaled up at least six times, but it should be emphasised that to do this it is necessary to modify the procedure. The volumes of many solvents, at various stages, cannot be multiplied by six for practical reasons. Consequently, protein concentrations will vary and differential separations by solvents will take place at different solvent concentrations.

Throughout this procedure there are many practical points which may not at first sight seem important, but which seriously affect the purity of the final products. The reasons for many of the steps in the procedure may also not seem obvious and, therefore, in order not to break the continuity of the method, numbers have been inserted at appropriate points and detailed explanations given afterward against the relevant number. A schematic diagram of the preparative method is given in Fig. 1.

A. Details of the Procedure

All operations are carried out at $4°$ unless stated otherwise in the text.

Calf thymus is obtained directly after slaughter of the animal and frozen on solid carbon dioxide. It should be stored at about $-80°$ until required. The gland is then warmed to $0°$ in ice, and membranes and connective tissue are removed. After mincing, 50 gm are taken and homogenized (Note 1) for 2 minutes at top speed in 700 ml of $0.14 M$ NaCl adjusted to pH 5.0 using

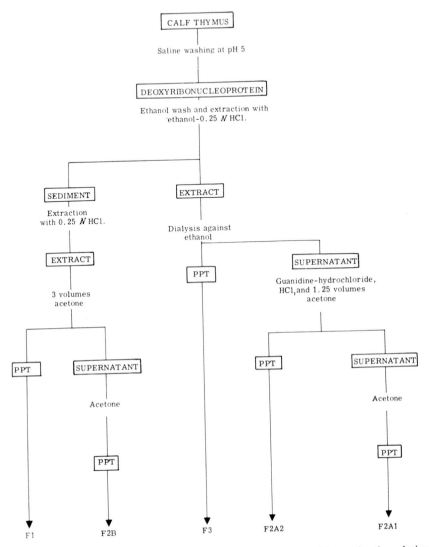

Fig. 1. A schematic diagram showing the isolation of all five histone fractions during one preparation.

N-HCl. The pH of the first homogenate is also adjusted to 5.0 by the careful addition of more N-HCl (note 2) and then centrifuged in an M. S. E. Mistral centrifuge at 1500 g for 30 minutes. The supernatant is rejected and the sediment washed again in a similar manner but only blending for 1 minute and centrifuging at 1500 g for 20 minutes. At this stage it should no longer be necessary to adjust the pH of the homogenate to 5.0 since the pH 5.0

saline should be sufficient to maintain it. The sediment is then washed four times more in a similar manner but only blending for 30 seconds, and centrifuging at 1500 g for 15 minutes. The sediment remaining after the pH 5.0 saline washings is then blended at half speed (Note 1) for 30 seconds, with 700 ml of 90% ethanol and centrifuged at 1500 g for 15 minutes (Note 3). The supernatant is rejected and the sediment placed in a 1-liter wide-necked polyethylene bottle containing 400 ml of ethanol–1.25 N HCl (4:1, v/v) and approximately 1500 (400 ml) mixed glass beads from 4–8 mm in diameter. The bottle is then placed on a Winchester shaker (Note 4) and shaken overnight (approx. 18 hours.) The contents of the pot, excluding the glass beads (Note 5), are then centrifuged at 1500 g for 15 minutes and the sediment extracted twice more in the same way, but using only 200 ml portions of ethanol–1.25 N HCl (4:1, v/v) and extracting for only 3 hours each time. The combined supernatants (780–800 ml) are clarified by filtering through a No. 4 sintered glass funnel and placed in dialysis bags and dia-lyzed against 4.5 volumes of ethanol overnight at room temperature (22°) with stirring. The dialysis is then repeated against a fresh 4.5 volumes of ethanol for 5 hours (Note 6). The precipitate which forms is F3 histone and is recovered by low-speed centrifugation, washed twice in 100 ml of the solution against which it had been dialyzed (Note 7), and then washed three times in acetone (Note 8) and finally dried under vacuum. The vacuum is released several times during this procedure and the powder broken up with a glass rod. The product should be a fine white powder which is completely and easily soluble in water.

The supernatant remaining after the precipitation of F3 which contains mainly F2A1 and F2A2 is then dialyzed again against the same solution, but with the addition of another 2 volumes (approx. 1600 ml depending on volume of extract) of ethanol, for 4 hours with stirring. It is then removed from the dialysis bag and kept in the inner cold room (−10°) overnight. After warming to room temperature (22°) the small precipitate that forms (approx. 40 mg) is removed by centrifugation and rejected (Note 9). The supernatant is then filtered through a No. 4 sintered glass funnel (Note 10). Solid guanidine hydrochloride (44 gm) is then added to the supernatant (Note 11) still at room temperature, together with 8.6 ml of concentrated HCl with rapid stirring. When all the guanidine hydrochloride is dissolved the solution is again clarified by filtering through a No. 4 sintered glass funnel. The volume of solution should now be about 660 ml since a volume decrease occurs during the dialysis stage. Next 1.25 volumes of acetone (825 ml) are added slowly with rapid stirring and the solution allowed to stir for a further 10 minutes. The precipitate which forms is F2A2 and is removed by low-speed centrifugation. It is washed twice in a solution similar to that from which it was precipitated (Note 12), then three times in acetone and dried under vacuum as described previously.

The supernatant which contains histone F2A1 is clarified again if necessary by filtering through a No. 4 sintered glass funnel and an equal volume (approx. 1500 ml) of acetone added. The precipitate which forms is F2A1 and is recovered by centrifugation, washed three times in acetone, and dried under vacuum as described above.

The sediment which remained after the extraction of histones F3, F2A2, and F2A1 from the saline-washed nucleoprotein using ethanol–1.25 N HCl (4:1, v/v) contains histones F1 and F2B and is treated as follows.

A further extraction of the sediment with 200 ml of ethanol–1.25 N HCl (4:1, v/v) is carried out for 1.5 hours in a similar manner to the previous three; in this case, however, the supernatant after centrifugation is rejected. The sediment is then extracted in the same manner with 200 ml of 0.25 N HCl for 18 hours, followed by two more extractions using 100 ml of 0.25 N HCl for 2 hours each, in order to extract histones F1 and F2B. The combined supernatants (approx. 390 ml) are clarified by filtration through a No. 4 sintered glass funnel, raised to room temperature, and 3 volumes (1170 ml) of acetone are added to precipitate F1. The precipitate is recovered by low-speed centrifugation and then washed in 100 ml of 0.25 N HCl–acetone (1:3, v/v) at room temperature, then three times in acetone and dried under vacuum as described previously.

The supernatant containing F2B is filtered through a No. 4 sintered glass funnel and 2 volumes (780 ml) of acetone are added. The precipitate that forms after stirring rapidly for a few minutes is F2B and is recovered by centrifugation at low speed, washed three times in acetone, and dried as described previously.

The yield of the various histone fractions from 50 gm, of calf thymus is 200–250 mg of each fraction. More recently, however, we have been carrying out preparations on cow thymus and have found an increased yield per weight of tissue. For 50 gm of cow thymus we have consistently obtained yields of up to 300 mg for each fraction.

At this stage the purity of the fractions can be judged by the methods discussed in Section X. If they are then required in a more highly purified state, or if the preparation has inadvertently gone wrong, then the methods described in Sections III–VI can be employed for their further purification.

B. Detailed Comments on the Procedure

NOTE 1

The blender is a Kenwood domestic blender Cat. No. A 956A. Two speeds are available, a top speed at approximately 12,000–15,000 rpm and a lower speed that we call half-speed but which is approximately 5000–6000 rpm.

NOTE 2

Saline (0.14 M NaCl) washing at pH 7 is not recommended since there are many proteolytic enzymes in the homogenate with maximum activities at about this pH. Indeed proteolytic activity can be demonstrated in the isolated histone fractions with a maximum activity at about pH 7.5 (*11*). At pH 3.5 such activity is much less and we have carried out some experiments in which we have saline-washed at pH 3.5. These preparations gave high yields and good-quality histone fractions. However, at pH 3.5 ribonucleoproteins are precipitated, together with other macromolecules. The possibility of contamination of the histone fractions with ribosomal and other proteins must therefore be increased. Even so, such a contamination cannot be detected by polyacrylamide gel electrophoresis, but this is probably due to the large excess of histones over other proteins in calf thymus. With other tissues, saline washing at pH 3.5 would probably give much more contamination. Saline washing at pH 5.0 is therefore a compromise situation and may be changed depending on the contamination that can be expected.

NOTE 3

This washing in 90% ethanol is essential since the next stage in the preparation is an extraction of the sediment with 80% ethanol–HCl. The single washing with 90% ethanol insures that the aqueous sediment is in approximately 80% ethanol when it is extracted at the next stage. If this is omitted then the first ethanol–HCl extraction will be diluted with saline and histones F1 and F2B may be extracted. If it is washed with too much ethanol then the extraction of F3 may not be as efficient.

NOTE 4

This method of extraction is essential for quantitative removal of F3, F2A1, and F2A2 without removing F1 and F2B. More vigorous means of extraction (e.g., by blending) invariably result in the removal of some F1 and F2B, presumably caused by local overheating. Less vigorous forms of extraction (e.g., by stirring) result in the incomplete extraction of F3, F2A2, and F2A1, which will then be extracted by the HCl along with histones F1 and F2B. The Winchester shaker used was purchased from Griffin and George Ltd., London; it has a stroke of 4 cm and shakes at 300 strokes/minute.

NOTE 5

Practically, this is carried out easily by pouring the contents of the bottle into a polyethylene funnel which just holds back the glass balls.

NOTE 6

This part of the procedure has caused problems with many workers in the past. Histone F3 precipitates as the chloride (at this pH and concentration of protein) at about 96% ethanol. It is therefore fairly obvious that ethanol with an appreciable water content will never reach this value and will never precipitate F3. On the other hand, if the concentration goes much above this value then histone F2A2 will begin to come down (see also note 9). In the past, we found this procedure very variable but, be careful control of the ethanol used, the concentration of protein in the dialysis bag, and the temperature and the mechanics of the dialysis, this procedure has been standardized. The volumes of ethanol–HCl used for extraction give the correct concentration of total protein (approx. 1 mg/ml) when starting with 50 gm of calf thymus. However, when working with different quantities of thymus or with different tissues this must be adjusted accordingly. The ethanol used should be checked for its content of water, even if it is labeled as anhydrous. In our experience, this is extremely variable, presumably due to absorption of moisture from the air. The dialysis should be carried out so that the contents of the bags are constantly moving to prevent local excessive concentrations of ethanol on the inner surface of the bags which would then precipitate F2A2. In our laboratory the dialysis is carried out in a 10-liter polyethylene bucket, using a magnetic stirrer to keep the ethanol solution and the dialysis bags moving. The magnetic stirrer is covered by a polyethylene cage so that it cannot hit the dialysis bags which become very brittle in the ethanol. Care should be taken when handling the bags after the dialysis because they are easily cracked.

NOTE 7

It is often forgotten that a protein precipitated from a solution containing other proteins and then separated by centrifugation is still wet with the solution containing the other proteins, and that if such a precipitate is washed with acetone then all proteins are precipitated with a consequent cross-contamination. All protein precipitates should, ideally, be washed in a solution identical to that from which they are precipitated. In the case of F3 this is done most easily by using the solution against which it has been dialyzed. The temperature should not change during the washing.

NOTE 8

The last acetone wash should be taken from a bottle that does not have an appreciable content of water, rather than, for example, a laboratory wash bottle. Acetone in laboratory wash bottles often contains sufficient water to cause difficulty in drying proteins.

Note 9

Histone F3 begins to precipitate at about 96% ehtanol and histone F2A2 marginally later. If the details given under Note 6 for the dialysis stage are not rigidly adhered to then cross contamination of F3 with F2A2, or F2A2 with F3, is possible. This stage is designed to prevent such cross-contamination and results in the rejection of approximately 40 mg of histone. There is always a danger however when rejecting a small portion in this manner that a non representative sample is being lost. For example, if the histone molecules are differentially modified by acetylation or phosphorylation, then they could be precipitated at slightly different solvent concentrations. This criticism of course applies equally to all methods that do not give a completely quantitative yield.

Note 10

When supernatants are decanted from protein precipitates they often contain a trace of that precipitate in a finely divided form. Filtration through a No. 4 sintered glass funnel at this stage is essential since the addition of guanidine hydrochloride and HCl at the next stage would resolubilize any precipitate and it would not be removed by the subsequent filtration which is usually only necessary because of the poor quality of guanidine hydrochloride (see also Note 11).

Note 11

Without guanidine hydrochloride and HCl, F2A1 aggregates and is difficult to separate from F2A2. However, we have found that guanidine hydrochloride is extremely variable in quality, and quite recently a number of our preparations were ruined because of exceptionally bad batches from the suppliers. If this reagent is at all suspect then it can be recrystallized easily from hot ethanol.

Note 12

This is most easily carried out by saving the solution against which these proteins were dialyzed, adding guanidine hydrochloride (6.6 gm/100 ml) and concentrated HCl (1.3 ml/100 ml), followed by 1.25 volumes of acetone. Such a solution should wash out the soluble F2A1 but not solubilize the F2A2.

III. Purification of F1

F1 is soluble in 5% (v/v) perchloric acid (0.74 N), and this has formed the basis of one method for its preparation (5) (see Section VIII,A). The procedure can be adapted, however, for the further purification of F1, as follows. The sample is dissolved in 0.01 N HCl at about 4 mg/ml and perchloric acid (PCA) is added up to 5% (0.74 N). The solution is stirred rapidly for 1 hour and any precipitate removed by high-speed centrifugation. The supernatant is then filtered throught a No. 4 sintered glass funnel and, if necessary, the filtration is repeated as the solution must be clear at this stage. F1 is then recovered from the clear supernatant by the addition of 0.25 of a volume of N-HCl followed by 7 volumes of acetone. The precipitate is then washed three times in acetone and dried under vacuum as described previously.

Alternatively, column chromatographic methods can be useful at this stage, and F1 can be eluted before the other histone fractions by using guanidinium chloride elution from IRC 50 (12) or from carboxymethyl cellulose using acetate buffer at pH 4.2 (13).

These procedures do not of course separate the subfractions of F1 which only differ from one another by a few amino acids, and reference should be made to papers by Cole and his colleagues for the detail of these separation techniques (14–16).

To remove nonhistone impurities that may be present if other methods are used to prepare F1 (e.g., PCA extraction), column chromatography on carboxymethyl cellulose at pH 8.5 is effective (5).

If care has not been taken to limit proteolysis during the preparation, then minor bands having a higher mobility than F1 are seen on polyacrylamide gel electrophoresis. These bands are a good measure of the quality of the preparation as a whole and if present the whole preparation of all the fractions should be regarded as suspect. In our experience these bands often indicate that the thymus has deteriorated during storage. Stellwagen *et al* (17) have indicated the ease with which F1 is degraded. Oliver *et al.* (10) have suggested, however, that even if such degradation has occurred a pure F1 can be obtained by chromatography on Amberlite CG-50.

IV. Purification of F2B

If the ethanol–HCl extraction of the histones F3, F2A1, and F2A2 has been carried out exhaustively as described in the main method above, then the only possible contamination of F2B would be with histone F1. This can therefore be purified by a process similar to that for the purification of F1, but with the emphasis on the precipitate and the necessity to wash it free of histone F1.

The sample is dissolved in 0.01 N HCl at about 4 mg/ml and PCA is added to 5% (0.74 N). The solution is stirred rapidly for 1 hour at 4° and the precipitate removed by centrifugation. The precipitate is then washed three times by mixing thoroughly with 4 volumes of 5% PCA; any lumps are broken up with a glass rod. The final precipitate is then washed with 4 volumes of acidified acetone (200 ml acetone–0.1 ml concd HCl), and then three times in acetone and dried under vacuum.

Other methods which can be used for separating minor components from F2B are chromatography on carboxymethyl cellulose (18), preparative polyacrylamide gel electrophoresis (19), Bio Gel P 60 exclusion chromatography (20), and Bio Gel P 100 exclusion chromatography (10).

More recently (21) F2B has been purified by preparative electrophoresis in a polyacrylamide gel slab at pH 2.7. This method of purification is of course applicable to all the partially purified fractions but gives only relatively small yields.

V. Purification of F3

F3 is probably the most difficult histone fraction to obtain in a pure form by selective extraction techniques only. The most frequent contaminant is histone F2A2 which tends to coprecipitate with it during the ethanol dialysis if this is not controlled extremely carefully (Note 6) However, F2A2 can be removed by recycling the material through the same procedure, in the following way. The impure F3 (100 mg) is dissolved in 30 ml of distilled water, and when completely dissolved 10 ml of 5 N HCl is added with rapid stirring, followed by 160 ml of ethanol. This solution should be clear and if not should be clarified by centrifugation or filtration through a No. 4 sintered glass funnel. This solution is then dialyzed against 7 volumes of ethanol for 17 hours at 4 °C as described previously (Note 6). The precipitate that forms is F3 and is washed twice with the solution outside the dialysis bag, then three times in acetone, and then dried under vacuum.

Alternatively partially purified F3 can be chromatographed on Bio-Rex 70 which is the method used by Hooper and Smith (22) to purify F3 for sequence analysis.

Another method which has been used successfully subsequent to a selective extraction with ethanol–HCl is exclusion chromatography on Sephadex G-100 (23). Highly purified fractions were obtained by running the F3 through the columns twice.

An alternative method of considerable interest exploits the fact that histone F3 is the only histone fraction to contain cysteine (24). F3 is attached

reversibly to an organomercurial Sepharose by its thiol groups and is thus
separated from the other histones. It can then be eluted with cysteine or
β-mercaptoethanol. This method may also be applied to whole unfractionated histone.

F3 and F2A2 also can be separated easily on Sephadex G-100 if the F3
is first oxidized to the dimer by incubating it in 6 M guanidinium chloride
at pH 8.3 at 37° for 18 hours (8).

VI. Purification of Histones F2A1 and F2A2

If the methods recommended above are carried out there should be no
cross-contamination between F2A1 and F2A2. The separation is an easy one
with a good plateau after the precipitation of F2A2, during which no further
protein is precipitated with the addition of more acetone. If the precipitates
are washed as recommended, no other bands are visible by polyacrylamide
gel electrophoresis even when overloaded (7). However, since all problems
cannot be foreseen, and calf thymus is an ideal tissue for working out preparative procedures that are not always easily applicable to other situations,
methods for the further purification of these two fractions are given.

A. F2A1

If this fraction is obtained contaminated with F2A2, and this can occur
if insufficient acetone has been added to precipitate all the F2A2, then this
can be remedied by recycling through the acetone precipitation stage. However, this is a little difficult since the protein was initially extracted in ethanol–1.25 N HCl (4:1, v/v) and subsequently dialyzed against ethanol.
The easiest procedure is to dissolve the protein in a solution made up by
preparing 40% (w/v) guanidinium chloride in water, adjusting to pH 7, and
then adding 3 volumes of ethanol. This can then be acetone-precipitated in
ways as described previously (7). A separation of F2A1 and F2A2 can also
be obtained by exclusion chromatography on Bio Gel P 100 (10) or by the
methods developed by Starbuck et al. (23).

B. F2A2

This fraction will only be contaminated with F2A1 if a considerable excess
of acetone has been added to precipitate F2A2, or if the F2A2 precipitate
has not been washed to remove the solution containing F2A1 prior to its

final washing in acetone. Contamination with F3 may occur if insufficient dialysis against ethanol has occurred in the previous stage. To remove F3 a redialysis against ethanol as described under the further purification of F3 will suffice. In this case of course the supernatant is required, and the F2A2 can be recovered by precipitation with acetone. F2A1 is removed easily by the three methods given above (7,23,10).

VII. Gel Exclusion Chromatography for the Preparation of All Five Histones

Although very useful for the purification of fractions prepared on a large scale by other methods, gel exclusion chromatography is not selective enough to separate all five histone fractions during one run. However, a recent publication by Böhm et al. (24) describes how a two-stage procedure can resolve whole histone into its five main components. This is an interesting development, and if such a technique is used in conjunction with the large-scale preparative techniques, there should be no difficulty in preparing the five histones in a pure form in relatively large quantities by these methods.

VIII. The Preparation of Individual Histone Fractions

At this stage it should be pointed out that the methods given so far, together with the further purification procedures, are to enable all five histone fractions to be prepared at the same time from the same chromatin sample. With most bulk separations of proteins using differential precipitation methods, the point at which one protein is precipitated leaving another in solution is often extremely variable, depending on the relative concentrations of the proteins, temperature, purity and state of hydration of the solvent, the presence of other ions, and in some cases one suspects even the phases of the moon! Although attempts have been made to limit these variables, inevitably when one is attempting to obtain five very similar proteins from the same sample some compromises need to be made. In the preparation detailed in Section II, certain overlapping fractions are discarded. However, if the requirements are for one or two histone fractions specifically, then the methods can be modified considerably by sacrificing other unwanted fractions, and in the case of F1, for example, another method entirely is more suitable. With the exception of the specific method for histone F3 described by Ruiz-Carrillo and Allfrey (25) and the selective extraction of F2A1 and F2A2 (7), the procedures adopted for the specific isolation of F2B, F3, F2A2, and F2A1 are all modifications of the main method given in Section II. In each case the other fractions are sacrificed in favor of the selected histone. The details for the preparation of individual fractions are given below.

A. F1

If F1 only is required then without doubt the perchloric acid (PCA) extraction method (5—Method I) is the easiest and gives a very pure product. It is to be preferred to the composite method for all five fractions since there is much less chance of proteolytic degradation, and it is a simple procedure that can be completed easily in one day.

First 100 gm of calf thymus is saline-washed as described above. The sediment obtained from the last washing is then homogenized with 400 ml of 5% PCA (0.74 N) at top speed for 2 minutes and the mixture centrifuged for 30 minutes at 1100 g. The sediment is then extracted twice more in the same manner but using 200 ml of PCA. The combined supernatants are then clarified by filtering through a No. 4 sintered glass funnel, and trichloroacetic acid is added to a final concentration of 18% (w/v) (1.1 M). The precipitate is recovered by centrifugation, washed once in acidified acetone (200 ml of acetone + 0.1 ml of concd HCl), three times in acetone, and dried under vacuum. The yield of F1 histone is from 400 to 450 mg.

Alternatively, the F1 can be recovered from the PCA extract by the addition of HCl and acetone, which also provides a method for the simultaneous separation and isolation of some non histone chromatin proteins that are co extracted with histone F1 under certain conditions (26).

There are two practical points concerned with this preparation.

(1) When centrifuging the trichloroacetic acid (TCA) precipitate in glass bottles, under certain conditions it seems to disappear much to the distress of the worker. However, the precipitate becomes visible again at the bottom of the centrifuge bottle when the acidified acetone is added. The acid (HCl) is essential since the TCA complexes of histones are soluble in acetone–water mixtures.

2. There has been some confusion about the value "5% PCA" (0.74 N) in our laboratory and other workers may have had the same problem. The figure 0.74 N was obtained by titration because of possible ambiguity. The PCA used is obtained as a 60% w/w solution. This is diluted 1:11 with water. Strictly then the solution used should be given thus; PCA solution (60% w/w)–H$_2$O (1:11, v/v).

B. F2B

The ethanol–HCl extraction of chromatin described in Section II to remove F3, F2A1, and F2A2 must be carried out exhaustively (at least four times). F1 and F2B are then extracted from the residue as described, using 0.25 N HCl. Then 3.5 volumes of acetone are added to precipitate all the F1 and possibly a little F2B. The precipitate is removed by centrifugation and the supernatant clarified by filtration through a No. 4 sintered glass funnel. F2B is finally precipitated using an excess of acetone, washed three times in acetone, and dried under vacuum.

C. F3

The preparative method for all fractions as described in Section II is followed as far as the extraction of the chromatin with 80% ethanol–HCl. Only two extractions are necessary since most of the F3 is extracted and contamination of the remaining protein is not important. The dialysis against ethanol can also be restricted for the same reason to one dialysis against 7 volumes of ethanol. The precipitate must be washed in the solution against which it was dialyzed in the same way as described in the main preparation. It is then washed three times in acetone and dried under vacuum.

D. F2A2

The extraction of the chromatin with ethanol–HCl is modified in the same way as described for F3 above. The dialysis against ethanol, however, must be exhaustive, i.e., twice against 7 volumes at room temperature. This insures the complete precipitation of F3 together with a little F2A2. The precipitate is removed by centrifugation and the supernatant clarified. Guanidine hydrochloride and HCl are then added as described in the main method, but F2A2 is precipitated with only 0.75–1 volume of acetone. It is then washed twice in a solution similar to that from which it was precipitated, three times in acetone, and dried under vacuum.

E. F2A1

The modifications of the preparation for F2A1 are similar to those described above for F2A2 until the precipitation with acetone. F2A2 is then removed from the mixture by precipitation with 1.5 volumes of acetone, together with a little F2A1. The precipitate is removed by centrifugation, the solution clarified, and F2A1 precipitated by adding an equal volume of acetone. It is then recovered, washed in acetone, and dried in the usual manner.

Alternatively, F2A1 and F2A2 can be specifically extracted from chromatin by ethanol–guanidine hydrochloride mixtures and the individual fraction isolated in a pure form as described previously (7).

IX. The Preparation of the Nucleated Erythrocyte Specific Histone F2C

In the inactive chromatin of nucleated erythrocytes a unique histone fraction designated 5 was discovered by Neelin and Butler (27). This fraction, later designated F2C by Hnilica (28), appears to partially replace the histone fraction F1 in the erythrocyte chromatin, and it is now thought to play some part in the final repression of the DNA (29,30).

This protein and its interactions with DNA are therefore of considerable interest, and two main methods have been developed for its preparation in large quantities, both of which give highly purified products.

A. Method of Murray, Vidali, and Neelin (20)

This method describes in detail the preparation of chromatin from chicken blood and the subsequent extraction of histones by careful titration with acid. If histone F2C only required the chromatin is serially extracted with H_2SO_4 at pH 2.8, 2.15, 2.15, and 1.95 as described in the paper, and this protein rejected. The chromatin is then further extracted at pH 1.9, and the F2C that is removed is recovered by precipitation with 2.5 volumes of ethanol. The amino acid analysis,and the starch gel electrophoretic pattern given, indicate that a very pure fraction can be obtained by this method. Over 600 mg can be isolated from 4 liters of blood.

B. Method of Johns and Diggle (31)

This method is based on the fact that after histone F1 is removed from chicken erythrocyte chromatin by washing in 0.5 M NaCl, histone F2C can be specifically extracted using 5% PCA. The PCA extract is then treated with concentrated TCA solution to a final concentration of 0.6 M and the precipitated F2C washed in acidified acetone, acetone, and dried. The yield is approximately 200 mg from 300 ml of blood. This method also gives a highly purified product as judged by amino acid analysis and polyacrylamide gel electrophoresis.

X. Characterization of Products

It is relatively easy to carry out a biochemical preparation, such as that given for the isolation of the five histones, and to finish up with five tubes containing white powder. Unfortunately this is the point at which many workers feel very satisfied and proceed to the next stage of their work. It must be stressed, however, that because of the variability inherent in this type of biochemical preparation, characterization of the products is equally as important as their preparation. This may seem to some workers an unnecessary and obvious comment, but the author has found on numerous occasions that because a preparative method has been carried out of the best of a given person's ability, then that person has automatically assumed that the products must be pure. All workers are fallible and frequently our preparations go wrong for no apparent reason. I would stress therefore that before any histones are used they should be characterized by the following methods.

A. Total Amino Acid Analysis

Amino acid analysis should be carried out on all samples and compared with the corresponding analyses of the purified fractions given here (Table I) or elsewhere. This will give a general indication of the purity of the histone but will not indicate much less than 5% cross-contamination with other fractions. It will also give an estimate of the protein content of the sample if an accurately known amount is taken for hydrolyses.

B. N-Terminal Amino Acid Analysis

This gives a useful indication of the purity of a sample. However, it is possible to have two proteins with the same N-terminal amino acid, and small associated peptides can give a complex pattern which bears no relationship to the small quantitative contamination. Histone F2B has a proline N-terminal group and histone F3 has an alanine N-terminal group. The other histones all have an acetyl N-terminal group. For quantitative N-terminal group analysis the methods of Phillips and Johns (11) and Phillips (32) can be used, but more rapid and suitable for much smaller quantities (5×10^{-9} mol) is the qualitative method of Gray (33) which is recommended.

C. Polyacrylamide Gel Electrophoresis

This is probably the best and certainly the most aesthetically pleasing method for determining the purity of a given histone fraction. The methods

TABLE I

The Amino Acid Composition of Purified Histone Fractions[a]

	F1	F2B		F2A2		F2A1		F3		F2C
Asp	2.5	5.0	(4.8)	6.6	(6.9)	5.2	(4.9)	4.2	(4.1)	1.7
Thr	5.6	6.4	(6.4)	3.9	(3.8)	6.3	(6.9)	6.8	(6.8)	3.2
Ser	5.6	10.4	(11.2)	3.4	(3.0)	2.2	(2.0)	3.6	(3.8)	11.9
Glu	3.7	8.7	(8.0)	9.8	(9.2)	6.9	(5.9)	11.5	(11.3)	4.3
Pro	9.2	4.9	(4.8)	4.1	(3.8)	1.5	(1.0)	4.6	(4.5)	4.7
Gly	7.2	5.9	(5.6)	10.8	(10.7)	14.9	(16.7)	5.4	(5.3)	5.3
Ala	24.3	10.8	(10.4)	12.9	(13.0)	7.7	(6.9)	13.3	(13.5)	16.3
Val	5.4	7.5	(7.2)	6.3	(6.1)	8.2	(8.8)	4.4	(4.5)	4.2
Cys/2	0.0	0.0	(0.0)	0.0	(0.0)	0.0	(0.0)	1.0	(1.5)	0.0
Met	0.0	1.5	(1.6)	0.0	(0.0)	1.0	(1.0)	1.1	(1.5)	0.4
Ile	1.5	5.1	(4.8)	3.9	(4.6)	5.7	(5.9)	5.3	(5.3)	3.2
Leu	4.5	4.9	(4.8)	12.4	(12.2)	8.2	(7.8)	9.1	(9.0)	4.7
Tyr	0.9	4.0	(4.0)	2.2	(2.3)	3.8	(3.9)	2.2	(2.3)	1.2
Phe	0.9	1.6	(1.6)	0.9	(0.8)	2.1	(2.0)	3.1	(3.0)	0.6
Lys	26.8	14.1	(16.0)	10.2	(11.4)	11.4	(10.8)	10.0	(9.7)	23.6
His	0.0	2.3	(2.4)	3.1	(3.0)	2.2	(2.0)	1.7	(1.5)	1.9
Arg	1.8	6.9	(6.4)	9.4	(9.2)	12.8	(13.7)	13.0	(13.5)	12.4
N-terminal group	Acetyl	Proline		Acetyl		Acetyl		Alanine		Acetyl

[a] The analyses are given as moles per 100 moles of all amino acids recovered, and no corrections have been made for hydrolytic losses. Figures in parentheses have been calculated from the amino acid sequences. The lysine value includes ε-N-methyl lysine.

used are numerous [see Table 1.6, (34)] and, depending on the pH and ionic strength of the gels, the mobility of the different fractions relative to one another can vary considerably. In some cases the order of migration of the five histones is quite different (35,36).

Polyacrylamide gel should not be used in isolation for characterizing histones since under a given set of conditions it is possible for contaminating proteins to have mobilities similar to those of the histones. In the acetic acid system we use, for example, globin runs exactly with histone F1.

The method we use routinely was developed by us in 1967. It uses a 20% polyacrylamide gel in 0.9% acetic acid. This gives a good resolution of bands except for histones F2B and F3 which have similar mobilities. In 1969 Panyim and Chalkley modified this acetic acid gel system by raising the pH, reducing the concentration of the gel, and adding urea. This effectively caused the separation of histones F2B and F3, and if this system is used in long gels (36), a good separation of all five histones can be obtained. Indeed many of the histone modifications are also resolved. These methods will give an indication of the purity of a fraction on as little as 10 μg of sample, but if more quantitative and accurate data are required then it is necessary to run the sample at various concentrations. Small amounts of other proteins (1%) can easily be detected if 100 μg of sample are applied to the gel system we use; and if quantitative methods are applied (9) to the stained bands then an accurate estimate of contamination can be obtained.

We are currently developing a method that separated all five fractions easily on a short gel without using urea. In this method the 0.9 N acetic acid is replaced by citric acid which improves the separation considerably.

Two new methods are now available for discriminating between arginine-rich and lysine-rich histones in polyacrylamide gel. The first is a differential destaining technique which washes away histones F1, F2B, and F2A2 and leaves only the arginine-rich histones F3 and F2A1 (39).

The second is a differential color staining technique which stains the lysine-rich histones F1 and F2B red and the other histone fractions blue. This method can be very useful in giving an indication of the amino acid composition of fractions separated on gels before they are actually isolated and analyzed (40).

XI. Storage of Histones

The histone fractions prepared and dried as described in this chapter should be white powders easily soluble in water. They should be stored in a vacuum desiccator over silica gel at room temperature. Samples stored on the laboratory shelf are slowly degraded by traces of proteolytic enzymes which are ubiquitous (6). Also histone F1 slowly forms β structures if stored in the presence of moisture and slowly becomes insoluble. Cold room or deep-freeze storage is not recommended because of the condensation of moisture on the sample each time it is opened at room temperature. Samples stored as recommended have been kept in the author's laboratory for over 10 years without any apparent changes in properties.

ACKNOWLEDGMENTS

The author's work described in this chapter has been supported by grants from the Medical Research Council and the Cancer Research Campaign. I also wish to thank Susan Forrester and Robert Nicolas for their skillful technical assistance in preparing histone fractions. I am also grateful to Butterworths, London, for permission to use much of my chapter on "Fractionation and Isolation of Histones," *Nuclear Components*, edited by G. Birnie.

REFERENCES

1. Philips, D. M. P. ed., "The Histones and Nucleohistones." Plenum, New York, 1971.

2. Elgin, S. C. R., Froehner, S. C., Smart, J. E., and Bonner, J., *Adv. Cell. Mol. Biol.* 1, 1 (1971).

3. Hnilica, L. S., "The Structure and Biological Function of Histones." Chemical Rubber Publ. Co., Cleveland, Ohio, 1972.

4. Bradbury, E. M., *Ciba Found. Symp. Struct. Funct. Chromatin* 28 1. (1975).

5. Johns, E. W., *Biochem. J.* 92, 55 (1964).

6. Phillips, D. M. P., and Johns, E. W., *Biochem. J.* 94, 127 (1965).

7. Johns, E. W., *Biochem. J.* 105, 611 (1967).

8. Sanders, L. A., and McCarty, K. S., *Biochemistry* 11, 4216 (1972).

9. Johns, E. W., *Biochem. J.* 104, 78 (1967).

10. Oliver, D., Sommer, K. R., Panyim, S., Spiker, S., and Chalkley, R., *Biochem. J.* 129 349 (1972).

11. Phillips, D. M. P., and Johns, E. W., *Biochem. J.* 72, 538 (1959).

12. Rasmussen, P. S., Murray, K., and Luck, J. M., *Biochemistry* 1, 79 (1962).

13. Johns, E. W., Phillips, D. M. P., Simson, P., and Butler, J. A. V., *Biochem. J.* 77, 631 (1960).

14. Kinkade, J. M., Jr., and Cole, R. D., *J. Biol. Chem.* 241, 5790 (1966).

15. Bustin, M., and Cole, R. D., *J. Biol. Chem.* 243, 4500 (1968).

16. Bustin, M., and Cole, R. D., *J. Biol. Chem.* 244, 5286 (1969).

17. Stellwagen, R. H., Reid, B. R., and Cole, R. D., *Biochim. Biophys. Acta* 155, 581 (1968).

18. Johns, E. W., *Eur. J. Biochem.* 4, 437 (1968).

19. Macpherson, A., and Murray, K., *Biochim. Biophys. Acta* 104, 574 (1965).

20. Murray, K., Vidali, G., and Neelin, J. M., *Biochem. J.* 107, 207 (1968).

21. Martinage, A., Sautiere, P., Kerckaert, J., and Biserte, G., *Biochim. Biophys. Acta* 420, 37 (1976).

22. Hooper, J. A., and Smith, E. L., *J. Biol. Chem.* 248, 3255 (1973).

23. Starbuck, W. C., Mauritzen, C. M., Taylor, C. W., Saroja, S., and Busch, H., *J. Biol. Chem.* 243, 2038 (1968).

24. Böhm, E. L., Strickland, W. N., Strickland, M., Thwaits, B. H., Van der Westhuizen, D. R., and Von Holt, C., *FEBS Lett.* 34, 217 (1973).

25. Ruiz-Carrillo, A., and Allfrey, V. G., *Arch. Biochem. Biophys.* 154, 185 (1973).

26. Sanders, C., and Johns, E. W., *Biochem. Soc. Trans.* 2, 547 (1974).

27. Neelin, J. M., and Butler, G. C., *Can J. Biochem Physiol.* **39**, 485 (1961).
28. Hnilica, L. S., *Experientia* **20**, 1 (1964).
29. Dick, C., and Johns, E. W., *Biochim. Biophys. Acta* **175**, 414 (1969).
30. Sotirov, N., and Johns, E. W., *Exp. Cell Res.* **73**, 12 (1972).
31. Johns, E. W., and Diggle, J. H., *Eur. J. Biochem.* **11**, 495 (1969).
32. Phillips, D. M. P. *Biochem. J.* **86**, 397 (1963).
33. Gray, W. R., Methods in Enzymology, Vol. 25: Enzyme Structure, Part B (C. H. W. Hirs and S. M. Timasheff, eds.), p. 121. Academic Press, New York, 1972.
34. Johns, E. W., *in* "Histones and Nucleohistones" (D. M. P. Phillips, ed.), p. 1. Plenum, New York, 1971.
35. Shepherd, G. R., and Gurley, L. R., *Anal Biochem.* **14**, 356 (1961).
36. Panyim, S., and Chalkley, R., *Arch. Biochem. Biophys.* **130**, 337 (1969).
37. Johns, E. W., *J. Chromatogr.* **42**, 152 (1969).
38. Johns, E. W., and Forrester, S., *J. Chtomatogr.* **55**, 429 (1971).
39. Barrett, I. D., and Johns, E. W., *J. Chromatogr.* **75**, 161 (1973).
40. Barrett, I. D., and Johns, E. W., *Biochem. Soc. Trans.* **2**, 136 (1974).

Chapter 12

Fractionation of Histones on Molecular Sieve Matrices

CLAUS von HOLT and WOLF F. BRANDT

*Department of Biochemistry, C.S.I.R.–Chromatin Research Unit,
University of Cape Town, South Africa*

I. Introduction

Methods for the isolation and fractionation of histones are being governed by two predominant properties of these nuclear proteins inherent to their biological function: their strong positive charge and their tendency to aggregate. Any strategy designed to purify the histones has to take into account these two properties in addition to those related to their molecular weight.

The decision on which method to use to purify histones will depend primarily on the aims of the investigation. These may call for two basically different methods: (1) fractionation of *all* the histones present in the type of chromatin under investigation, and (2) purification of *one* particular histone. Though for both major methodological areas well-proven techniques are available which in most cases, after suitable modifications, probably will lead to the desired purification of one histone or fractionation of the set of histones from a particular organism or tissue, no simple methodology can be prescribed that will cure all histone isolation problems. In our experience success in the isolation and purification of histones depends not only on the exploitation of their chemical properties but also largely on conditions only indirectly related to the chemical nature of histones. For example, as the ratio of histones to total cellular proteins decreases it becomes more difficult to get pure histones. Also the presence of extra- or intranuclear macromolecules which may interact with histones to result in complexes with drastically altered chemical properties will complicate the purification of histones. As the genome decreases in size it becomes more difficult to separate the histones from the cellular proteins. In addition, the more metabolically active the nuclei, the higher the concentration of nonhistone proteins in the nucleus (*1*) and the more difficult it becomes to purify the histones. The nonhistone protein in the nucleus may amount to approximately 500 different protein species (*2*). The presence of extra-nuclear polyanions like the sulfated polysaccharide in *Physarum poly-cephalum* (*3*) will require special modifications of existing methods at the various stages of histone purification (*4*) because even traces of such com-pounds will drastically affect the properties of histones. The uncertainty which exists in the literature as to the occurrence of particular histones in unicellular eukaryotes (*5*) may partially be due to the above-outlined difficulties.

II.   Isolation of Nucleoprotein and Dissociation of Histones from DNA

A.   Preparation of Nucleoprotein

Purified nuclei as a source of nucleoprotein will facilitate the subsequent histone fractionation. However, in many investigations, particularly in the field of comparative histone biochemistry, this will be difficult to achieve either because suitable methods to purify nuclei have not yet been developed, or if available, the yields may be too low. If in such cases one wishes to proceed to the preparation of a nucleoprotein without prior purification of nuclei, the advantages of higher yields of crude chromatin have to be weighed against the disadvantages of a likely contamination of nucleoprotein with extranuclear macromolecules which may become complexed to the chroma-tin in the course of its preparation.

If no specific consideration needs to be given to the topography of histones within the chromatin complex, then the procedures employed for nucleo-protein isolation may involve conditions that lead to a temporary dissocia-tion of histones from the DNA. If organic solvents are to be used care should be taken to remove traces of aldehydes from the solvents; otherwise, subsequent structural investigations on the histones will be hampered by blocked NH_2 groups. Histones for chromatin reconstitution studies are best isolated as histone complexes under nondenaturing conditions.

Basically three types of chromatin purification methods have been developed. The first consists of a sequence of washing procedures during which the main component of chromatin, nucleohistone, is being kept insoluble in 0.14 M NaCl, whereas the bulk of the cytoplasmic and ribo-somal proteins adhering to the outer nuclear membrane are being solubilized by buffers of suitable composition and ionic strength with or without the addition of nonionic detergents like Triton X-100 or Nonidet. Usually no special attempt is made to remove inner nuclear membrane components. Probably most chromatin preparations are heavily contaminated with fragments of nuclear membranes or some of their components. The second principle applied in the preparation of chromatin consists in the solubilization of the main components of chromatin under conditions which lead to the dissociation of the DNA–protein complex at high NaCl molarity (approx. 2 M NaCl) [for a review on factors affecting the solubility of nucleohistone see Fredericq (6)] with subsequent precipitation at low NaCl concentrations (0.14 M NaCl) of a DNA–protein complex predominantly nucleohistone in nature. Third, chromatin can be reversibly expanded into a gel in distilled water (6) and reprecipitated by increasing the NaCl molarity to approximately 0.14 M NaCl. These three principles of chromatin purification are being applied in various combinations to produce chromatin from nuclei or cell homogenates as a starting material for histone isolation. In view of the methods applied for its isolation chromatin is chemically an ill-defined nucleic acid–protein complex whose detailed composition—in particular with respect to its protein part—will depend on the isolation history.

B. Dissociation of Histones from DNA

Histones may be dissociated from DNA for the purpose of fractionation of the total complement or the purification of individual histones in three ways:

1. Extraction of total histones with dilute mineralic acid. This was introduced by A. Kossel in 1884 [references cited in Johns (7) and Hnilica (8)].

2. Dissociation of histones from DNA with salt [references reviewed in Fredericq (6)].

3. Selective extraction with mineralic acid and acid–ethanol mixtures (9, 10).

Any of these three methods will result not only in the solubilization of histones but also in that of other chromatin proteins. Their nature again may vary depending on the chromatin source and may possibly cause problems in the subsequent fractionation and purification of histones.

1. EXTRACTION OF HISTONES WITH DILUTE MINERALIC ACID

Nucleoprotein is suspended (or if necessary blended) (10 mg/ml) in an excess of 0.25 N H_2SO_4 or 0.25 N HCl and shaken for 1 hour at +4 °C. The concentration of acid may be raised to 0.4 N to keep the extraction volume smaller. The suspension is centrifuged at 10,000 g for 15 minutes and the supernatant, containing histones, is decanted for further processing. The precipitate is reextracted with the same volume of 0.25 N H_2SO_4 or 0.25

N HCl and recentrifuged. The two supernatants are combined and extensively dialyzed against distilled water before freeze-drying. It is essential that even traces of free H_2SO_4 are removed before freeze-drying; otherwise charring of the protein occurs. If the crude histone is to be recovered by acetone or ethanol precipitation without prior dialysis, H_2SO_4 should be used for extraction because of the low solubility of histone sulfates in organic solvents (7).

2. DISSOCIATION OF HISTONES FROM DNA WITH NaCl

Deoxyribonucleoprotein may be dissociated into protein and DNA at 2 M NaCl. The resulting solution is extremely viscous. This makes large-scale separation of the histones from DNA on molecular sieve columns (11) impractical. However, the DNA can be completely removed from the salt-dissociated DNA–protein solution as the protamine–DNA complex (12) (Section III, E).

III.   Fractionation of Histones on Molecular Sieve Matrices

A.   Theoretical Considerations

A fraction containing the total complement of histones either as hydrochlorides or sulfates, if not contaminated seriously with other proteins, neutral polysaccharides, or polyanionic polysaccharides, can be fractionated on molecular sieve matrices like Bio-gel or Sephadex. To minimize the aggregation of histones dilute HCl (0.02 N) has been introduced by Cruft (13) as a solvent for the fractionation of histones on molecular sieves. Even under such conditions the separate elution of all the individual histones is not possible, and molecular sieve fractionation of histones with dilute HCl has been mainly used to subfractionate the arginine-rich group or lysine-rich group of histones [for a review of literature, see Hnilica (8)].

The behavior of histones on molecular sieves is abnormal (14); usually histone elution volumes correspond to those of compounds with molecular weights several times that of histone (14). It has been suggested that the abnormal elution volumes of histones are the result of an extended chain conformation together with a conformational rigidity due to prolyl residues (14). Variation in the ionic strength of NaCl leads to conformational changes which in turn may cause histone–histone interaction (15). Either of these two changes may contribute to the fractionation of the histones on molecular sieve matrices in dilute acid in the presence of NaCl (16, 17) and may be the underlying reason for the apparently larger molecular weight of histones. Should aggregation be the cause then it must be postulated that such aggregates, if they are formed under the conditions of separation (0.02 N HCl–0.05 M NaCl), are mainly homoaggregates, because the histone complements of calf, chicken, trout, and sea urchin can be separated almost completely into their component histones in that solvent (Section III,C).

However, other types of interaction between histones and Bio-Gel (for example) have to be taken into account. A random sample of fresh Bio-Gel on titration reveals the presence of a weakly acidic group with a pK' of 5.6 (0.1 M equivalent per gram of dry matrix). This carboxyl group, probably arising from partial hydrolysis of the matrix amide group, may vary in concentration depending on the use of the column. Bonilla (18) reported that basic proteins are being retained on Bio-Gel columns and can only be eluted by increasing the NaCl concentration. This adsorption is obvious in the case of a basic sperm protein with 50 mol % arginine from the mollusc *Patella granatina* (19). Whereas the histone complement also present in the chromatin from these cells is eluted in various fractions from Bio-Gel with 0.02 N HCl–0.05 M NaCl, the basic, arginine-rich protein is only removed from the column with 0.1 N HCl. This indicates the operation of a process involving interaction between the guanidinium group of arginine and the carboxyl group of the Bio-Gel matrix. Such an interaction may partially determine the elution order of calf thymus and chicken histones from Bio-Gel which correlates with the arginine content of the histones (Section III.B). Histone H4, which has highest arginine content, is most strongly retained. Similar observations have been made by Dixon (20). The sea urchin histones H2B$_{(1)}$ and H2B$_{(2)}$ have a higher arginine content (21,22) than the H2B histones from other organisms and thus have a larger elution volume (Fig. 2) in spite of the increased molecular weight. The effect of increasing the molarity of NaCl on the elution profile (Section III,B) is compatible with the assumption of the operation of an ion-exchange and/or hydrogen-bonding mechanism.

From these results it is obvious that the fractionation and purification of histones on molecular sieve matrices is not governed solely by the molecular weight of the histones. Additional properties can be exploited to further the purification. Noncovalent interactions between the arginine residues of the proteins and the carboxyl groups of Bio-Gel, as well as conformational changes due to the solvent and additional solutes, will affect the elution volumes. In addition, histone–histone interaction leading to homo- and/or heteroaggregates may modulate the separation pattern which under ideal conditions should be determined by the Stokes radius of the disaggregated molecules only. The reversible, covalent dimerization of histone H3 through cysteine offers another handle to manipulate the elution pattern.

Though no satisfactory understanding exists at this stage as to the precise nature of the interaction between histones, the Bio-Gel matrix, and the solvent, the histone complement can nevertheless be successfully fractionated on Bio-Gel columns into its components (16,17,25,26). In general we found the separation power of Bio-Gel superior to that of Sephadex.

B. Experimental Conditions

Separation of histones can be achieved on columns (Bio-Gel and Sephadex) between 80–150 cm in length with diameters varying between 1.5 and 14 cm, the amount of sample being approximately 15–20 mg/cm² of column cross-section area. Because of the low content of aromatic amino acids in histones concentrations in the eluates are monitored between 206 and 230 nm.

The best separation of fractions is achieved at temperatures between +20 and +25°C. Figure 1 shows the effect of increasing NaCl concentration in 0.02 N HCl on the elution pattern of calf thymus histones from a Bio-Gel column (16).

Optimum separation is achieved at NaCl concentrations of 0.05 M and 0.1 M (Fig. 1). Only histones H2A and H3 remain unresolved. Their separation is possible in a sodium acetate–bisulfite buffer of pH 5.1 (Section III,C). The chicken erythrocyte histones, including the erythrocyte-specific histone H5, can be separated by a similar method (17) (Section III,C).

C. Two-Step Procedures for the Fractionation of Histones

On the basis of these separations the following conditions can be chosen (see also Figs. 2 and 3).

1. FIRST COLUMN RUN

Sample: Dissolved in 8 M urea–1% mercaptoethanol–0.02 N HCl.

Bio-Gel P-60 column: Varying diameter depending on sample size (15–20 mg/cm² column cross-section area); recommended length approximately 90 cm.

Eluant: 0.02 N HCl–0.02 N NaN₃(pH 1.7)–0.05 M NaCl, temperature +20° to +25°C. Fractions containing H1 or H2B or H4 (calf histones) and, in the case of chicken erythrocyte histones, H1 or H5 or H2A or H4 are cut according to the OD profile (Fig. 2), pooled, dialyzed at +4°C against distilled water, and finally freeze-dried. Fractions containing a mixture of H2A and H3 (calf histones) or H3 and H2B (chicken erythrocyte histones) are combined, dialyzed against distilled water at +4°C, the freeze-dried. These mixtures are subsequently processed in a second column run under different conditions.

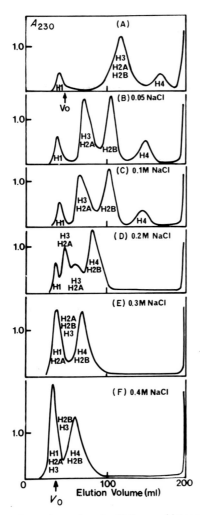

FIG. 1. Gel exclusion chromatography of calf thymus histones. Column: Bio-Gel P-60 (1.5 × 150 cm) samples were dissolved in 8 M urea–1% mercaptoethanol and left overnight at +4°C. Eluant: 0.02 N HCl–0.02% sodium azide with various concentrations of NaCl. Flow rate: 6 ml/cm²/hour; fraction volume: 1.2 ml. Data from Böhm et al. (16).

2. SECOND COLUMN RUN

Sample: H2A–H3 (calf) or H2B–H3 (chicken) mixture from first run dissolved in eluant (see below).

Sephadex G-100 column: minimum length 90 cm, various diameters according to sample size (15–20 mg/cm² column cross-section area).

Eluant: 0.05 M sodium acetate–0.005 M sodium bisulfite (pH 5.1), tem-

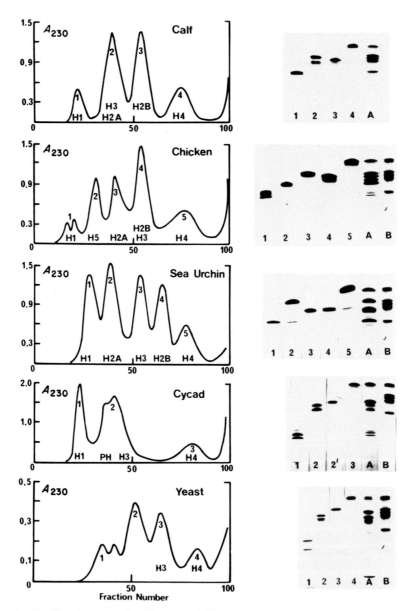

FIG. 2. Fractionation of acid-extracted histones from various organisms on Bio-Gel columns. Samples: 10–15 mg protein/cm² of column cross-section were dissolved in 6 M urea –1% mercaptoethanol. Eluant: 0.02 N HCl–0.05 M NaCl except yeast histones which were eluted with 0.01 N HCl. The elution profiles from various columns (1.5 × 100 cm to 2.5 × 150 cm) have been normalized by aligning the elution volumes of mercaptoethanol. Polyacrylamide gel numbers correspond to column fractions, gels A to acid-extracted histones, and gels B to calf thymus histones. Data from Refs. *16,17,21,23,24.*

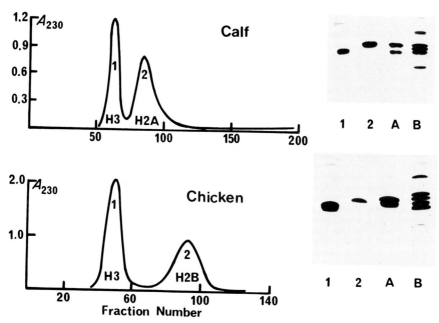

FIG. 3. Fractionation of histone pairs from calf thymus (H3–H2A) and chicken erythrocytes (H3–H2B). Column: Sephadex G-100, 1.5 × 150 cm. Samples: Fraction 2 from calf thymus histones (Fig. 2) and fraction 4 from chicken erythrocytes (Fig. 2) freeze-dried and dissolved in eluting buffer, stored at +4°C overnight. Eluant: 0.05 M sodium acetate–0.005 M sodium bisulfite (pH 5.1), fraction volume 1.2 ml and 1.3 ml respectively. The numbering of polyacrylamide gels corresponds to fraction numbers; gels A: unseparated pairs; gels B: acid-extracted histones from calf thymus and chicken erythrocytes, respectively. Data from Refs. 16,17.

perature +20 to +25°C. The sodium bisulfite is added as a protease inhibitor (27). Fractions containing H3 or H2A (calf thymus) and H3 or H2B (chicken erythrocytes) (Fig. 3) are combined, dialyzed against distilled water at +4°C, and subsequently freeze-dried.

If the histone fractions isolated in the two column runs are contaminated with overlapping, neighboring histones, then a repeat of the procedure and/or more discriminate cutting of the fractions will result in electrophoretically pure histones. In the case of chicken histones H2A and H5 the slight cross-contamination can be removed by rechromatographing the fractions in 0.02 NHCl in the absence of NaCl on a Bio-Gel P-60 column. Under these conditions H2A elutes later, whereas the H5 position is not affected. Total histones isolated from cycad pollen, yeast, sea urchin sperm (Fig. 2), and trout (26) have been fractionated in a similar fashion with varying resolution. No separation of the acetylated histone species from the nonacetylated can be achieved in this system.

The elution volume of a histone in itself is not a particularly specific criterion as to the identity of that histone. In the absence of a satisfactory theory explaining the interaction among histones, molecular sieve matrix, and solvent, considerable caution has to be exercised in the interpretation of data from molecular sieve fractionation. In addition, in most cases of histone fractionation on molecular sieve columns impure extracts are being processed which contain uncontrolled amounts and types of either cytoplasmatic and/or nuclear nonhistone proteins possibly interacting with histones. This may well result in the formation of complexes with more or less pronounced differences in elution properties. This may be the reason why rat liver histones only separate into three fractions (Hl, H2A-H2B-H3, and H4) under conditions where these histones from sea urchin sperm, calf thymus, and chicken erythrocyte are much better separated (Fig. 2). The danger of taking only one of the physicochemical or chemical properties of a histone fraction as sufficient proof of identity becomes obvious in the case of two histones isolated from sea urchin sperm (*Parechinus angulosus*) (*28,29*). Both elute as one fraction between H3 and H4 (Fig. 7). The fraction has proline at the N-terminus (*28,29*), just as histone H2B$_{(calf)}$ (*30*) has. The extractability in John's procedure corresponds to H2B histone but the electrophoretic mobility is identical to that of histone H3. This fraction contains two different histones homologous to histone H2B$_{(calf)}$ (*28,29*) (see Section IV,D). Therefore neither extractability criteria nor retention volumes, electrophoretic mobility, or the N-terminal group allow the unequivocal identification of a histone. A similar situation pertains to yeast histones (*23*) and to the two histones Phl and PH2 from cycad pollen which coelute with the arginine-rich histone H3 but are very dissimilar from the latter in that they are lysine-rich histones (*24*). Also one of the proteolytic degradation products of histone H4, present in histone extracts, elutes at a position indistinguishable from that of histone H4 (*31*).

D. Removal of Proteolytic Degradation Products from Histones H3 and H4

Proteolytic histone degradation products accompany the histones in various amounts through extraction and purification procedures. Nucleoprotein on storage, even at -20°C, is being degraded by proteolytic enzymes which preferentially attack histones Hl, H3, and H4 (*31*). Histone H3 is cleaved at the C-terminal site of lysine residue 23 and histone H4 in a homologous region C-terminal of lysine residue 16. The proteolytic H4 fragment (residue 17 through 102 of histone H4) can contaminate H4 isolated via chromatography on Bio-Gel P-60. Both the H3 and H4 fragments are also present in the arginine-rich histone fraction (*31*) extracted with 0.25 N HCl in 80%

ethanol in John's procedure (9). Whereas the H3 fragment (residue 24 through 135 of histone H3) clearly separates from histone H3 (Bio-Gel P-60 in 0.02 N HCl–0.05 M NaCl), the fragment from histone H4 elutes with a volume indistinguishable from that of H4. The separation of the H4 fragment from histone H4 requires rechromatography of the H4 fraction from column run 1 on Bio-Gel P-60 in 0.02 N HCl–0.12 M NaCl (Fig. 4). The H3 fragment can be further purified through dimerization in the same way as described for the intact parent molecule (see also Section IV,E).

E. Fractionation of Histones into Histone–Histone Complexes and Histone Hl

Physicochemical studies on histones aiming for insight into their natural arrangement in the chromatin complex have been hampered by the state of denaturation in which one finds the histones after they have been in contact during their extraction with perchloric, hydrochloric, or sulfuric acid, mixtures of hydrochloric acid and ethanol, or dilute HCl on molecular sieve columns. Such treatment leaves the histones largely in a random coil structure (32) making them unsuitable components in, e.g., chromatin reconstitution studies unless suitable arrangements are made to allow their renaturation (33).

Through a combination of salt dissociation from DNA at pH 5, removal of the DNA as the protamine complex from the solution of dissociated chromatin, and subsequent molecular sieve chromatography and ammonium sulfate precipitation, the histones can be fractionated into complexes consisting of histone H3 and H4, histone H2B and H2A, and histone Hl (12). The histone H3–H4 complex sediments with an S value (sedimentation coefficient) of 2.6 under non-denaturing conditions at pH 7 (33,34). There is increasing evidence that the H3–H4 as well as the H2B–H2A complex are the natural forms in which the histones are complexed in chromatin. Cross-linking experiments result in covalently linked H3–H4 histones (35), and cross-linking with tetranitromethane covalently binds the tyrosines in histone H2A and H2B (36). Considering the molecular dimensions of tetranitromethane the H2A and H2B molecules in chromatin must be very closely associated indeed. Since the findings by Kornberg and Thomas (35) of the suitability of histone complexes for chromatin reconstitution, the use of histone complexes for physicochemical and biological investigations related to the topography of histones in chromatin becomes of increasing importance.

Commercial protamine sulfate, used for the DNA precipitation, is frequently contaminated with histones. To remove the contaminants a 2% (w/v) protamine solution prepared at room temperature in distilled water is cooled to +4°C. The bulk of the protamine precipitates as a sticky preci-

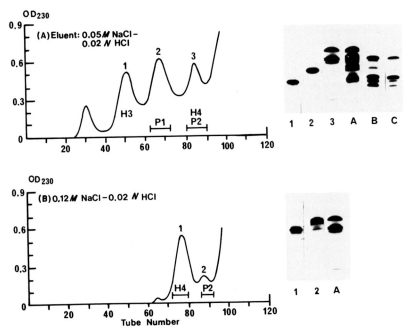

FIG. 4. Separation of proteolytic degradation products from undegraded histones H3 and H4. Column: Bio-Gel P-60, 2.5 × 100 cm. Sample: (A) Arginine-rich histone fraction from John's extraction (1960) of partially degraded cycad pollen nucleoprotein. (B) Rechromatography of column fraction 3. Samples were dissolved in 8 M urea–1% mercaptoethanol. Eluant: (A) 0.05 M NaCl–0.02 N HCl; (B) 0.12 M NaCl–0.02 N HCl. (A) Polyacrylamide gel numbering corresponds to column fractions. Gel A: ethanolic HCl extract of partially degraded nucleoprotein; gels B and C: acid extract of partially degraded and undegraded nucleoprotein respectively. (B) Polyacrylamide gel 1: column fraction 1 (histone H4); gel 2: proteolytic degradation product of histone H4 (column fraction 1); gel A: sample applied to column. From Brandt et al., with permission of the Federation of European Biochemical Societies (31).

pitate which is essentially free from histones. The material is freeze-dried. For column sizes and sample loads see Section III,C.

Deoxyribonucleoprotein (see Section II,A) is dissolved (4 mg/ml) in a solution of 2 M NaCl–0.005 M sodium bisulfite–0.05 M sodium acetate (pH 5.0). The bisulfite is included as a protease inhibitor (27). Equal volumes of the deoxyribonucleoprotein solution and a protamine solution (same solvent, 20 mg/ml) are mixed and dialyzed against 0.15 M NaCl–0.005 M sodium bisulfite–0.05 M sodium acetate (pH 5.0) at +4°C till the precipitation of the DNA–protamine complex is complete (UV monitoring of the supernatant at 260 nm). After centrifugation the supernatant is concentrated either through a UM 10-membrane (Amicon) or freeze-dried after dialysis against distilled water.

The concentrated histone–protamine solution or the freeze-dried preparation dissolved in 0.005 M sodium bisulfite–0.05 sodium acetate (pH 5.0) is applied onto a Sephadex G-50 column and developed with 0.005 M sodium bisulfite–sodium acetate (pH 5.0) into two fractions, histones and protamine (Fig. 5). The histone fraction can be either concentrated or after dialysis freeze-dried as before.

The concentrated histone solution or the reconstituted histone solution is applied to a Sephadex G-100 column and eluted as before with sodium bisulfite–sodium acetate to result in two main fractions (Fig. 6). Fraction 1 consists of histones H4, H3, and H1 and fraction 2 of histones H2A and H2B. Histones H4 and H3 are precipitated from fraction 1 with ammonium sulfate at 70% saturation. The three fractions histone H1, mixture of histones H3 and H4, and mixture of histones H2A and H2b (the lysine-rich histone fraction, the arginine-rich fraction, and the slightly lysine-rich fraction respectively) are dialyzed against distilled water and freeze-dried or dialyzed against suitable buffer solution for further processing.

IV.   Purification of Individual Histones

A.   General Considerations

The purification of histones on molecular sieve matrices in dilute acid in the presence or absence of NaCl is particularly successful in conjunction with the selective procedures of extraction as developed by Johns et al. (9).

Gel electrophoresis (37) is generally used as a sensitive and rapid method to establish the purity of histone fractions. Considering the relative mobilities of the various histones and the resulting degree of separation, approximately 5 μg of any single histone are being applied to the gel. Staining with Amido black will allow the detection by visual inspection of about 0.5 μg as a very faint band. Coomassie blue will increase the sensitivity by a factor of about 2–5. A few such justvisible protein fractions, present in a histone preparation, can therefore still constitute a considerable contamination and make such histone preparation unsuitable for certain structural, immunochemical, or radiochemical investigations. For such purposes column fractions should be cut to contain only the center part and should be recycled through the fractionation procedure (see also Fig. 7).

A frequent source of contamination are the Bio-Gel columns if they have been used before for protein, in particular histone fractionation. Small amounts of basic proteins are always retained on the Bio-Gel, and they may gradually accumulate and coelute in subsequent experiments. These re-

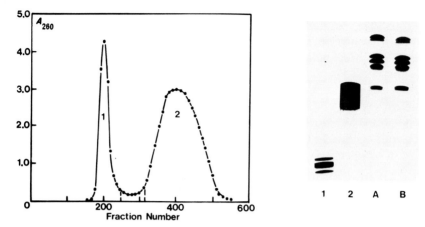

FIG. 5. Exclusion chromatography of a histone–protamine mixture. Column: Sephadex G-50, 14 × 70 cm. Sample: 900 mg histones and protamine (excess remaining from DNA precipitation) in 250 ml. Eluant: 0.05 M sodium acetate–0.005 M sodium bisulfite (pH 5.0). Flow rate: 700 ml/hour, fraction volume 20 ml. Numbering of polyacrylamide gels corresponds to column fractions: gel A and gel B to protamine-displaced and acid-extracted histones respectively. Gels 1 and 2 were subjected to electrophoresis for 1 hour only instead of 3.5 hours. From van der Westhuyen and von Holt (12), with permission of the Federation of European Biochemical Societies.

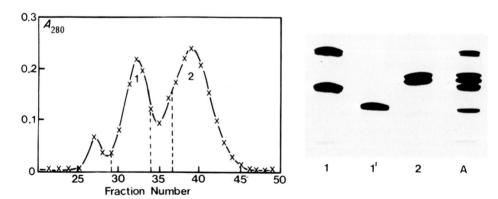

FIG. 6. Exclusion chromatography of protamine-displaced histone from calf thymus. Column: Sephadex G-100, 2.5 × 90 cm. Sample: 50 mg histones, dissolved in eluting buffer. Eluant: 0.05 M sodium acetate–0.005 M sodium bisulfite (pH 5.0). Flow rate: 15 ml/hour; fraction volume 6 ml. Polyacrylamide gel 1: precipitate from column fraction 1 (histones H3 and H4) at ammonium sulfate saturation 0.7; gel 1′: supernatant (histone H1); gel 2: column fraction 2 (histones H2A and H2B); gel A: calf thymus histones. From (12), with permission of the Federation of European Biochemical Societies.

tained proteins can be removed if the columns are "scrubbed" with 0.1–0.2 N HCl. However, such treatment may change the properties of the Bio-Gel matrix by cleaving amide bonds and thereby increasing the ion-exchange properties of Bio-Gel. This can be monitored by titration of representative samples of Bio-Gel.

B. Histone Hl

Histone Hl is eluted with the outer volume from a Bio-Gel P-60 column (Section III,C). It is usually clearly separated from the later eluting histone fractions and generally is not contaminated with other histones or proteins of lower molecular weight, but it may be contaminated with proteins of equal or higher molecular weight. In addition, histone Hl is heterogeneous, and it can be fractionated into a number of Hl species with different sequences.

We found that the histone Hl from sea urchin sperm (*Parechinus angulosus*) consists of a single polypeptide chain only which can be extracted with 5% perchloric acid (*10*) from nucleoprotein as a sequentially pure protein (*42*).

C. Histone H2A

This histone is efficiently purified through recycling in the two-column procedure (Section III,C). Contamination with histone H3 may be removed through dimerization (Section IV,E) if the source of histone stems from organisms evolutionarily older than rodents (see Section IV,C). For the removal of histone H5 contamination see Section III,C.

D. Histone H2B

Histone H2B isolated either via the two-column procedure (see Section III,C) or the John's fractionation (*9*) is frequently contaminated with varying amounts of histone H3. This H3 contamination in histone H2B fractions can best be removed through oxidation to the histone H3 dimer if the histone H3 contains a single cysteine, with subsequent separation of the dimer from the histone H2B on a Sephadex G-100 column (Section IV,D). Three oxidation–separation cycles usually remove the histone H3 contamination (Fig. 7).

Histone H2B from sperm of the sea urchin *Parechinus angulosus* elutes as a single fraction clearly separated from histone H3 (Bio-Gel P-60 or P-30, see Fig. 2) On gel electrophoresis this fraction has a mobility nearly identical to histone H3 and shows a double band reminiscent of histone acetylation of histone H3 or H4. If subjected to chromatography in urea with a NaCl gra-

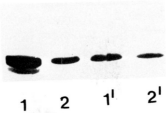

FIG. 7. Purified histone H2B from chicken and crocodile erythrocytes. Histone H2B has been prepared according to Oliver *et al.* (*38*) from total histone and purified on Bio-Gel P-60 (see also Fig. 2) (gels 1 and 2). In gel 1 (chicken erythrocyte histone H2B) approximately triple the amount of histone has been applied to make the histone H3 contamination visible. Histone H3 contaminations have been removed via dimerization (gels 1′ and 2′) as described in Section IV,D. The histone H3 content was assessed via quantitative determination of the alanine end-group (histone H3) and proline end-group (histone H2B) as described previously (*39,40*). Histone H3 content before iodozobenzoate treatment: 15% (gels 1 and 2); after iodozobenzoate treatment: 2.5% (gels 1′ and 2′). From van Helden *et al.* (*41*).

dient on *O*-(carboxymethyl)cellulose (CM-cellulose), the histone H2B separates into two distinct components with slightly different electrophoretic mobilities (Fig. 8). These two proteins, histone $H2B_{(1)Parechinus}$ and histone $H2B_{(2)Parechinus}$, have different though homologous primary structures (*21, 22,28,29*) and are both homologous to calf thymus histone H2B. Though the CM-cellulose–urea system is very effective in separating the sea urchin histones H2B$_1$ and H2B$_2$, we found that the chromatography of the histone complement on CM-cellulose in 6 *M* urea with a salt gradient (*43*) does not

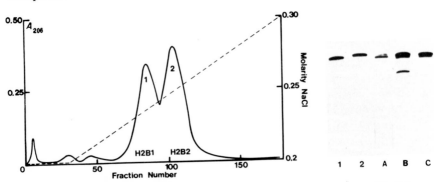

FIG. 8. CM-cellulose chromatography of sea urchin sperm histone fraction H2B. Column: CM-cellulose (2.5 × 30 cm) equilibrated with eluting buffer. Sample: 100 mg histone H2B extracted according to Johns *et al.* (*9*) and further purified on Bio-Gel P-60 (see also Fig. 2). Eluant: 0.05 *M* sodium acetate–6 *M* urea (pH 4.5). Polyacrylamide gels 1 and 2: column fraction 1 (histone H2B$_{(1)}$, and column fraction 2 (histone H2B$_{(2)}$; gel B: histone H2B–containing fraction (*9*); gel C: histone H2B fraction from Bio-Gel (Fig. 2). From Strickland *et al.* (*42*).

result in a clear separation into the single histones but in fractions containing a multiplicity of N-terminal amino acids (43), indicating probably histone–histone interaction due to the increasing salt gradient.

E. Histone H3

Though the purification of histone H3 on a preparative scale is possible in a two-step column procedure (Section III,C), the purification is preferentially achieved through a combination of selective extraction (9), dimerization through its cysteine residue, and subsequent molecular sieve fractionation (44). However, this very efficient method can be applied only for the isolation of histone H3 containing a single cysteine residue. Histone H3's from animals higher in the evolutionary scale than rodents have two cysteine residues in their amino acid sequence (45) and on oxidation preferentially form an intramolecular dimer and a series of higher polymers which cannot be easily separated.

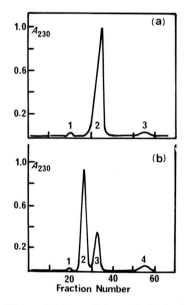

FIG. 9.   Purification of histone H3 dimer. Column: Sephadex G-100, 15 × 100 cm. Sample: 800 mg of crude chicken erythrocyte histone H3 (method of Johns *et al.* (9) containing histone H2A contaminant before (a) and after (b) oxidation with *o*-iodozobenzoate. Eluant: 0.01 N NCl–0.02% sodium azide. Fraction 2 in chromatogram (a): crude histone H3 monomer; fraction 2 in chromatogram (b): histone H3 dimer; fraction 3 in chromatogram (b): histone H2A. From Brandt and von Holt (44), with permission of the Federation of European Biochemical Societies.

Typically crude histone H3 prepared by method 2 of Johns (9), which is still contaminated with histone H2A, is oxidized at pH 7 and O °C in 8 M urea to the dimer with o-iodosobenzoate equivalent to its cysteine content, determined spectrophotometrically with p-chloromercyribenzoate (46). The urea solution in 0.1 N HCl is left prior to the addition of histone H3 for 1 hour to decompose cyanate and is then adjusted with Tris to pH 7. To avoid the blockage of N-terminal amono acids with cyanate, urea solutions should contain Tris as one of the buffer components to trap any cyanate (47). The H3 dimer can be eluted with 0.01 N HCl separately from histone H2A on a Sephadex G-100 column (Fig. 8) to result after reduction with excess of 2-mercaptoethanol in pure H3 monomer and its acetylated species (Figs. 9 and 10). This method has been successfully applied to the purification for sequence determinations of histone H3 from chicken (39, 40), shark (48), sea urchin, cycad, and mollusc (49)—all histones with only one cysteine in their sequence. Histone H3 from rat liver also can be purified in this fashion.

F.   Histone H4

Histone H4 from a variety of organisms generally can be purified easily on Bio-Gel P-60 in 0.01 N HCl–0.05 M NaCl (Fig. 11 and Section III,C).

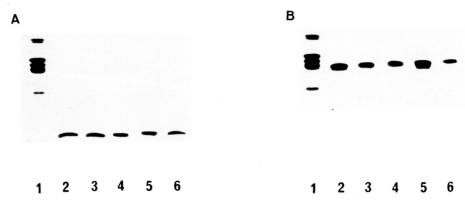

FIG. 10.   Gel electrophoresis of histone H3 from various organisms. Gel series A and B: histone H3 dimers and monomers respectively. Gel 1: acid-extracted calf thymus histones; 2: chicken erythrocytes (*Gallus domesticus*); 3: shark erythrocytes (*Paroderma africanum*); 4: sea urchin (*Parechinus angulosus*); 5: mollusc sperm (*Patella granatina*); 6: cycad pollen (*Encephalartos caffer*). The histones were purified via dimerization of either the histone H3–containing fraction from John's selective extraction procedure (9) or the histone H3–containing fraction from exclusion chromatography (Figs. 3 and 4). The histone H3 is acetylated to a different extent in the samples (slower-moving subfractions of the monomers). From Brandt *et al.* (49), with permission of the Federation of European Biochemical Societies.

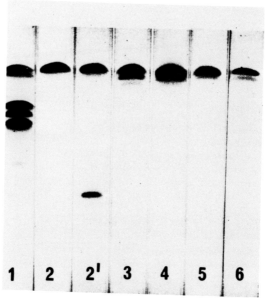

FIG. 11. Gel electrophoresis of histone H4 from various organisms. Histones were extracted according to John's procedure (9) and further purified on Bio-Gel P-60 (Fig. 2). Gel 1: calf thymus histones; 2: sea urchin sperm, reduced; 2': sea urchin sperm, partially oxidized; 3: mollusc sperm; 4: chicken erythrocytes; 5: shark erythrocytes; 6: cycad pollen. See also Fig. 10. From Strickland *et al.* (28,29).

Histones H4, isolated from a number of lower animals, have been reported to contain cysteine (50). For the sea urchin histone H4 (*Parechinus angulosus*) it has been shown that threonyl (Thr) residue 73 has been replaced by cysteine (28,29). In subjecting crude histone fractions from echinoderms to iodosobenzoate oxidation to produce H3 dimers the possible formation of mixed dimers between histone H3 and histone H4 has to be considered.

ACKNOWLEDGMENTS

Work from this laboratory was supported by grants from the Council for Scientific and Industrial Research (C.S.I.R.) and the University of Cape Town Research Committee to one of the authors (C.v.H.). We thank Mrs. D. M. Loureiro, Mrs. C. Lee, Mrs. M. Morgan, and Mr. E. Lee for help in the preparation of the manuscript.

REFERENCES

1. Bonner, J., Dahmus, M. E., Fambrough, D., Huang, R. C., Marushige, K., and Tuan, D. Y. H., *Science* **159**, 47 (1968).
2. Peterson, J. L., and McConkey, E. H., *J. Biol. Chem.* **251**, 548 (1976).
3. McCormick, J. J., Blomquist, J. C., and Rusch, H. P., *J. Bacteriol.* **104**, 1110 (1970).
4. Mohberg, J., and Rusch, H. P., *Arch. Biochem. Biophys.* **134**, 577 (1969).

5. Elgin, S. C. R., and Weintraub , H., *Ann. Rev. Biochem.* **44**, 725 (1975).
6. Fredericq, E., *in* "Histones and Nucleoproteins" (D. M. P. Phillips, ed.), p. 136. Plenum, New York, 1971.
7. Johns, E. W., *in* "Histones and Nucleohistones" (D. M. P. Phillips, ed.), p. 2. Plenum, New York, 1971.
8. Hnilica, L. S., "The Structure and Biological Functions of Histones," p. 13. CRC Press, Cleveland, Ohio, 1972.
9. Johns, E. W., Phillips, D. M. P., Simson, P., and Butler, J. A. V., *Biochem. J.* **77**, 631 (1960).
10. Johns, E. W., *Biochem. J.* **92**, 55 (1964).
11. Loeb, J. E., *Biochim. Biophys. Acta* **157**, 424 (1968).
12. van der Westhuyzen, D. R., and von Holt, C., *FEBS Lett.* **14**, 333 (1971).
13. Cruft, H. J., *Biochim. Biophys. Acta* **54**, 611 (1961).
14. Phillips, D. M. P., and Clarke, M., *J. Chromatogr.* **46**, 320 (1970).
15. Boublik, M., Bradbury, E. M., and Crane-Robinson, C., *Eur. J. Biochem.* **14**, 486 (1970).
16. Böhm, E. L., Strickland, W. N., Strickland, M., Thwaits, B. H., van der Westhuyzen, D. R., and von Holt, C., *FEBS Lett.* **34**, 217 (1973).
17. van der Westhuyzen, D. R., Böhm, E. L., and von Holt, C., *Biochim. Biophys. Acta* **359**, 341 (1974).
18. Bonilla, C. A., *Anal. Biochem.* **32**, 522 (1969).
19. Strickland, W. N., Brandt, W. F., and von Holt, C., unpublished observation.
20. Dixon, G. H., *Ciba Found. Symp., Struct. Funct. Chromatin* **28**, 267 (1975).
21. Strickland, W. N., Strickland, M., Brandt, W. R., and von Holt, C., *Eur. J. Biochem.* in press.
22. Strickland, M., Strickland, W. N., Brandt, W. F., and von Holt, C. *Eur. J. Biochem.* in press.
23. Brandt, W. F., and von Holt, C., *FEBS Lett.*, in press.
24. Brandt, W. F., and von Holt, C., *FEBS Lett.* **51**, 84 (1975).
25. Sung, M. T., and Dixon, G. H., *Proc. Natl. Acad. Sci. U.S.A.* **67**, 1616 (1970).
26. Candido, P. M., and Dixon, G. H., *J. Biol. Chem.* **247**, 3868 (1972).
27. Panyim, S., Jensen, R. H., and Chalkley, R., *Biochim. Biophys. Acta* **160**, 252 (1968).
28. Strickland, W. N., Strickland, M., Brandt, W. F., Morgan, M., and von Holt, C., *FEBS Lett.* **40**, 161 (1974).
29. Strickland, M., Strickland, W. N., Brandt, W. F., and von Holt, C., *FEBS Lett.* **40**, 346 (1974).
30. Iwai, K., Ishikawa, K., and Hayashi, H., *Nature (London)* **226**, 1056 (1970).
31. Brandt, W. F., Böhm, L., and von Holt, C., *FEBS Lett.* **51**, 88 (1975).
32. Bradybury, E. M., and Crane-Robinson, C., *in* "Histones and Nucleo-histones" (D. M. P. Phillips, ed.), p. 85. Plenum, New York, 1971.
33. D'Anna, J. A., and Isenberg, I., *Biochem. Biophys. Res. Commun.* **61**, 343 (1974).
34. Lewis, P. N., *Biochem. Biophys. Res. Commun.* **68**, 329 (1976).
35. Kornberg, R. D., and Thomas, J. O., *Science* **184**, 2419 (1974).
36. Martinson, H. G., and McCarthy, B., *Biochemistry* **14**, 1073 (1975).
37. Panyim, S., and Chalkley, R., *Arch. Biochem. Biophys.* **130**, 337 (1969).
38. Oliver, D., Sommer, K. R., Panyim, S., Spiker, S., and Chalkley, R., *Biochem. J.* **129**, 349 (1972).
39. Brandt, W. F., and von Holt, C., *Eur. J. Biochem.* **46**, 407 (1974).
40. Brandt, W. F., and von Holt, C., *Eur. J. Biochem.* **46** 419 (1974).
41. Van Helden, P., Strickland, W. N., Brandt, W. R., and von Holt, C., unpublished observation.
42. Strickland, W. N., Schaller, H., Strickland, M., and von Holt, C., *FEBS Lett.* **66**, 322 (1976).

43. Kobayashi, Y., and Iwai, K., *J. Biochem.* **67**, 465 (1970).
44. Brandt, W. F., and von Holt, C., *FEBS Lett.* **14**, 338 (1971).
45. Panyim, S., Sommer, K. R., and Chalkley, R., *Biochemistry* **10**, 3911 (1971).
46. Benesch, R., and Benesch, R. E., *Methods Biochem. Anal.* **10**, 43 (1962).
47. Henschen, A., and Edman, P., *Biochim. Biophys. Acta* **263**, 351 (1971).
48. Brandt, W. F., Strickland, W. N., and von Holt, C., *FEBS Lett.* **40**, 349 (1974).
49. Brandt, W. F., Strickland, W. N., Morgan, M., and von Holt, C., *FEBS Lett.* **40**, 167 (1974).
50. Subirana, J. A., *FEBS Lett.* **16**, 133 (1971).

Chapter 13

Chromatographic Fractionation of Histones

THOMAS G. SPRING[1]

Department of Biophysical Sciences, University of Houston,
Houston, Texas

AND

R. DAVID COLE

Department of Biochemistry, University of California, Berkeley,
Berkeley, California

I. Introduction

A major focus of cell biology, biochemistry, and biophysics today is the elucidation of the structure, function, and metabolism of those proteins associated with DNA in the nucleus of the eukaryotic cell. One reason for rapid progress in this area has been the availability of suitable methodology for fractionation of nuclear proteins, both on an analytical scale and a preparative one. This chapter will deal only with the chromatographic fractionation of the histones, the basic, nuclear proteins characterized by a high content (approximately 20–30 mol%) of lysine plus arginine and an absence of tryptophan. Several reviews of histone methodology have been published (*1–3*); the reader is referred to them for additional information on the isolation of cell nuclei, extraction methods for nuclear proteins, and electrophoretic and chromatographic separations.

The histones usually fall into five major classes though there are occasionally special forms such as one associated with avian erythrocytes. These histone classes have been given various designations based on the method of separation, or the amino acid composition (*2*). The nomenclature which will be used here is the most recent one (see chapter by Bradbury in this volume).

Although the histone classes are distinct and separable, each class may represent a mixture of proteins that differ in amino acid sequence and/or in the level of phosphorylation, acetylation, or methylation. For example, sub-

[1] *Present address*: Abbott Laboratories, Diagnostic Division, Abbott Park, North Chicago, Illinois.

fractions of H1 have been separated chromatographically (4) and have been demonstrated to differ in amino acid sequence (5); their chromatographic resolution is not due to differences in the level of phosphorylation or complexing with RNA (6). The subfractions of H1 are the result of multiple molecular forms (7) present within a single organism which differ in proportions as cell or tissue phenotype differs (8–11) and which respond differently to changes in physiological state (12). The subfractions are not simply the result of polymorphisms present in the population (5) as have been detected in H2A, H3, and H5 (13). Subfractions of the histones H2A, H2B, H3 and H4 are more commonly due to differences in the level of acetylation of certain lysine residues or differences in the level of methylation of certain lysine or arginine residues. These subfractions are usually not detected by chromatographic separation techniques, but are sometimes seen in high-resolution gel electrophoresis (14) or in amino acid sequencing studies (15). Since both the acetyl and methyl groups show turnover, the level of histone acetylation or methylation will depend on the physiological state of the tissue at the time of histone extraction. The metabolism of the acetyl and methyl groups is most conveniently studied using radioactive tracers and measuring their incorporation into the various histone fractions after electrophoretic or chromatographic separation. Phosphorylation of histones can also introduce microheterogeneity, and since phosphate groups often show a high rate of turnover, the level of histone phosphorylation is highly dependent on the physiological state of the tissue and/or the phase of the cell division cycle. The level of histone phosphorylation has been studied by measurement of ^{32}P incorporation into histones separated by high-resolution gel electrophoresis (16,17). By this technique the phosphorylated subfractions can be separated, due to differences in net charge on the histones. Ion exchange methods do not generally afford clean separation of the differentially phosphorylated subfractions, although some resolution has been reported using very shallow gradients (18). In the resolving systems studied to date, little attention has been given to the distinction between the more commonly studied O-phosphoryl groups on serine and threonine residues (alkaline-labile phosphates) and the less commonly studied N-phosphoryl groups on lysine and histidine residues (19) (acid-labile phosphates), which are not stable in the usually acidic conditions of electrophoresis or histone extraction.

II.   General Considerations in Choice of Methods

There are presently two general chromatographic methods applicable to histones: gel filtration and ion exchange. The historical development of these techniques and examples of their application can be found elsewhere (1).

Neither technique, even with more recent improvements, is capable of completely separating all histone classes in a single column run. Both techniques can be used for the preparative isolation of pure histone fractions (as would be required for determining amino acid composition and for sequencing) or for analytical measurements on separated histones (e.g., measuring the incorporation of radioactive phosphate, acetate, methyl groups, or amino acids into individual histones in metabolic studies). Cross-contamination of some histone fractions with nonhistone proteins can occur in either method, but this contamination can usually be detected and can be reduced by proper isolation of nuclei and histones (20). More will be said about this in the following section. The methods selected for presentation below represent two well-tested and widely used techniques which should be generally applicable to histones from any biological source. The methods will cover the chromatographic separation of a complete mixture of histones, such as would be obtained by an acid extraction of nuclear material. Resolution of H1 histone subfractions will also be covered. Other separation techniques utilizing fractional precipitation will not be covered. Large-scale fractional precipitation techniques can be usefully combined with chromatographic purification where large quantities of homogeneous histones are required (e.g., in amino acid sequencing studies).

III. Gel Filtration Chromatography

A. General Remarks

Since gel filtration is technically quite simple and requires no specialized elution solvents, it may be the preferred technique where speed and convenience are primary considerations. Numerous investigators have used gel filtration to fractionate histones. Methods have been reported with the polyacrylamide gel filtration materials Bio-Gel P-10 (21), P-30 (22), P-60(23), and P-100 (24), and with beaded dextran materials Sephadex G-75 and G-100 (25). All of these techniques utilized an acidic pH (0.01–0.02 M HCl) to minimize the aggregation of histones seen at high pH and to suppress the ionization of COOH groups which are present in low amounts in the gel filtration material, thus preventing ion exchange interference. The low pH has the added advantage of preventing proteolysis by an endogenous nuclear protease (known to be extractable from chromatin) which has a pH optimum above pH 7 (26). Low pH has the possible disadvantage (for investigators interested in the native structure of histones) of denaturing what is thought to be the native, aggregated state of histones and possibly removing nitrogen-bound phosphoryl groups (19). A method for the isolation of histones under more nearly "native" conditions and chromatography on Sephadex

G-100 at pH 5 isolates what are thought to be specific aggregates of H2A and H2B as well as specific aggregates of H3 and H4 (27,28). At low pH, where aggregation effects are minimized, histones still chromatograph anomalously (25). The molecular weights (13) of the calf thymus histones (H1 = 21,500; H2A = 14,004; H2B = 13,774; H3 = 15,324; and H4 = 11,282) would predict an elution order of H1, H3, H2A, H2B, and H4 from a gel filtration column; the observed order is 1, 2A, 2B, 3, and 4. In addition the histones show elution positions that are inconsistent with their known molecular weight range when globular, nonbasic proteins are used as calibrating markers. Histones chromatograph as if they were proteins of molecular weight 22,000–125,000 on Bio-Gel P-100 (25). The reason for this is that the histones are highly extended polypeptides at low pH and that elution position is a function of size and shape rather than molecular weight alone. Another important variable in determining the elution position of the histones is the concentration of NaCl used in the elution buffer for the column. Using a Bio-Gel P-60 column and an elution buffer of $0.02~M$ HCl–0.02% NaN$_3$(pH 1.7), Böhm et al. (23) showed that the elution volumes of all the histones were smaller when the concentration of NaCl in the eluant was increased in the range $0.0–0.4~M$ (23). The explanation is that salt-induced (presumably hydrophobic) aggregation of histones occurs even at pH 1.7. However, this phenomenon can be exploited to optimize the resolution of the various histone fractions since all the histones do not aggregate to the same extent in salt. This method has been applied to the separation of the six histone species in chicken erythrocyte nuclei and is reported in detail below (29). The method resolves histones H1, H2A, H4, and H5 on a Bio-Gel P-60 column, and resolves the resulting H2B–H3 mixture on a second column of Sephadex G-100. With minor variations, it ought to be a suitable general method for histone fractionation.

Techniques similar to, but not identical with this one have been used on Bio-Gel columns to separate histones and protamines from fish gonads (21), pea and Arbacia punctulata sperm histones (24), and tadpole liver histones (22), all with similar results. With histones from different biological sources, the resolution may not be optimal under the conditions specified below, but experimentation with the NaCl concentration should provide a way of optimizing resolution, based on the data of Böhm et al. (23).

B.   Procedure for Bio-Gel P-60

Deoxyribonucleoprotein is isolated from fresh chicken blood according to the method of Murray et al. (30), with some modifications. Briefly the procedure calls for washing the erythrocytes at 4 °C with $0.14~M$ NaCl–0.01 M trisodium citrate, and then lysing them in a 0.6% (w/v) saponin solution in

0.14 M NaCl followed by repeated washing of the nuclei in the saline–citrate until the reddish hemoglobin color has been removed. The nuclear wash solution also contains 0.05 M NaHSO$_3$ as a protease inhibitor (26). It has also been useful to include 0.5% (v/v) Triton X-100 in the nuclear wash buffer in order to facilitate the removal of cytoplasmic proteins (31). The nuclei are generally lysed by this procedure, and the resulting deoxyribonucleoprotein pellet can be further freed of nonhistone proteins by dispersing it at 4°C in 2 M NaCl–0.05 M NaHSO$_3$–0.05 M Na acetate (pH 5.0) and then reprecipitating the nucleohistone complex by reducing the concentration of NaCl to 0.3 M by dilution with the 0.05 M NaHSO$_3$–0.05 M Na acetate buffer (pH 5.0). Histones can be extracted directly from this pellet with 0.4 M H$_2$SO$_4$, followed by dialysis against distilled water (4°C) and lyophilization. Although this procedure is used by the authors of the original method, other extraction and concentration methods for histones should be applicable here, as long as the histones are undegraded and free from contamination by nonhistone proteins. The histones are dissolved in fresh 8 M urea–1% (v/v) 2-mercaptoethanol and stored overnight at 4°C prior to chromatography in order to achieve complete disaggregation and complete conversion of H3 from the disulfide cross-linked dimer to the sulfhydryl monomer. Chromatography is performed at room temperature on a column of Bio-Gel P-60 (Bio-Rad Laboratories, Richmond, California), equilibrated in 0.02 M HCl–0.05 M NaCl–0.02% (w/v) NaN$_3$ (pH 1.7). A typical procedure began with a 64-mg sample of chicken erythrocyte histone dissolved in 4 ml of urea–mercaptoethanol, which was chromatographed on a 2.5 × 160 cm column of Bio-Gel P-60 at a flow rate of 5.5 ml/cm^2/hour, a pressure head of 40 cm, and a fraction volume of 3.5 ml. Protein is monitored by measurement of the 230-nm absorbance of each fraction versus a blank. Histones are eluted in the order H1, H5, H2A, H2B, H3, and H4 (see Fig. 1). Resolution will depend on the sample size and purity, column dimensions, flow rate, and age of the column. The authors recommend repacking the column after 6–8 runs to maintain resolution. Other factors which reduce resolution are: (1) overloading of the column with histone; (2) contamination of the histone sample with nonhistone proteins, which usually run near the void volume of the column and contaminate H1; and (3) degradation of the histone sample by proteolysis prior to chromatography, which generates heterogeneous peptides that can contaminate one or more fractions. Generally if the starting material is sufficiently free of nonhistone proteins and degradation products, then H1, H5, and H4 can be obtained in pure form by this column step. H2A is sometimes contaminated with H5 and can be further purified by rechromatography on Bio-Gel P-60 in 0.02 M HCl–0.02% (w/v) NaN$_3$ (no salt). Under these conditions, the H2A is less aggregated and elutes at a greater volume, while H5 is unchanged in elution position.

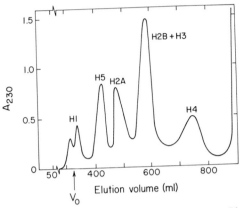

FIG. 1.   Chromatography of chicken erythrocyte whole histone on Bio-Gel P-60 in 0.02 M HCl–0.05 M NaCl–0.02% NaN$_3$ (pH 1.7). Samples were dissolved in fresh 8 M urea–1% 2-mercaptoethanol and stored overnight at 4 °C before application. Column dimensions, 2.5 × 160 cm; sample weight, 64 mg; sample volume, 4 ml; fraction volume, 3.5 ml. The strong ultra-violet (UV) absorption towards the end of the inner volume is due to the elution of urea and mercaptoethanol. From van der Westhuyzen *et al.* (*29*), with permission of North-Holland Publ. Co.

C.   Further Resolution on Sephadex G-100

The resolution of H2B and H3 is performed (*29*) at room temperature on a column of Sephadex G-100 (Pharmacia Fine Chemicals, Piscataway, New Jersey) equilibrated and run in 0.05 M Sodium acetate–0.05 M NaHSO$_3$ (pH 5.1). Under these conditions, the H3 is more aggregated than H2B and a clean separation is achieved. The authors report dissolving the lyophilized H2B–H3 mixture (8.5 mg) recovered from the Bio-Gel P-60 column in 1.0 ml of pH 5.1 buffer and storing overnight at 4 °C. The sample was chromato-graphed on a 1.6 × 90 cm column at a flow rate of 3.7 ml/cm²/hour, collec-ting 1.3 ml fractions. Protein was again monitored by 230-nm absorbance (see Fig. 2). The reported recovery of protein from the Bio-Gel column was 83% and from the Sephadex column, 78% (based on the recovery of pooled fractions shown in Figs. 1 and 2).

IV.   Ion Exchange Chromatography

A.   General Remarks

Chromatographic fractionation of histones by ion exchange is generally done on weak cation exchange materials containing fixed COOH groups, such as carboxymethyl cellulose or Amberlite IRC-50. These fractionations

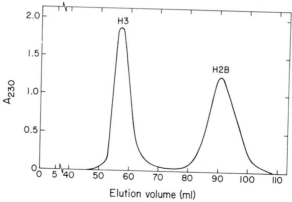

FIG.2. Chromatography of chicken erythrocyte H3 and H2B mixture on Sephadex G-100 in 0.05 M sodium acetate–0.005 M sodium bisulfite (pH 5.1). Column dimensions, 1.6 × 90 cm; sample weight, 8.5 mg; sample volume, 1 ml; fraction volume, 1.3 ml. From van der Westhuyzen et al. (29), with permission of North-Holland Publ. Co.

are necessarily run at a pH near neutrality (in order for the carboxylate groups of the ion exchange material to be in the ionized form) and eluted with a salt gradient. The main problem with this kind of fractionation is the tight binding and aggregation of histones under these conditions. This problem has been overcome in the Amberlite IRC-50 system by the use of guanidinium chloride (GuCl) as an eluting salt. The protein-denaturing properties of the GuCl are sufficient to overcome much of the aggregation and to allow reasonable, although not perfect, resolution of the histones. Although it is less convenient in many ways, Amberlite IRC-50 gives better resolution than does carboxymethyl cellulose except when H3 is the only histone whose complete resolution is critical. The method described below is essentially the Amberlite system of Luck et al. (32) with technical modifications and additional applications.

B. Preparation of UV-Transparent Guanidine

One of the annoying problems with this technique is that GuCl can interfere with the measurement of protein in the effluent from the column by the usual colorimetric procedures (33) or by A_{230} measurements (the A_{280} of histones is low and quite variable due to the absence of tryptophan and low levels of tyrosine). A turbidity method for measuring protein in 1.1 M trichloroacetic acid is commonly used instead and has been described (32). The disadvantage of this method is that it is relatively insensitive and destructive, often requiring a significant portion of a fraction just for the protein measurement. This problem can be overcome by preparing or puchasing

UV-transparent GuCl so that A_{230} can be measured. This method is both sensitive and nondestructive. For investigators who are making limited use of this method, purchase of purified GuCl may be more practical, despite the high cost. A reasonable source of such GuCl has been Sigma Chemical Co., St. Louis, Missouri (Grade I guanidine $\cdot$ HCl, catalog G4505). The UV background of this material is sufficiently low for A_{230} measurements, although the background may depend on the batch. An alternate method for purifying GuCl starting with practical-grade Gu_2CO_3 is described below.

Day 1. One kilogram of guanidinium carbonate (Gu_2CO_3, Matheson, Coleman, and Bell, practical grade) is stirred for 30 minutes with 2 liters of distilled water in a 4-liter flask and then put in a 37°C bath for 2–4 hours. Not all of the particles dissolve, so the suspension is filtered through two layers of filter paper on a Buchner funnel and then the filtrate is stirred overnight with 110 gm of activated charcoal. (At this point 2 gallons of absolute ethanol should be chilled to 4°C.)

Days 2 and 3. The charcoal suspension is filtered through four layers of filter paper and a layer of moist Celite. To the clear filtrate is added 6 liters of chilled absolute ethanol, with stirring. The Gu_2CO_3 is precipitated by this procedure, and the precipitation process is allowed to proceed for 2 days at 4°C before isolation.

Day 4. The precipitated Gu_2CO_3 is collected by filtration on a Buchner funnel, and the precipitate is washed with 2 liters of cold absolute ethanol and air-dried overnight. The yield of dry precipitate is 500–900 gm.

Day 5. The precipitate is slurried with a small quantity of distilled water in a beaker and concd HCl is added with stirring (approximately 1 ml of HCl/gm of precipitate) until the pH of the solution reaches 1. The solution is left overnight to allow complete evolution of CO_2.

Day 6. The pH of the solution is adjusted to 6.8 with a concentrated NaOH solution, and the final volume is measured (it should be 1–2 liters and the concentration of GuCl should be 40–50%, w/v). Add solid $NaH_2PO_4 \cdot H_2O$ and Na_2HPO_4 to a final concentration of 0.05 M each (total phosphate concentration, 0.1 M) and adjust the pH, if necessary, to 6.8 with concd NaOH or HCl. The solution is then Millipore-filtered and stored in the cold as a concentrated GuCl solution. The A_{222} of the concentrate should be less than 0.35 and the A_{230} less than about 0.18. The absorbance of the solutions depends on the temperature, with the room temperature reading being about 10% higher than the reading at 4°C. The concentration of GuCl is measured by refractive index, using standard solutions of guanidine HCl made up in 0.1 M phosphate buffer (pH 6.8) for comparison. The refractive index versus % GuCl (w/v) is a linear function up to about 60% (2). Stock solutions of lower concentrations $GuCl-PO_4$ are made by dilution with 0.1 M $NaPO_4$

buffer, pH 6.8. Upon standing, these stock solutions sometimes develop a precipitate. In this case, the solution should be filtered before using.

C. Preparation of Amberlite IRC-50 Columns

Amberlite IRC-50 (CG-50) 200–400 mesh (Mallinkrodt Chemical Works) or Bio-Rex 70, 10 meq/gm, 200–325 mesh (Bio-Rad Laboratories, Richmond, California) is suspended in water and allowed to settle; fine particles are decanted and the resin is sequentially washed as previously described (2) with 2 M HCl, H_2O, 2 M NaOH, H_2O, 2MHCl, H_2O, 2 M NaCl; the resin is titrated to pH 7 with NaOH. After each wash, the resin is suction-filtered. Finally it is resuspended in a starting buffer (7 or 8.5% GuCl depending on the particular resolution required). The column is packed with resin in starting buffer and is washed with several column volumes of starting buffer prior to sample addition. If starting samples of histone contain much nonhistone protein or degradation products it is best to start at 7% GuCl to elute most of these contaminants clearly before the first histone; this takes almost an extra day compared with starting the elution gradient at 8.5% GuCl.

D. Fractionation of a Total Histone Mixture

Histones are usually isolated by acid extraction as described above or elsewhere. Recommended column size for a preparative column (50–100 mg total histone) is 2.5 × 60 cm. Elution is with a 700-ml linear gradient of 8–13% GuCl in 0.1 M NaPO$_4$ (pH 6.8), followed by 100 ml of 40% GuCl, run at a flow rate of 30–40 ml/hour. For analytical columns (2–6 mg), a 0.6 × 60 cm column is eluted with 50 ml of a linear 8–13% GuCl gradient, followed by 10 ml of 40% GuCl at a flow rate of 5 ml/hour. The columns can be reused after equilibrating with 8% GuCl; repacking is necessary only when the flow rate becomes too slow. Samples are dissolved in 0.2–2 ml of 8% GuCl buffer and centrifuged, if turbid, prior to application to the column. Under these conditions most nonhistone contaminants appear in the breakthrough peak (see Fig. 3), followed by several degradation products (34) if present, and then by H1 (sometimes partially fractionated into subfractions), H2A–H2B (only partially resolved) and H3–H4 (unresolved or only partially resolved by the steep rise from 13 to 40% GuCl).

In certain rare situations peaks are seen to elute between H2B and H3. In the case of pea histones (35) a peak in this region was stated to contain a mixture of histones. In the case of wheat germ histones (36), H3 and H4 were found in this position in addition to their usual position in the 40% GuCl wash. The wheat germ histones are apparently associated with phospho-

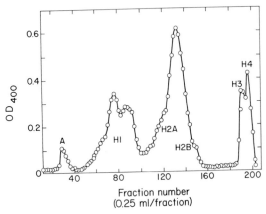

FIG.3.   Fractionation of acid-extracted histones of calf thymus chromatin by column chromatography on Amberlite CG-50; 0.6 × 55 cm; 3.0 mg of histone were used. Protein concentration in the effluent fractions was determined by optical density (OD) at 400 nm of the turbid solutions resulting when the 0.26-ml fractions were mixed with 1.1 M trichloroacetic acid in a total volume of 1.56 ml. From Bonner et al. (1).

lipids since incubation with phospholipase C caused all the H3 and H4 to elute in their normal positions. A special family of Hl-like histones found in sea cucumbers also elutes between H2B and H3 (37). Of course, contaminants might be found in any part of the chromatogram but the only particularly suspect regions are the initial peak and the last peak where the 40% GuCl breaks through (20). This can be seen in Fig. 4 where chromatograms are compared for histones isolated from highly purified nuclei and heavily contaminated nuclei (the latter was an acidic extract from material that was essentially a homogenate of the entire tissue).

E.   Further Fractionation of H1 Subfractions

Histone 1 is most readily isolated from nuclear material by extraction at 4°C with 5% perchloric or 5% trichloroacetic acid (2). By using a shallower gradient of GuCl than just described and a reduced flow rate, Kinkade and Cole (4) were able to show that Hl can be subfractionated on the Amberlite IRC-50 system. Using, e.g., a 2.3 × 52 cm column and a gradient of 7–14% GuCl (total volume of 3400 ml), 100 mg of calf thymus Hl fraction could be fractionated at a flow rate of 10 ml/hour. Depending on the exact starting point of the gradient and minor variations of flow rates these columns require 3 to 6 days. Increasing the flow rate to 30 ml/hour or more, gave much poorer resolution. Careful pooling of fractions and rechromatography allowed isolation of some subfractions which were pure enough for amino

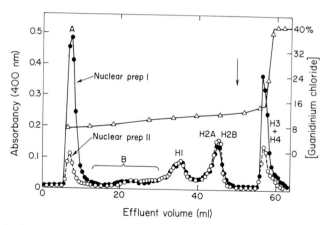

Fɪɢ.4. A chromatographic comparison of acid-soluble proteins extracted from nuclei of different purities. The vertical arrow indicates the point at which elution with 40% guanidinium chloride was begun. Protein was determined on aliquots of 0.3 ml according to Luck *et al.* (*32*), and the concentration of guanidinium chloride (triangles) by refractometry. The acid-soluble proteins applied in each case were from nuclear preparations containing 2.2 mg of DNA; preparation I (closed circles) from very crude nuclei; preparation II (open circles) from purified nuclei. From Stellwagen and Cole (*20*), with permission of the American Society of Biological Chemists, Inc.

acid sequencing (*7*), although not all fractions are equally resolved. The load applied to the column can be increased, or decreased, if the height of the column, the volume of the elution gradient, and the flow rate are increased or decreased proportionally. This results in a fairly constant elution time for the lysine-rich histones. The resolution can be improved only a little by using a shallower $8.5-13 \pm 0.5\%$ gradient, and care must be taken that no degraded histone is present in the sample, since it will tend to elute with the early Hl peaks (*4*). Analytical columns have also been run using a 1.2×15 cm column and 180 ml of a $8.5-14\%$ GuCl gradient at 1 ml/hour (*38, 12*). These columns can be regenerated by washing with 40% GuCl buffer and then recquilibrating with 7 or 8.5% GuCl buffer. Typical results are shown in Fig. 5.

It should be reemphasized that the subfractions of Hl separated by this chromatography system are distinct molecules with different amino acid sequences (*7*) and are not the result of multiple phosphorylation or complexing with RNA; this point has been demonstrated by chromatography of rabbit thymus Hl before and after treatments with either alkaline phosphatase or ribonuclease (*6*). The phosphorylation of Hl subfractions *in vivo* and *in vitro* has been detected by incorporation of ^{32}P into the subfractions separated on Amberlite IRC-50 columns. The change in charge due to enzy-

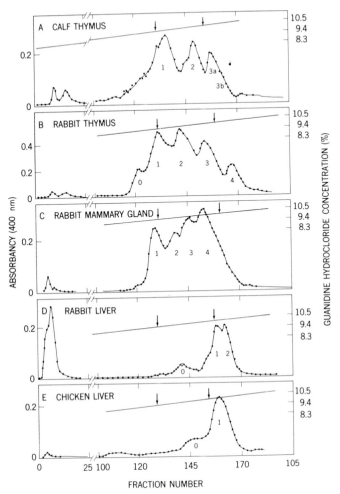

FIG.5. Chromatography on Amberlite IRC-50 of H1 histones derived from various tissues. (A) 80 mg of calf thymus; (B) 80 mg of rabbit thymus; (C) 80 mg of rabbit mammary gland; (D) 22 mg of rabbit liver; (E) 26 mg of chicken liver; The chicken liver extract was chromatographed on a 2.3 × 15 cm column; all others on a 2.3 × 30 cm column. The arrows point to the 9.4% and 9.9% guanidine hydrochloride concentration, as determined by refractive index. The numbered peaks are all H1 histone. From Bustin and Cole (8), with permission of the American Society of Biological Chemists, Inc.

mic phosphorylation at the cyclic adenosine monophosphate (cAMP)-sensitive site does not appreciably affect the elution position of the subfractions (39), although multiple phosphorylation has been reported to decrease the retention of the subfractions substantially (18), and phosphorylation at different sites might affect the chromatography differently.

V.   Special Histones and Special Problems

The application of the above chromatographic method for the fractionation of histones other than the usual Hl, H2A, H2B, H3 and H4 has also been reported. Greenaway and Murray (40) reported the fractionation of H5 from chicken erythrocytes on an Amberlite IRC-50 column into two subfractions, which were later shown by sequence analysis to differ by a single amino acid substitution (arginine for glutamine); this apparently is the result of a genetic polymorphism. An unusual histone (labeled histone T) was isolated by Wigle and Dixon (41) from trout testis and was also similar to trout Hl in being extractable with 5% trichloroacetic acid. The purification of this histone by ion exchange was performed on a 3 × 30 cm column of carboxymethyl cellulose (CM-52, Whatman) eluted with a 2-liter gradient from 0.15 to 0.75 M LiCl in 0.01 M lithium acetate (pH 5), at a flow rate of 100 ml/hour. Histone T was not found to contain subfractions and was successfully fractionated and purified by this technique.

Although other chromatographic techniques have been described for the fractionation of histones, none allows complete resolution of all histone fractions in a single column run. Other chromatographic systems work well in specific cases, but the two methods described here represent two well-tested versatile methods which are capable of handling most separation problems. Until such time as better chromatographic separation methods become available, they will probably be the methods of choice.

REFERENCES

1. Bonner, J., Chalkley, C. R., Dahmus, M., Fambrough, D., Fujimura, F., Huang, R. C., Huberman, J., Jensen, R., Marushige, K., Ohlenbusch, H., Olivera, B. and Widholm, J. *in* "Methods in Enzymology," Vol. 12; Nucleic Acids, Part B (L. Grossman and K. Moldave, eds.) p. 65. Academic Press, New York, 1968.

2. Hnilica, L. S., "The Structure and Biological Function of Histones." Chem. Rubber Publ. Co., Cleveland, Ohio, 1972.

3. O'Malley, B. W., and Hardman, J. G., eds., "Methods in Enzymology," Vol. 40: Hormone Action, Part E, Nuclear Structure and Function. Academic Press, New York, 1975.

4. Kinkade, J. M., Jr., and Cole, R. D., *J. Biol. Chem.* **241**, 5790 (1966).

5. Kinkade, J. M., Jr., and Cole, R. D., *J. Biol. Chem.* **241**, 5798 (1966).

6. Evans, K., Hohmann, P., and Cole, R. D., *Biochim. Biophys. Acta* **221**, 128 (1970).

7. Rall, S. C., and Cole, R. D., *J. Biol. Chem.* **246**, 7175 (1971).

8. Bustin, M., and Cole, R. D., *J. Biol. Chem.* **244**, 5286 (1968).

9. Kinkade, J. M., Jr., *J. Biol. Chem.* **244**, 3375 (1969).

10. Nelson, R. D., and Yunis, J. J., *Exp. Cell Res.* **57**, 311 (1969).

11. Panyim, S., and Chalkley, R., *Biochem. Biophys. Res. Commun.* **37**, 1042 (1969).

12. Hohmann, P., and Cole, R. D., *Nature (London)* **223**, 1064 (1969).

13. Elgin, S. C. R., and Weintraub, H., *Ann. Rev. Biochem.* **44**, 725 (1975).

14. Ruiz-Carrillo, A., Wangh, L. J., Littau, V. G., and Allfrey, V. G., *J. Biol. Chem.* **249**, 7358 (1974).

15. DeLange, R. J., and Smith, E. L., *Ann. Rev. Biochem.* **40**, 279 (1971).
16. Balhorn, R., and Chalkley, R., in *"Methods in Enzymology,"* Vol. 40: Hormone Action, Part E, Nuclear Structure and Function (B. W. O'Malley and J. G. Hardman, eds.), p. 138. Academic Press, New York, 1975.
17. Sung, M., and Smithies, D., *Biopolymers* **7**, 39 (1969).
18. Gurley, L. R., Walters, R. A., and Tobey, R. A., *J. Biol. Chem.* **250**, 3936 (1975).
19. Chen, C., Smith, D. L., Bruegger, B. B., Halpern, R. M., and Smith, R. A., *Biochemistry* **13**, 3785 (1974).
20. Stellwagen, R. H., and Cole, R. D., *J. Biol. Chem.* **243**, 4452 (1968).
21. Sung, M., and Dixon, G. H., *Proc. Natl. Acad. Sci. U.S.A.* **67**, 1616 (1970).
22. Morris, S., and Cole, R. D., unpublished results.
23. Böhm, E. L., Strickland, W. N., Strickland, M., Thwaits, B. H., van der Westhuyzen, D. R., and von Holt, C., *FEBS Lett.* **34**, 217 (1973).
24. Sommer, K. R., and Chalkley, R., *Biochemistry* **13**, 1022 (1974).
25. Phillips, D. M. P., and Clarke, M., *J. Chromatogr.* **46**, 321 (1970).
26. Bartley, J., and Chalkley, R., *J. Biol. Chem.* **245**, 4286 (1970).
27. van der Westhuyzen, D. R., and von Holt, C., *FEBS Lett.* **14**, 333 (1971).
28. Kornberg, R. D., and Thomas, J. O., *Science* **184**, 865 (1974).
29. van der Westhuyzen, D. R., Böhm, E. L., and von Holt, C., *Biochim. Biophys. Acta* **359**, 341 (1974).
30. Murray, K., Vidali, G., and Neelin, J. M., *Biochem. J.* **107**, 207 (1968).
31. Stellwagen, R. H., Reid, B. R, and Cole, R. D., *Biochim. Biophys. Acta* **155**, 581 (1968).
32. Luck, J. M., Rasmussen, P. S., Satake, K., and Tsvetikov, A. N., *J. Biol. Chem.* **233**, 1407 (1958).
33. Lowry, O. H., Rosebrough, N. J., Farr, A. L., and Randall, R. J., *J. Biol. Chem.* **193**, 265 (1951).
34. Blobel, G., and Potter, V., *Science* **154**, 1662 (1966).
35. Fambrough, D. M., Fujimura, F., and Bonner, J., *Biochemistry* **7**, 575 (1968).
36. Fazal, M., and Cole, R. D., unpublished results.
37. Phelan, J. J., Subirana, J. A., and Cole, R. D., *Eur. J. Biochem.* **31**, 63 (1972).
38. Stellwagen, R. H., and Cole, R. D., *J. Biol. Chem.* **243**, 4456 (1968).
39. Langan, T. A., Rall, S. C., and Cole, R. D., *J. Biol. Chem.* **246**, 1942 (1971).
40. Greenaway, P. J., and Murray, K., *Nature (London), New Biol.* **229**, 233 (1971).
41. Wigle, D. T., and Dixon, G. H., *J. Biol Chem.* **246**, 5636 (1971).

Chapter 14

Cytochemical Quantitation of Histones

NIRMAL K. DAS

Department of Cell Biology, College of Medicine,
University of Kentucky,
Lexington, Kentucky

AND MAX ALFERT

Department of Zoology,
University of California, Berkeley,
Berkeley, California

I. Introduction

Attempts to identify histones at the individual cell level go back to Caserosson (*1*) who thought that "diamino acid" proteins or histones had ultraviolet absorption spectra different from other cellular proteins.[1] However, later studies by Mirsky and Pollister (*2*) failed to show such differences in the absorption spectra between histones and nonhistone proteins. Pollister and Ris (*3*) distinguished histones from nonhistone proteins on the basis of the solubility of the former in sulfuric acid. They applied Millon's reaction for tyrosine to cell preparations in the presence of sulfuric acid or trichloroacetic acid (TCA). Since histones are soluble in sulfuric acid, but not in TCA, the decrease in staining intensity in the presence of sulfuric acid was used as an indirect measure of cellular histones. Anionic and cationic dyes have also been used to stain cellular proteins at different pH values (*4*). The quantitative and qualitative studies of histones at the individual cell level have been aided substantially with the discovery, by Alfert and Geschwind (*5*) that the anionic dye acid fast green binds selectively to nuclear histones at an alkaline pH. In this article we will describe the alkaline fast green (AFG)

---

[1]Caspersson made this observation in 1941 and discussed it in 1950 (*1*).

staining technique and its application for quantitative measurements of nuclear histones. In addition, a few selected cytochemical methods, which have been used separately or in conjunction with the AFG technique, will be described to illustrate the usefulness of such techniques in studies on histone patterns during growth and development. Other cytochemical staining methods for histones are discussed in an excellent review article by Bloch (6).

II. AFG Staining Procedure

The AFG staining procedure which has been worked out by Alfert and Geschwind (5) is as follows:

1. Fix plant and animal tissues or cultured cells in 10% neutral formalin[2] for a period of at least 1 hour. Wash fixed materials thoroughly, preferably overnight, in running cold tap water to remove formalin.

2. Dehydrate and embed tissue pieces in paraffin; prepare paraffin sections.

3. Hydrolyze deparaffinized sections or cultured cell preparations in 5% TCA for 15 to 20 minutes at 90 °C to remove deoxyribonucleic acid (DNA). Wash slides 3 times, 10 minutes each, in 70% ethanol and rinse several times in distilled water to remove TCA.

4. Place slides for 1 to 2 minutes in 0.1 M Tris-HCl or 0.1 M sodium barbitol buffer at pH 8.0–8.2. Stain slides for 30 minutes at room temperature in 0.1% solution of acid fast green (National Aniline Division) made in the above buffer.

5. Wash stained slides for 5 minutes in distilled water, followed by a quick rinse in 95% ethanol. Dehydrate, clear, and mount slides.

The precedure described above is suitable for staining chromosomal histones of various types of cells. However, it might occasionally be necessary to change the staining and washing schedule. Bloch and Godman (7) have modified this procedure by replacing HCl with TCA in all steps of the Feulgen procedure to avoid loss of histones. Later studies (8, 9) suggested that little or no histone is removed from formalin-fixed nuclei during the conventional HCl–Feulgen staining procedure (10). Nuclei can therefore be first stained with Feulgen, mapped, and measured for DNA content. They are subsequently restained with AFG and measured again for histone content.

[2] Formalin solution, which is usually acidic, becomes neutral if left overnight in a bottle containing marble chips.

III. Specificity of AFG Staining

AFG staining is affected by the type of fixative used (5). Fixation of cells in 10% neutral formalin is suitable for preservation of histones and of cell morphology as well. Cells can also be fixed in methanol or ethanol for staining of nuclear histones. However, these fixatives cause considerable shrinkage of cells and may lead to loss of (some) histone during hydrolysis. The chromosomal staining specificity is abolished in cells fixed in divalent metal ion-containing fixatives (such as Zenker's or Susa's which contain mercuric chloride). Such metal ions could link carboxyl groups of proteins to the acid dye (5). Acetic acid–alcohol (Carnoy's) fixative, commonly used in cytology, also affects the specificity of AFG staining, perhaps because of the shift in the isoelectric point favoring acid dye binding (5), and may also lead to loss of histones.

The AFG stainability of nuclear basic proteins depends on the prior removal of DNA by hot TCA or deoxyribonuclease (DNase). This frees guanidino groups of arginine and the epsilon amino groups of lysine of basic proteins to react with acid dye ions (5).

The fast green staining at the empirically chosen pH of 8.0–8.2 is selective for chromosomal basic proteins, which have a higher isoelectric point than most other cellular proteins. Little or no stain is seen in the nucleolus or cytoplasm of cells (Figs. 1 and 2). However, at a low acid pH, all cellular proteins, including cytoplasmic and nucleolar proteins, stain (Figs. 3 and 4).

The specificity of AFG for basic proteins has been determined in model experiments. Alfert and Geschwind (5) observed that among various known proteins spotted on filter papers, basic proteins, histones, and protamines all stain well with AFG, while acidic proteins do not stain. Methylated albumin and basic proteins, lysozyme, and cytochrome c all give a positive reaction to AFG. However, such proteins do not interfere with AFG staining of nuclear histones because of their relatively low concentrations in most cells. Although ribosomal basic proteins have not been tested in this model experiment, the presence of these proteins in relatively high concentrations in the cytoplasm and the nucleolus of some cells could lead to AFG staining of these cellular compartments (11–15).

Formalin-fixed onion root tip cells, mouse liver or kidney cells, etc. demonstrate specific nuclear histone staining (Fig. 1). Such cell preparations are routinely used as standards in AFG staining of new materials. The specificity of histone staining in fixed cells is also ascertained by subjecting cell preparations to the procedure of deamination or acetylation, both of which affect the epsilon amino groups of the lysine residue of histones (6). Both of these procedures abolish AFG stainability of somatic or adult histones

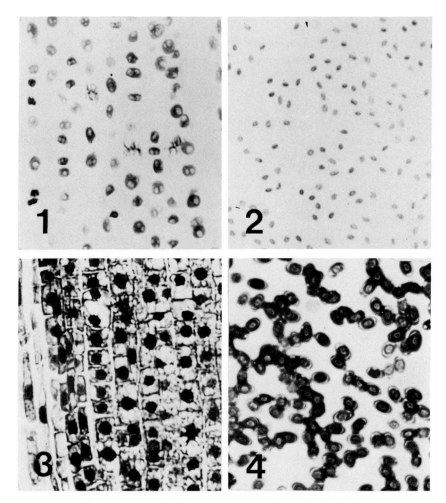

FIGS. 1–4. Photomicrographs of animal and plant cells stained with fast green at pH 8 (Figs. 1 and 2) and at pH 2.2 (Figs. 3 and 4). Figures 2 and 4 are blood smears of the lizard *Sceloporus occidentalis*, prefixed in absolute methanol to prevent hemolysis; Figs. 1 and 3 are sections of onion root tips. Reproduced from Alfert and Geschwind (5). × 550.

which contain an overall high lysine residue. The AFG staining of protamines or arginine-rich sperm nuclear proteins is not affected by deamination or acetylation (*6,16*).

The deamination and acetylation procedures (*6*) are described below.

DEAMINATION

1. Prepare 10% solutions of sodium nitrite and acetic acid. Chill to ice-bath temperature.
2. Mix equal volumes of these solutions and incubate slides, from which DNA has been removed by hot TCA or DNase hydrolysis, for 45 minutes (3 changes, 15 minutes each) at ice-bath temperature.
3. Wash slides thoroughly in cold distilled water and stain with AFG.

ACETYLATION

1. Remove DNA from cell preparations by hot TCA or DNase hydrolysis, and wash slides thoroughly in water.
2. Dehydrate slides through 70%, 95%, and 100% ethanol.
3. Incubate slides in acetic anhydride containing 1% acetic acid for 60 minutes at 60 °C.
4. Bring slides through ethanol to distilled water and stain with AFG.

IV. Quantitation of Histones of Individual Nuclei

A. Cytophotometry of AFG-Histones

Earlier biochemical analysis suggested that the DNA–histone ratio in various cells remains constant (17,18). This observation was supported by cytophotometric measurements of Feulgen-DNA and AFG-histones of nuclei of animal and plant cells.

Measurements (19) of Feulgen-DNA and AFG-histones of individual nuclei show that histones maintain a constancy similar to that of DNA in rat thyroid and mouse kidney nuclei under different experimental conditions (Table I) (20). In both plant and animal cells, the DNA–histone ratio of the same individual nuclei remains constant throughout the whole interphase, including the DNA-synthetic or the S-phase (6,7,20,21) (see also Figs. 5–7). These results have suggested that DNA and histones are synthesized concurrently.

B. Autoradiographic Study of Histone Synthesis

We have also used autoradiography, in combination with Feulgen-DNA cytophotometry, to study the temporal relationship between DNA and histone synthesis in primary spermatocytes of a marine worm *Urechis caupo*

TABLE I

VOLUME, DNA, AND PROTEIN CONTENT OF NUCLEI FROM MAMMALIAN TISSUES UNDER VARIOUS EXPERIMENTAL CONDITIONS[a,b]

		Modal volume (μm^3)	Average total protein content	Histone content (Mean ± SE)	DNA content (Mean ± SE)
Rat liver	2n	120	10.6	28 ± 2	54 ± 1
	4n	240	20.2	56 ± 2	106 ± 2
	8n	480	38.4	125 ± 6	199 ± 4
Rat thyroid (2n)	Hypophysect.	39	3.2	27 ± 1	52 ± 1
	Control	78	6.0	27 ± 1	53 ± 1
	Propylthiouracil-treated	110	7.6	26 ± 1	52 ± 1
Mouse kidney (2n)	Collecting duct	60	6.0	33 ± 1	51 ± 1
	Prox. convoluted tubule: castrate	102	10.1	31 ± 2	50 ± 1
	Prox. convol. tub.: castr. + androgen	144	13.2	31 ± 2	53 ± 1

[a] Reproduced from Alfert (20) with the permission of the publisher.
[b] DNA in terms of Feulgen dye and histone in terms of alkaline fast green stainability, in arbitrary units of dye content per nucleus. Total protein content, in arbitrary units, calculated from the concentrations of tyrosine mercurial (Millon reaction) in nuclear sections, taking into account the modal nuclear volumes. The karyometric determinations were done on series of 100 to 200 nuclei, and the photometric measurements were done on 20 to 30 nuclei for each value reported in the table. For technical details see Alfert and Geschwind (5).

(22). Autoradiographic demonstration of the time of histone synthesis during the cell cycle is usually difficult, because labeled basic amino acids used in these studies are incorporated in both histones and nonhistone proteins and also because only a small fraction of the total proteins labeled in growing cells at any time is histone (23,24). However, materials, such as primary spermatocytes of Urechis, which are limited in their growth potential to ensuing meiotic division, are suitable for autoradiographic studies of histone patterns during the premeiotic interphase. The following procedures have been used in our study:

1. For 30 minutes expose the coelomic fluid of male Urechis, containing free-floating spermatocytes at different stages of maturation, to 50 μCi/ml of L-arginine-^{3}H (sp act 1.4 Ci/mmol).

2. Wash cells 2 to 3 times, at ice-bath temperature, in cell-free and nonradioactive coelomic fluid.

3. Smear cells on slides, expose to formalin vapor (to preserve cell shape), air-dry, and postfix in 10% neutral formalin.

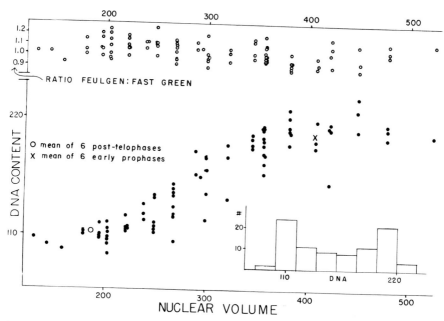

FIG. 5. The relation between volume (μm³), DNA (Feulgen), and histone (fast green) content in intermitotic nuclei of growing onion root tips. Lower right: A frequency distribution of DNA values typical of a mitotically active tissue. Middle: Individual measurements of DNA content plotted against nuclear volume. Upper: Individual DNA–histone ratios of the same nuclei, plotted against nuclear volume. Reproduced from Alfert (20) with the permission of the publisher.

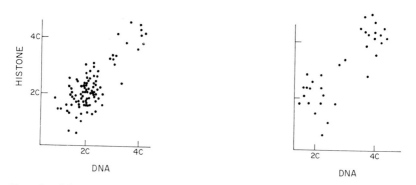

FIGS. 6 and 7. Relative amounts of DNA plotted against relative amounts of histone for the same individual nuclei of 1-day-old rat liver (Fig. 6, *left*) and in growing cultures of rat fibroblasts (Fig. 7, *right*). Note the same DNA–histone content of nuclei at different stages of interphase. Reproduction from Bloch and Godman (7) with the permission of the authors and the publisher.

4. Stain slides with TCA- or HCl-Feulgen and coat slides with liquid emulsion (Kodak NTB2).

5. Develop slides, map cells, and record silver grains present over these cells.

6. Remove grains in 0.1% aqueous solution of potassium ferricyanide: place slides in this solution for 2 minutes and then in Kodak acid fixer (F–5) for 1 minute; all silver grains are found to be removed from processed emulsions after two to three such changes.

7. Wash, dehydrate, clear, and mount slides. Measure nuclear Feulgen-DNA content of mapped cells by the two-wavelength cytophotometric technique (25).

The combined Feulgen-DNA measurements and grain counts on the same individual cells show the rate of incorporation of [H³]arginine approximately doubles as cells enter the S-phase (Fig. 8). This rate remains constant throughout the whole S-phase; it decreases again as cells leave the S-phase. That the increased incorporation of [H³]arginine during the S-phase is due to synthesis of histones has been suggested from experiments in which most of the labeled protein of the S-phase cells was found to be acid-labile (22). Furthermore, no increase in incorporation of [H³]tryptophan was seen in cells at the S-phase (22). Histones lack tryptophan and they are acid-labile (6). The temporal relationship between DNA and histone syntheses has been demonstrated in several biochemical studies (23,24).

V.   Histone Patterns during Spermatogenesis and Early Embryogenesis

A.   Spermatogenesis

A transition from somatic histones to protamines or arginine-rich histones takes place during spermatogenesis in many organisms (26). The significance of the transition still remains unknown (26). Cytochemical techniques have been valuable for demonstrating such nuclear basic protein changes during spermatogenesis (6,16,26,27). Changes from histones to protamines in sperm nuclei of Rhine salmon were recognized by earlier biochemists (28,29). Alfert (16,27) showed later, by use of AFG staining, that the shift from somatic histones to protamines, or proteins rich in arginine, occurs at a late spermatid stage. Nuclear protamines (but not arginine-rich sperm nuclear histone) are lost during hot TCA hydrolysis used in AFG staining. Thus, the loss of AFG stainability of spermatid nuclei is

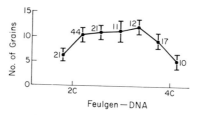

FIG. 8. Relative amounts of DNA plotted against the number of grains counted over the same cells. Cells were labeled for 30 minutes with [H³]arginine (50 μCi/ml), smeared, fixed, stained with Feulgen, and coated with emulsion. The procedure for removal of grains is described in the text; nuclear DNA content was measured by the two-wavelength method ($\lambda_1 = 580$ nm and $\lambda_2 = 595$ nm). Each point represents the mean and standard error of the grain number counted over the whole cells; number of cells is indicated on the left side of each point. Reproduced from Das and Alfert (22) with the permission of the publisher.

an indication of the replacement of typical nuclear histones by protamines in these nuclei (Fig. 9a and b).

Spermatid and sperm nuclei containing protamines can be stained with AFG (or also with eosin or bromphenol blue) after removal of DNA by mild picric acid hydrolysis (6). The picric acid procedure is described below:

1. Hydrolyze formalin-fixed cell preparations for 6 hours at 60 °C with a saturated solution of picric acid.

2. Wash slides in running tap water and also 70% ethanol to remove picric acid.

3. Stain slides with AFG without further hydrolysis in hot TCA.

The transition from typical histones to protamines or arginine-rich proteins during spermiogenesis can further be demonstrated by staining cells with AFG after acetylation or deamination (Fig. 10). As mentioned before, acetylation or deamination does not block AFG staining of arginine-rich proteins.

The increase in the arginine content in nuclear proteins during spermiogenesis in several organisms has also been demonstrated by the use of the Sakaguchi reaction for arginine (6,30). This procedure is as follows:

1. Bring formalin-fixed cell preparations to distilled water. Blot off water from slides and place them in a staining jar.

2. Mix freshly prepared solutions of 5 parts of barium hydroxide (filter before using) and 1 part of 1% sodium dichloro-α-napthol 1 (Eastman Organic Chemicals) dissolve in tertiary butanol or 70% ethanol. Pour this mixture in the jar, completely covering the slides. Stain for 10–15 minutes at room temperature.

3. Pour in 1% urea solution to allow the crust of barium carbonate, formed in the jar, to flow over the jar without coming in contact with cells.

4. Transfer slides rapidly through three changes of tertiary butanol and

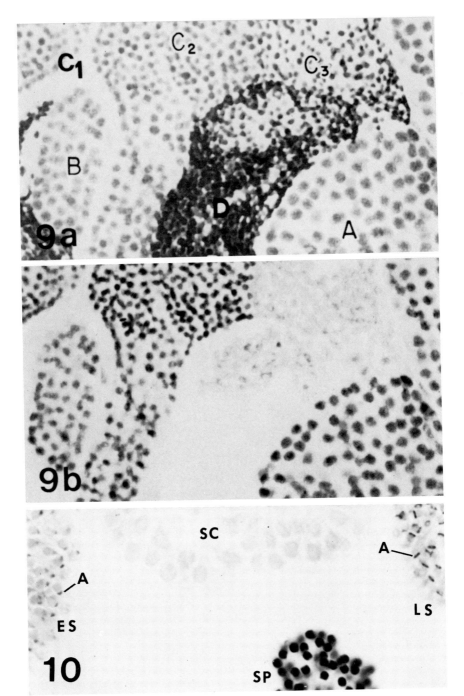

two changes of xylene. Both butanol and xylene contain 5% tri-*N*-butylamine (Eastman Organic Chemicals). Mount slides.

Quantitative changes in Feulgen-DNA and AFG-histones during spermiogenesis in a grasshopper have been studied by Alfert (*27*). The Feulgen–fast green ratio remains unchanged from the primary spermatocyte to the early spermatid stage (Fig. 11). However, a considerable decrease in the fast green stainability is seen at the midspermatid stage (Fig. 11). perhaps due to masking of histones by other proteins. This is followed by a sudden increase in the nuclear AFG stainability at the late spermatid stage (Fig. 11). That this increase is due to the replacement of typical somatic histones by a more arginine-rich protein is indicated by the Sakaguchi reaction (Table II).

B.   Early Embryogenesis

In an attempt to elucidate regulatory mechanisms of embryonic differentiation, cytochemical methods have been used to study embryonic histones of several organisms (*6,27,31–35*). The highly basic proteins of sperm nuclei of various organisms are no longer seen after fertilization; it has been suggested that these proteins are replaced by "cleavage" (*6*) or "juvenile" (*32*) histones, followed by the appearance of somatic or adult histones at gastrulation (*6*). Other studies suggest that the shift to adult histones takes place during or prior to early cleavage stages (*27,31,33–35*).

We have recently studied the AFG staining characteristics of nuclear basic proteins during postfertilization maturation divisions and early embryonic stages in *Urechis caupo* (*31*). *Urechis* sperm nuclei contain protamine-type protein, as shown by cytochemical and biochemical methods (*36*). These sperm nuclei normally do not stain with AFG due to the loss of protamine during hot TCA hydrolysis. However, within 10–15 minutes after the entrance into the egg cytoplasm, the sperm nuclei stain well with AFG (Figs. 12 and 13). Both the male and female pronuclei and nuclei of embryos

FIG. 9.   (a) Feulgen-stained salmon testis cells. (b) The same cells restained by alkaline fast green (AFG) after extraction with hot TCA. Note that nuclei of primary spermatocytes (A), secondary spermatocytes (B), and spermatids at early (C_1) and middle (C_2) stages of condensation are well stained with AFG. Nuclei of spermatids at later stages of condensation (C_3) and mature sperm (D) do not stain with AFG. Reproduced from Alfert (*16*) with the permission of the publisher. ×850.

FIG. 10.   Cluster of *Urechis caupo* primary spermatocytes (SC), early spermatids (ES), late spermatids (LS), and mature sperm (SP) stained with alkaline fast green (AFG) after removal of DNA by picric acid followed by acetylation. Note that acetylation does not block the AFG stainability of mature sperm nuclei which contain arginine-rich protamine; the AFG stainability of acrosomal (A) basic protein is also not affected by acetylation. Reproduced from Das *et al.* (*36*).

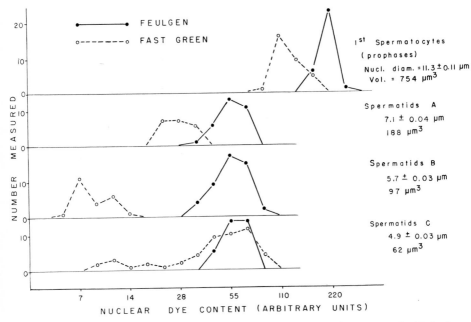

FIG. 11. Distribution of nuclear Feulgen-DNA and alkaline fast green stainable basic proteins of a grasshopper (*Chortophaga viridifasciata*): primary spermatocytes, early spermatids (A), midspermatids (B), and late spermatids (C). Reproduced from Alfert (*27*) with the permission of the publisher.

TABLE II

INCREASE IN THE ARGININE–TYROSINE RATIO DURING SPERMATOGENESIS IN
THE GRASSHOPPER, *Chortophaga viridifasciata*[a]

	Sakaguchi E^b_{510} (Arginine)	Million E^b_{490} (Tyrosine)	Arg/Tyr	Average decrease in (arginine) after hot TCA
1° Cyte	.297 ± .007	.185 ± .008	1.61	16.5%
Tid B	.302 ± .006	.181 ± .007	1.67	14.5%
Sperm	.392 ± .007	.131 ± .004	2.99	24%

[a] Reproduced from Alfert (*27*) with the permission of the publisher.
[b] Each value represents mean and standard error of 20 measurements of optical density in 4-μm sections.

from early cleavage to later stages of development also stain well with AFG (Figs. 14 and 15). Acetylation blocks stainability at all stages of development. These results suggest that the transition from protamine to adult histones occurs very soon after fertilization in *Urechis*. Such observations are sup-

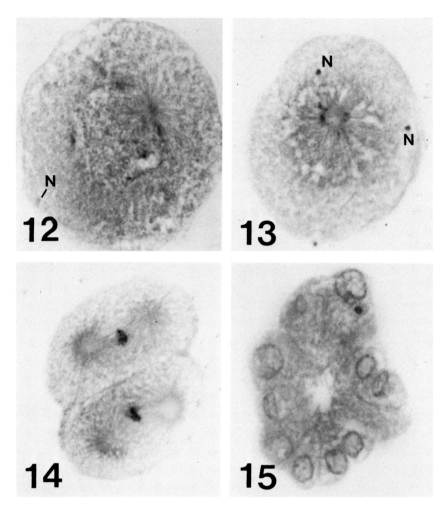

FIGS. 12–15. AFG-stained sections (10 μm) of *Urechis caupo* embryos. Five minutes (Fig. 12), 15 minutes (Fig. 13), 1½ hours (Fig. 15) after fertilization. Note that the sperm nucleus (N) is only lightly stained 5 minutes after fertilization (Fig. 12); it is well stained 10 minutes later (Fig. 13). Nuclei at the 2-cell (Fig. 14) and 64-cell (Fig. 15) stages are also well stained. Reproduced from Das *et al.* (*31*) with the permission of the publisher. ×800.

ported by biochemical analyses which show that all five major histones are present in embryos at the cleavage, blastula, and gastrula stages (Fig. 16). The absence of qualitative differences in histone patterns at various early developmental stages in *Urechis* would seem to argue against the involvement of histones alone in the control of patterns of RNA synthesis during embryogenesis.

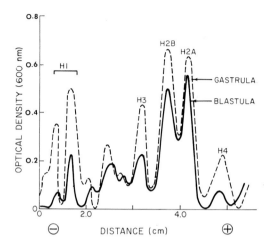

Fig. 16. Densitometer tracing profiles of Coomassie Blue–stained acid-soluble nuclear proteins of *Urechis caupo* embryos. Proteins were separated by sodium dodecylsulfate–polyacrylamide gel electrophoresis [see Weber and Osborn (37)]. H1 = very lysine-rich histone; H2B and H2A = slightly lysine-rich histones; H3 and H4 = arginine-rich histones. Calf thymus histones have been used as markers for the identification of these five major histones. Note that all major acid-soluble nuclear histones from blastula and gastrula embryos are qualitatively the same; similar nuclear histone patterns (not shown) were also noted in embryos at the early cleavage stage.

VI. Summary and Conclusion

Cytochemical methods described in this article have been valuable tools for studying quantitative and qualitative nuclear histone changes in individual cells during growth and development. Information concerning nuclear DNA–histone constancy in proliferating and nonproliferating cells has been obtained. It has been shown that transition from typical somatic histones to more basic nuclear proteins in several organisms occurs in late spermiogenesis. The time of normal chromosomal DNA–histone reorganization during embryogenesis in various organisms has been ascertained. However, cytochemical methods are mainly limited to detection and analysis of the content of nuclear basic proteins and their overall composition; they do not provide information about chemical modifications and quantitative or qualitative modulations of individual histone fractions, which could influence nuclear activity. Therefore, coordinated cytochemical and biochemical studies are necessary for the understanding of the role of histones in nuclear function.

ACKNOWLEDGMENTS

These studies were supported partly by General Research Support Grants (RR 05373–4), NIH, and partly by grants from the National Science Foundation.

REFERENCES

1. Caspersson, T., "Cell Growth and Cell Function." Norton, New York, 1950.
2. Mirsky, A. E., and Pollister, A. W., *J. Gen. Physiol.* **30**, 117 (1946).
3. Pollister, A. W., and Ris, H., *Cold Spring Harbor Symp. Quant. Biol.* **12**, 147 (1947).
4. Singer, M., *Int. Rev. Cytol.* **1**, 211 (1952).
5. Alfert, M., and Geschwind, I. I., *Proc. Natl. Acad. Sci. U.S.A.* **39**, 991 (1953).
6. Bloch, D. P., *in* "Chemistry and Cytochemistry of Nucleic Acids and Nuclear Proteins" (M. Alfert, H. Bauer, C. V. Harding, and P. Sitte, eds.), Vol. V-3, Section d, p. 1. Springer-Verlag, Berlin and New York, 1966.
7. Bloch, D. P., and Godman, G. C., *J. Biophys. Biochem. Cytol.* **1**, 17 (1955).
8. Mattingly, Sister A., *Exp. Cell Res.* **29**, 314 (1963).
9. Das, N. K., Siegel, E. P., and Alfert, M., *J. Cell Biol.* **25**, 387 (1965).
10. Stowell, R. E., *Stain Technol.* **20**, 45 (1945).
11. Talepores, P., *J. Histochem. Cytochem.* **7**, 322 (1959).
12. Alfert, M., and Goldstein, N. O., *J. Exp. Zool.* **130**, 403 (1955).
13. Gifford, E. M., Jr., and Tepper, H. B., *Am. J. Botany* **49**, 706 (1962).
14. Horn, E. C., and Ward, C. L., *Proc. Natl. Acad. Sci. U.S.A.* **43**, 776 (1957).
15. Das, N. K., and Alfert, M., *J. Cell Sci.* **12**, 781 (1973).
16. Alfert, M., *J. Biophys. Biochem. Cytol.* **2**, 109 (1956).
17. Mirsky, A. E., and Ris, H., *Nature (London)* **163**, 666 (1949).
18. Vendrely, R., and Vendrely, C., *in* "Chemistry and Cytochemistry of Nucleic Acids and Nuclear Proteins" (M. Alfert, H. Bauer, C. V. Harding, and P. Sitte, eds.), Vol. V-3, Section c, p. 1. Springer-Verlag, Berlin and New York, 1966.
19. Pollister, A. W., Swift, H., and Alfert, M., *J. Cell. Comp. Physiol.* **38**, Suppl. 1, 101 (1951).
20. Alfert, M., *Union Int. Sci. Biol., Ser. B, Symp. Fine Struc. Cells* **21**, 157 (1955).
21. Rasch, E., and Woodard, J. W., *J. Biophys. Biochem. Cytol.* **6**, 263 (1959).
22. Das, N. K., and Alfert, M., *Exp. Cell Res.* **49**, 51 (1968).
23. Prescott, D. M., *J. Cell Biol.* **31**, 1 (1966).
24. Robbins, E., and Borun, T. W., *Proc. Natl. Acad. Sci. U.S.A.* **57**, 409(1967).
25. Patau, K., *Chromosoma* **5**, 341 (1952).
26. Bloch, D. P., *Genetics* **61**, Suppl. 1 (1969).
27. Alfert, M., *Colloq. Ges. Physiol. Chem.* **9**, 73 (1958).
28. Miescher, F., "Die histochemischen und physiologischen Arbeiten." Vogel, Leipzig, 1897.
29. Kossel, A., "The Protamines and Histones." Longmans Green, London, 1928.
30. Deitch, A. D., *in* "Introduction to Quantitative Cytochemistry" (G. L. Weid, ed.), p. 327. Academic Press, New York, 1966.
31. Das, N. K., Micou-Eastwood, J., and Alfert, M., *Dev. Biol.* **43**, 333 (1975).
32. Das, C. C., Kaufman, B. P., and Gay, H., *J. Cell Biol.* **23**, 423 (1964).
33. Vaughan, J. C., *J. Histochem. Cytochem.* **16**, 473 (1968).
34. Moore, B. C., *Proc. Natl. Acad. Sci. U.S.A.* **50**, 1018 (1963).
35. Kopecny, V., and Pavlok, A., *J. Exp. Zool.* **191**, 85 (1975).
36. Das, N. K., Micou-Eastwood, J., and Alfert, M., *J. Cell Biol.* **35**, 455 (1967).
37. Weber, K., and Osborn, M., *J. Biol. Chem.* **244**, 4406 (1969).

Part D.   Fractionation and Characterization of Nonhistone Chromosomal Proteins. I

Chapter 15

The Isolation and Purification of the High Mobility Group (HMG) Nonhistone Chromosomal Proteins

GRAHAM H. GOODWIN AND ERNEST W. JOHNS

Division of Molecular Biology, Chester Beatty Research Institute,
Institute of Cancer Research–Royal Cancer Hospital,
London, England

I.   Introduction

Chromatin contains a group of nonhistone chromosomal proteins that are less firmly bound than the histones and can be extracted from the chromatin with 0.35 M NaCl (*1*). This nonhistone protein fraction is highly heterogenous (*2*); it is probably made up of structural proteins, nuclear enzymes, and possibly small amounts of regulatory proteins. The 0.35 M NaCl extract can be subdivided into two fractions based on their solubility in trichloroacetic acid (TCA) (*2*); (1) the many high-molecular-weight proteins which are insoluble in 2% TCA; and (2) the smaller number of lower-molecular-weight proteins which are soluble in 2% TCA. The former group we have termed the low mobility group (LMG) proteins and the latter the high mobility group (HMG) proteins because of their relative electrophoretic mobilities in polyacrylamide gels (*2*). We have been mostly concerned with the HMG proteins for they contain a number of interesting proteins with high contents of acidic and basic amino acids. These proteins were in fact originally discovered in perchloric acid extracts of chromatin and also in the histone

H2B fraction and were separated from the histone by carboxymethyl celluose (CM-cellulose) chromatography (3,4). Since then we have developed procedures for isolating fairly large quantities of three of the HMG nonhistone proteins, HMG1, 2, and 17 (5–9). [See Goodwin et al. (2) for the numbering of HMG proteins.]

The HMG proteins bind to DNA and histones (10,11), and they are probably chromatin structural proteins, though being present in smaller quantities than the histones they presumably have a more specific function.

One of the HMG proteins, HMG2, exhibits multiple forms which could be due to sequence microheterogeneity (as is the case with histone H1) or postsynthetic modifications such as methylation or acetylation. This article describes in detail the large-scale isolation of proteins HMG1, 2, and 17 and also the isolation of the HMG2 subfractions.

II.   Isolation of HMG Proteins

A schematic diagram for the isolation of proteins HMG1, 2, and 17 is given in Fig. 1.

A.   Isolation of Total HMG Proteins from Calf Thymus

The most convenient source for the large-scale preparation of nonhistone proteins is calf thymus, despite the fact that this tissue has a rather active protease(s) which partly degrades histone H1 and HMG protein causing a number of additional bands to appear in the 0.35 M NaCl extract (see below).

To isolate HMG proteins the thymus tissue (either fresh or stored for up to 1 month at $-20\,°C$) is blended with saline–ethylenediaminetetraacetic acid (EDTA) and the chromatin is washed several times with the saline–EDTA. The crude chromatin is extracted with 0.35 M NaCl and the LMG proteins and any contaminating nonchromosomal proteins precipitated with 2% TCA. After completely removing the precipitated material, the HMG proteins are recovered from the supernatant by acetone precipitation. To minimize the volume of acetone used here the precipitation is carried out at an alkaline pH—at pH 10 only 3 volumes of acetone are required to quantitatively precipitate the HMG proteins. The proteins are then converted to the chloride form by washing with acetone–HCl; this prevents aggregation of the proteins.

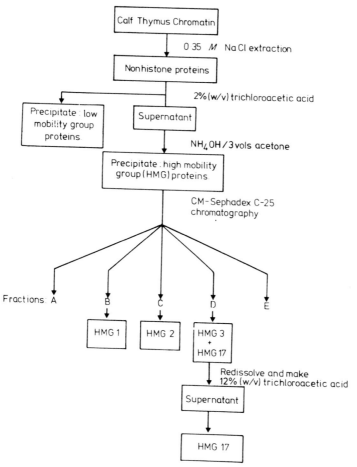

FIG. 1.   Schematic diagram for the isolation of HMG nonhistone chromosomal proteins.

METHOD

Operations up to the acetone precipitation step are carried out at 4 °C. One kilogram of minced calf thymus is divided into four portions and each is blended with approximately 700 ml of 0.075 M NaCl–0.025 M EDTA (pH 7.5) for 2 minutes at full speed (12,000 rpm) in a domestic blender (Kenwood, Kenmix). Each homogenate is passed through a domestic nylon sieve to remove connective tissue and then centrifuged for 30 minutes at 2000 g. The four sediments are grouped into two and each is blended with 700 ml of NaCl–EDTA for 1 minute at full speed. The homogenates are centrifuged and the crude chromatin pellets washed three more times with

NaCl–EDTA by blending for 30 seconds and centrifuging for 15 minutes. The pellet at this stage should be a cream color; if it is still pink it should be washed further.

The two chromatin sediments are then each extracted 3 times with 500 ml of 0.35 M NaCl (adjusted to pH 7 with 1 M NaOH) by blending for 1 minute at half speed followed by centrifugation at 2000 g for 15 minutes. The total extract (3 liters) is made 2% (w/v) TCA by the addition of 100% (w/v) TCA. The precipitate is removed by centrifugation at 2000 g for 15 minutes, and the supernatant is passed through a No. 4 sintered glass funnel (Sinta Glass, Gallenkamp, England). The filtrate should be absolutely clear. The filtrate is made 0.01 M with respect to β-mercaptoethanol before precipitating the HMG proteins in the following manner at room temperature. First 15 ml of 0.880 ammonia solution is added to each liter of solution (raising the pH to about 10) followed rapidly by the addition of 3 volumes of acetone. After collecting the protein precipitate by centrifuging at 2000 g for 15 minutes, it is washed twice with acetone–0.1 M HCl (6:1 v/v) and then 3 times with acetone before drying under vacuum. The average yield of total HMG protein is about 1 gm.

B.   Fractionation of Total HMG Protein by CM-Sephadex Chromatography: The Isolation of Proteins HMG1 and HMG2

The gradient-sievorptive chromatography principle described by Kirkegaarde (12) is employed here for the initial part of the chromatography. This allows the HMG protein to be loaded onto the column at a fairly high ionic strength (0.15 M NaCl), thereby reducing aggregation. Because the column is preequilibrated at a low ionic strength an inherent salt gradient is set up in the column after the sample is applied. The acidic material elutes at the excluded column volume, and the inherent salt gradient that follows elutes HMGl at about 0.1 M NaCl. The other HMG proteins, being more firmly bound than HMGl, are subsequently brought off the column by conventional ion-exchange chromatography on applying a linear salt gradient rising from 0.15 M NaCl. The method described here is a modification of that described originally (7) in that the ratio of ion-exchanger to protein has been increased and the protein is loaded onto the column at a lower ionic strength (0.15 M rather than 0.2 M NaCl). These changes have made the chromatography very reproducible, and 150–200 mg of each of the purified proteins HMG1 and HMG2 can be obtained from 2 gm of total HMG proteins.

METHOD

All operations were carried out at room temperature. Total HMG protein (usually 2 gm from the bulking of two of the above preparations) is dissolved in 10 ml of 7.5 mM sodium borate buffer (pH 8.8) containing 10 mM β-mercaptoethanol.[1] After readjusting the pH to about 8.8 with 1 N NaOH the solution is dialyzed overnight versus the 7.5 mM borate–mercaptoethanol buffer containing 0.15 M NaCl. (In this buffer and in the elution buffers described below containing NaCl, the NaCl is added to the 7.5 mM borate–mercaptoethanol (pH 8.8) buffer without readjusting the pH.) The dialyzed solution is clarified by centrifuging at 90,000 g for 30 minutes and, if necessary, filtering through a small No. 4 sintered glass funnel. The sample is loaded onto a 5 × 50 cm column (approx. 1 liter column volume) of CM-Sephadex C25 equilibrated with the 7.5 mM sodium borate buffer (pH 8.8) containing 10 mM β-mercaptoethanol. Then 400 ml of the 7.5 mM borate-mercaptoethanol buffer containing 0.15 M NaCl is then pumped through the column at a flow rate of 2 ml/minute. This is followed at the same flow rate by a linear salt gradient, the two chambers of the gradient-forming device each containing 1.6 liters of the 7.5 mM borate–mercapto-ethanol buffer containing 0.15 M NaCl and 2.0 M NaCl respectively. The elution profile is shown in Fig. 2. Protein is precipitated from the fractions A–E by acidifying to 0.1 N HCl and adding 6 volumes of acetone (fractions D and E are first diluted to a salt concentration of 0.35 M NaCl). The precipitates are collected by centrifugation at 2000 g for 10 minutes, washed once with acetone–0.1 N HCl (6:1, v/v), then washed 3 times with acetone before drying under vacuum.

The polyacrylamide gel electrophoretic analysis of the five fractions is shown in Fig. 3. Fraction A is acidic material (a mixture of proteins and

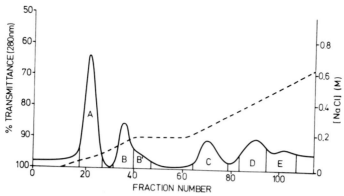

FIG. 2.   Elution profile of the CM-Sephadex C25 chromatography of total HMG protein. ——, Transmittance at 280 nm; ————, sodium chloride concentration. Twenty-milliliter fractions were collected.

[1] When fractionating smaller quantities of protein than described here it is advisable to omit the mercaptoethanol in the chromatography buffer so as to allow the detection of protein in the eluate by reading the absorption at 220–230 nm.

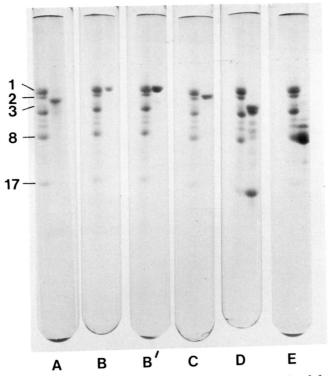

FIG. 3. Comparative polyacrylamide gel electrophoresis analysis of fractions A–E obtained by CM-Sephadex C25 chromatography of total HMG protein. The right-hand side of the gels shows fractions A–E as indicated; the left-hand side of each gel shows the total HMG protein.

nucleic acid). Fraction B is protein HMG1. The shoulder B' is possibly a modified HMG1 but it also contains another contaminating protein and is therefore collected separately. Fraction C is protein HMG2. Fraction D contains mainly HMG3 and HMG17. The latter protein can be obtained from this fraction by differential TCA precipitation (see Section II, C below). Fraction E is mainly HMG8. We now have some evidence that HMG3 and HMG8 are degradation products of HMG1 and histone H1 respectively; HMG3 is probably the N-terminal two-thirds of HMG1 and HMG8 the N-terminal half of H1 (J. M. Walker, G. H. Goodwin, and E. W. Johns, unpublished results).

The amino acid analyses of proteins HMG1 and HMG2 (fractions B and C) are given in Table I. Both proteins are characterized by the high contents of acidic and basic amino acids. Both proteins contain two cysteines which can form intramolecular disulfide bridges. The proteins can

TABLE I

AMINO ACID COMPOSITION (MOL%) AND N-TERMINAL
AMINO ACIDS OF PROTEIN HMG1, HMG2,
AND HMG17

Amino acid	HMG1	HMG2	HMG17
Asp	10.7	9.3	11.2
Thr	2.5	2.7	1.7
Ser	5.0	7.4	3.2
Glu	18.1	17.5	11.2
Pro	7.0	8.9	11.9
Gly	5.3	6.5	9.9
Ala	9.0	8.1	17.1
Val	1.9	2.3	2.2
Cys ($\frac{1}{2}$)	0.0	0.0	0.0
Met	1.5	0.4	0.0
Ile	1.8	1.3	0.1
Leu	2.2	2.0	1.2
Tyr	2.9	2.0	0.0
Phe	3.6	3.0	0.0
Lys	21.3	19.4	24.7
His	1.7	2.0	0.2
Arg	3.9	4.7	4.9
N-terminal amino acid	Gly	Gly	Pro

therefore exist in two forms (with or without disulphide bridge), and these are seen as closely running doublets on polyacrylamide gel electrophoresis (*13*).

It has been noticed that occasionally a small amount of pink color co-elutes with protein HMG2. This has been identified as a cytochrome. Its presence in the HMG preparation occurs if the chromatin is not sufficiently washed with saline–EDTA prior to the 0.35 *M* NaCl extraction. This material can be removed from the HMG2 protein by chromatography on CM-cellulose at pH 9 as in the preparation of protein HMG2 subfractions (see below) when it elutes at a higher salt concentration than HMG2.

C. Isolation of Protein HMG17

Protein HMG17 is soluble in 12% TCA while the other proteins in the CM-Sephadex fraction D are not, and hence it can be isolated by a simple TCA precipitation step.

METHOD

Fraction D protein is dissolved in water (4°C) at a concentration of about 15 mg/ml. Then 100% (w/v) TCA is added to a final concentration of 12% (w/v). The precipitate is removed by centrifuging at 1000 g for 10 minutes and the supernatant is clarified by filtering through a No. 4 sintered glass funnel. Protein HMG17 in the filtrate is precipitated by making the solution 0.2 M H$_2$SO$_4$ and adding 6 volumes of acetone precooled to −10°C. After washing the precipitate with acetone–HCl (0.5 ml concd. HCl to 200 ml acetone) and then several times with acetone, it is dried under vacuum.

Figure 4 shows the polyacrylamide gel electrophoretic analysis of the HMG17 protein. The amino acid analysis is given in Table I.

D. Isolation of Subfractions of HMG2

Isoelectric focusing of HMG2 reveals that it has four main subfractions focusing between pH 7 and 9. These subfractions can be isolated by CM-cellulose chromatography at pH 9 using a shallow salt gradient (8).

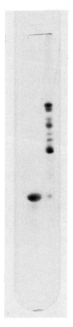

FIG. 4.   Comparative polyacrylamide gel electrophoresis of purified HMG17 (left-hand side) and total HMG protein (right-hand side). From Goodwin *et al.* (7).

METHOD

All operations are carried out at room temperature. First 150 mg of protein HMG2 are dissolved in 15 ml of 7.5 m*M* sodium borate buffer (pH 9) and readjusted to pH 9 with 1 *N* NaOH. The solution is dialyzed overnight versus 4 liters of the 7.5 m*M* borate buffer (pH 9). After centrifugation at 35,000 *g* for 1 hour the solution is applied to a 2.5 × 15 cm CM-cellulose column equilibrated with 7.5 m*M* borate buffer (pH 9). An 800-ml linear salt gradient, 0–0.15 *M* NaCl dissolved in the borate buffer, is pumped

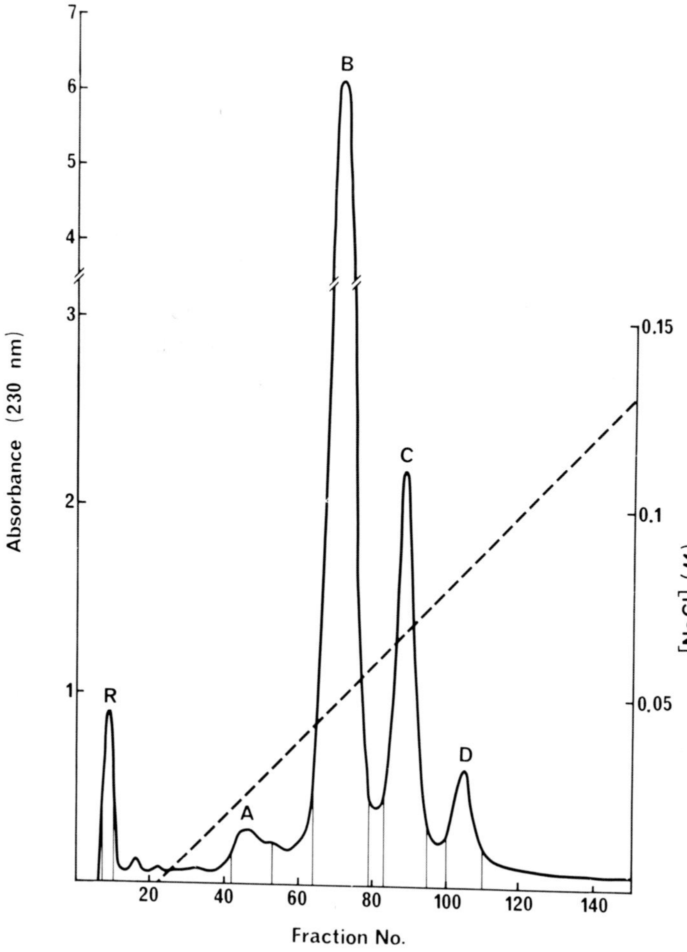

FIG. 5. The elution profile of the CM-cellulose chromatography of total protein HMG2. ——, Absorbance at 230 nm; ––––, NaCl concentration. Five-milliliter fractions were collected. From Goodwin *et al.* (*8*).

through the column at a flow rate of 0.8 ml/minute. Then 5-ml fractions are collected and the optical density at 230 nm is measured. The elution profile is shown in Fig. 5. Protein is precipitated from the fractions in the usual manner by acidifying to 0. 1 N HCl and adding 6 volumes of acetone.

The isoelectric focusing analysis of the total HMG2 and the five column fractions is shown in Fig. 6. The run-through peak (R) is a small amount of HMG1 contamination in the HMG2, and peaks A, B, C, and D are the four HMG2 subfractions eluting in order of increasing basicity.

III.   Alternative Procedures

As mentioned in Section I the HMG proteins can also be extracted from chromatin with 5% perchloric acid (PCA). This forms the basis for an alternative procedure for isolating total HMG and protein HMG1 (6). Briefly, 3.5 volumes of acetone and 0.03 volumes of concentrated HCl are added to the 5% PCA extract. This precipitates histone H1 which is removed by centrifugation leaving the HMG proteins in the supernatant. The HMG proteins are precipitated by the addition of another 2 volumes of acetone.

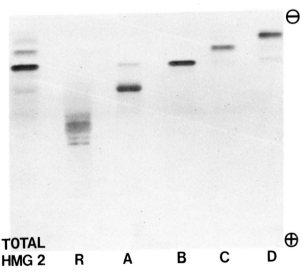

Fig. 6.   Isoelectric focusing of total protein HMG2 and fractions R, A, B, C, and D obtained by CM-cellulose chromatography of total protein HMG2. From Goodwin *et al.* (8).

The HMG1 can then be isolated from the mixture by a batch absorption method using CM-cellulose.

The isolation of total HMG from chromatins other than those of calf thymus and chicken erythrocyte is preferably carried out by 5% PCA extraction rather than by the 0.35 M NaCl extraction method. This is because chromatins with substantial amounts of light "euchromatin" (e.g., liver chromatin) tends to be partially solubilized by 0.35 M NaCl, resulting in the contamination of the HMG protein with substantial amounts of histone. The HMG protein prepared by the PCA extraction method can be fractionated by CM-Sephadex chromatography (Section II,B), and we have successfully isolated proteins HMG1 and HMG2 from calf liver chromatin by this procedure.

ACKNOWLEDGMENTS

The authors thank R. Nicolas and V. Wright for their technical assistance and also Dr. J. Walker and R. Smith for amino acid analyses. The authors also thank the editors and publishers of *Biochimica et Biophysica Acta* and *FEBS Letters* for allowing us to reproduce Figs. 4, 5, and 6. This investigation has been supported by grants to the Chester Beatty Research Institute (Institute of Cancer Research–Royal Cancer Hospital) from the Medical Research Council and the Cancer Research Campaign.

REFERENCES

1. Johns, E. W., and Forrester, S., *Eur. J. Biochem.* **8**, 547 (1969).
2. Goodwin, G. H., Sanders, C., and Johns, E. W., *Eur. J. Biochem.* **38**, 14 (1973).
3. Johns, E. W., *Biochem. J.* **92**, 55 (1964).
4. Johns, E. W., *Eur. J. Biochem.* **4**, 437 (1968).
5. Goodwin, G. H., and Johns, E. W., *Eur. J. Biochem.* **40**, 215 (1973).
6. Sanders, C., and Johns, E. W., *Biochem. Soc. Trans.* **2**, 547 (1974).
7. Goodwin, G. H., Nicolas, R. H., and Johns, E. W., *Biochim. Biophys. Acta* **405**, 280 (1975).
8. Goodwin, G. H., Nicolas, R. H., and Johns, E. W., *FEBS Lett.* **64**, 412 (1976).
9. Sanders, C., Ph.D. Thesis, Univ. of London (1975).
10. Shooter, K. V., Goodwin, G. H., and Johns, E. W., *Eur. J. Biochem.* **47**, 263 (1974).
11. Goodwin, G. H., Shooter, K. V., and Johns, E. W., *Eur. J. Biochem.* **54**, 427 (1975).
12. Kirkegaard, L. H., *Biochemistry* **12**, 3627 (1973).
13. Walker, J. M., Goodwin, G. H., and Johns, E. W., *Eur. J. Biochem.* **62**, 461 (1976).

Chapter 16

Identification of Contractile Proteins in Nuclear Protein Fractions

WALLACE M. LeSTOURGEON

Department of Molecular Biology,
Vanderbilt University,
Nashville, Tennessee

I. Introduction

Before studies on nuclear proteins other than histones became common-place, the contractile proteins actin and myosin were isolated from calf thymus nuclei (*1,2*). Also, cytologists, histologists, and ultrastructuralists had compiled a sizable body of evidence from immunological, enzymic, and ultrastructural studies that contractile proteins exist in nuclei *in vivo* [for an excellent review and theories of function, see Forer (*3*)]. To those beginning the early studies on the nonhistone proteins of chromatin the prevailing bias seemed to be that chromatin proteins regulate chromatin function, and studies were designed primarily to understand their mechanism of action. The previous evidence for nuclear contractile proteins was largely unknown to investigators probing genome function.

More recently Jockusch and colleagues (*4–7*) obtained biochemical and cytological evidence for actin and myosin in nuclei and nucleoli of *Physarum polycephalum*. In a confirming study, LeStourgeon *et al.* (*8*) purified and identified actin and myosin in nonhistone fractions during attempts to purify and characterize the major nonhistones of *Physarum* which underwent dramatic changes during cellular differentiation. In these studies it was demonstrated that actin increased in intranuclear concentration on the establishment of quiescent cell states in *Physarum* as well as in mammalian cells. In an associated study it was also shown that the intranuclear flux of actin during periods of cellular differentiation was due to intranuclear transport of preexisting cytoplasmic actin (9). Based on the apparent increase in actin during periods of heterochromatization it was suggested that contractile protein may function in this chromatin state transition (8, 10). In previously unpublished findings it has also been observed that myosin is specifically concentrated in nuclear membrane fractions. More recently others characterizing the nonhistones of various cell types have identified actin, myosin, and perhaps tropomyosin as major components in nonhistone protein fractions (11–14). Usually these proteins constitute several percent of total nonhistone protein. In addition to contractile proteins evidence has also been obtained that the cytoplasmic proteins α and β tubulin also exist in isolated interphase chromatin from rat liver (12). To date the study of Sanger (11) using fluorescently labeled heavy meromyosin (HMM) to specifically locate actin in unlysed cells is most definitive. Heavy meromyosin is the larger of two tryptic products of myosin and contains the globular head portions which specifically bind actin. In the Sanger study actin was confirmed in interphase nucleoli [originally suggested by Jockusch (5)], and during mitosis in rat kangaroo cells actin was discovered to be specifically concentrated at the kinetochore complex, centriolar regions, and in the mitotic spindle. These findings suggest an actin–myosin interaction in the force-producing mechanism for chromosome movement and may alone explain the presence of actin in nuclei. The interphase localization of actin in nucleoli is more obscure although in some eukaryotes microtubules polymerize from nucleation centers in nucleoli in early prophase (15) and an initial association between microtubules and actin might occur at this time.

In characterizing the protein components of subcellular structures it is not uncommon for investigators to assume that the presence of a protein in an isolated organelle is in itself evidence for an organelle-specific function. This assumption is not naive only if a particular protein can be shown to exist in but one subcellular structure prior to cell fractionation. When a protein can be routinely recovered in significant amounts from more than one subcellular compartment, unequivocal evidence for an integral function in a given organelle can only be the actual elucidation of that function. The recent recognition that the major cytoplasmic proteins actin, myosin, α and β tubulin, and perhaps tropomyosin are also major proteins of isolated interphase nuclei, nucleoli, chromatin, and metaphase chromosomes has served to remind investigators in the field of the above cautions. While it is true that considerable evidence (8, 11) argues against these proteins being cytoplasmic contaminants incurred during nuclear isolation, as yet no single experiment has proved it (excepting perhaps the Sanger study). Even if contamination is eliminated as the source of these components in interphase nuclei one can argue for permissible nuclear pores but no intranuclear interphase function. This argument might be made with respect to many nonhistone proteins in addition to those above. It is neither the purpose nor is it within the scope of this article to discuss all the data and the significance of nuclear contractile proteins. Rather it is the purpose of this article to describe methods for the rapid identification of contractile proteins in nuclear and chromatin protein fractions. The identification of contractile proteins in nonhistone protein fractions has been of value in providing internal markers, in helping to establish the fact that the major nonhistones apparently function in structural capacities, and in helping to focus attention on the more minor components as putative gene regulatory macromolecules.

II. Identification of Actin

Actin is a highly conserved globular protein with a molecular weight of 41,785 as determined through sequence analysis (16). Globular actin monomers (g-actin) self-associate in 0.6 M KCl–5 mM MgCl$_2$ (pH 6.8) to form a duplex helical coil (F-actin) about 55 Å in diameter and up to several microns in length. Actin appears to be ubiquitously distributed among all eukaryotes including plants, unicellular algae, and fungi (3, 8, 17–23). In muscle, a highly specialized tissue, F-actin filaments form the "thin filaments" of the sarcomere and are drawn past the "thick filaments" of associated myosin in the presence of ATP and CA^{2+}. Actin in nonmuscle cells is also involved in force-generating systems through mechanisms apparently similar to sarcomere action. For reviews and relevant papers dealing with contractile protein chemistry and function, see references 24–28.

Recently it has been reported that actin from heart muscle and nonmuscle cells (platelets) may be coded for by separate genes because at least one residue difference has been detected through sequence studies (29). However, all actins so far isolated comigrate in sodium dodecyl sulfate (SDS)–polyacrylamide gels (8), and except for minor sequence differences, actins generally appear identical. Actin can be readily identified due to its many unique properties; e.g., identical mobility in SDS–gel electrophoresis; polymerization in 0.6 M KCl–5 mM Mg^{2+} yielding characteristic filaments visible, after negative staining, in the electron microscope (22,30,31); stoichiometric binding and actomyosin precipitate formation with muscle myosin; characteristic and specific interaction with heavy meromyosin to form the very diagnostic decorated filaments or "arrowhead complexes" (30); highly conserved amino acid composition including the very rare and diagnostic residue N^t-methylhistidine (32–34); and in addition when added to purified myosin actin markedly stimulates the myosin ATPase in low ionic strength in the presence of Mg^{2+} (35–37).

Those suspecting actin in nuclear protein fractions may choose among several rather straightforward procedures for confirmation. The choice of procedure will depend somewhat on individual expertise, available equipment, and collaborative possibilities as outlined below.

A. Confirmation of Molecular Weight and N^t-Methylhistidine Content

Before lengthy projects are designed for actin confirmation the first objective should be to determine if a questionable protein comigrates in gels with an actin standard. This in itself is usually diagnostic, for in the author's experience using high-resolution thin-slab SDS gels,[1] proteins which have been found to migrate with skeletal muscle actin have been confirmed to be actin (chromatins examined include 3T3, CHO, HeLa, rat liver, *Physarum*, and 10T1/2 mouse embryo fibroblasts). Since contractile protein chemists are not rare and since actin can be easily purified in gram quantities, it is usually not difficult to obtain a few micrograms for comparative studies. If this is not possible an actomyosin precipitate suitable for comparative studies can easily be obtained through standard procedures (20,38,39). In SDS-containing high-resolution discontinuous gels[1] (8.75% acrylamide) actin migrates with an apparent molecular weight of 46,000 (8,19) with an R_f usually very near 0.609 (8).

If a particular protein is found to comigrate with an actin standard, the most rapid confirmation is the determination of N^t-methylhistidine since it is particularly rare but universally present in actins. Of course to obtain an amino acid analysis roughly 50 μg of purified protein must be obtained. As described in a second article by the author (see Chapter 26) this quantity of protein can easily be purified through preparative electrophoresis. The often-used procedure of cutting bands and reclaiming protein is acceptable but more time-consuming and considerably less efficient. The unusual residue N^t-methylhistidine can be identified using the automatic amino acid analyzer following classical procedures (40,41). However, the basics column of Beckman PA-35 resin must be roughly twice as long (about 20 cm)

---

[1] For details of high-resolution thin-slab SDS–gel electrophoresis, methods for purifying proteins through preparative gel electrophoresis, and several additional figures showing actin migration in gels see the author's second article in this volume, p. 387.

as routinely required for adequate resolution of lysine, histidine, and arginine (8). For details see Fig. 1.

Emphasis here is placed on the identification of N^L-methylhistidine in a protein that comigrates with an actin standard; however, actins from all species have essentially identical amino acid compositions and total residue composition should be compared for further confirmation.

B. Myosin Precipitation of Endogenous Actin and Actomyosin Formation

A second straightforward biochemical procedure for identifying actin in chromatin takes advantage of the binding specificity between actin and myosin. This procedure has been used for identifying actin in rat liver chromatin (12) as well as in nonmuscle cells (17, 19). Actin can be routinely extracted from chromatin with 1.0 M KCl–20 mM Tris-HCl (pH 7.0) (8, 12). Purified skeletal-muscle myosin can be added in excess and the clarified

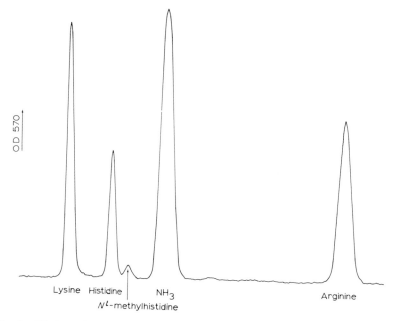

FIG. 1. Elution profile of the basic residues of actin purified from *Physarum* nuclei showing the position of N^t-methylhistidine. For resolution of this residue a 19 × 1.0 cm column of Beckman PA-35 resin was used. The column was eluted with 0.25 M citrate (pH 5.28) with a flow rate of 1.91 ml/minute. The composition and resolution of the basic residues of actin from mammalian cells are essentially identical. N^t-methylhistidine is present at 1 mol per mole of protein.

extract dialyzed to 0.05 M KCl in the presence of 5 mM MgCl$_2$. Actin present in the 1.0 M KCl extract will bind specifically to myosin and precipitate out of solution (actomyosin). In the author's experience substituting 10 mM ethylenediaminetetraàcetic acid (EDTA) for Mg^{2+} in the low-salt dialysis solution generally leads to a much cleaner precipitate of actomyosin since some nonhistones are insoluble in low salt in the presence of divalent cation. The soluble protein after dialysis can be precipitated with trichloroacetic acid (TCA), washed in methanol to remove TCA, solubilized in SDS sample buffer,[1] and analyzed on SDS gels. The disappearance of the putative protein which comigrates with actin from the soluble fraction and the appearance of this protein in the myosin precipitate is diagnostic for actin. If the weight ratio (dye binding in gels) of myosin to actin is roughly 4:1 in the precipitate, further evidence for actin can be obtained if the precipitate visually contracts "superprecipitates" in 5 mM ATP–5 mM Mg^{2+} (pH 6.8). Myosin alone will not superprecipitate under these conditions.

In some nuclei and chromatin (especially lower eukaryotes) enough myosin is present so that a native actomyosin precipitate will form spontaneously simply by dialyzing a clarified 1.0 M KCl extract into low salt [0.05 M KCl–20 mM Tris (pH 7.0)] (8). This might be a suitable first experiment since extraneous purified myosin is not required. The actomyosin precipitate is characteristic in the enrichment of high-molecular-weight myosin (about 200,000) and actin (46,000) (see Fig. 2). These precipitates should contract as above, and the myosin can almost be purified in one step by reextracting the precipitate with 0.6 M KCl. Sometimes it is necessary to add ATP and Mg^{2+} to aid actomyosin dissociation, but in the author's experience myosin can be resolubilized rather selectively from fresh precipitate simply by reextracting with 0.6 or 1.0 M KCl. Usually isolated chromatin from mammalian cells does not contain enough myosin to bring down significant quantities of endogenous actin, and only when sizable quantities of nuclei can be obtained (10–20 ml wet packed volume) is this procedure routinely successful. Actomyosin precipitates do not form from overly dilute solutions, and the 1.0 M KCl extraction buffer should be kept to about 2–3 times the volume of the nuclear pellet. If Triton X-100 or other detergents are used in nuclear isolation they should be rinsed out with water before extraction with 1.0 M KCl.

C.   Ultrastructure of Actin–HMM Complexes

If the electron microscope is routinely used by an investigator a third rapid confirmation of action is the identification of "arrowhead" complexes formed on reaction with heavy meromyosin (HMM) (22,30,42,43). HMM is the larger of two tryptic products of myosin containing the globular head

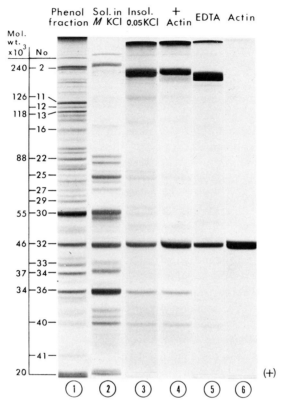

FIG. 2. Electrophoretic profiles showing the position of actin (46,000) and myosin (approx. 250,000) in total residual chromatin proteins and in various actomyosin preparations. Gel 3: phenol-soluble acidic nuclear proteins of *Physarum*; gel 4: a similar protein fraction to which 3 μg of rabbit muscle actin was added; gel 5: the 1.0 *M* KCl soluble proteins which form an insoluble complex in 0.05 *M* KCl–5 m*M* EDTA; and gel 6: purified rabbit muscle actin. The inconsistent mobilities evident in band 2 (myosin) are the result of varying amounts of protein per band. These variations occur most frequently in the very high-molecular-weight regions and are a result of the fact that in the contractile complexes about 4 times more myosin than actin is present.

portions of the myosin molecule which specifically bind actin and possess the ATPase activity of myosin. In this procedure HMM can be prepared following the procedure of Pollard *et al.* (*44*). It is then added to a clarified 1.0 *M* KCl extract of chromatin etc. (chromatin extract should be about 1 mg/ml total protein) and dialyzed to 0.6 *M* KCl–5 m*M* MgCl$_2$–5 m*M* MgCl$_2$–5 m*M* phosphate buffer (pH 6.8), fixed about 2 minutes in 1% aqueous formalin, and then negatively stained with 1% aqueous uranyl acetate (*45*). Other procedures may also be followed (*22,30*).

With the electron microscope negatively stained F-actin appears as long thin filaments about 55–65 Å in diameter which under ideal conditions appear as a duplex beaded coil. Since one molecule of HMM binds one G-actin monomer in the F-actin filament, the filaments of F-actin bind HMM so that the short tail portion of HMM stick away at angles from the long axis of the F-actin filament (about 120–150 Å in diameter). In negatively stained preparations these structures resemble a chain of arrowheads with projections spread about 350 Å apart and all arrowheads pointing in the same direction. Since nothing else but actin reacts with HMM to form these structures they are diagnostic for actin. However, further confirmation can be obtained by demonstrating that 5 mM ATP–5 mM MgCl$_2$ will dissociate the actin–HMM complexes (46).

D.   Enzymology and Other Procedures

The procedures described above for confirmation of actin in nuclear protein fractions are the most direct, unequivocal, and least time-consuming. However, further properties of actin and contractile proteins in general can be examined if desired. Among the more widely tested phenomena is the actin stimulation of the myosin ATPase and the determination of the presence of the regulatory proteins tropomyosin and troponin.

Purified native myosin is a Mg^{2+}-dependent ATPase. Microscale assay procedures have been designed for monitoring this activity (47), and generally the method of Martin and Doty (48) is used for determination of released inorganic phosphate. Adding actin stimulates by several fold (10–20 times) the myosin ATPase (21,35,36), and the phenomena can be used to further confirm actin in nuclear protein fractions. Experimentally, a putative purified actin or various chromatin fractions can be added to a myosin ATPase assay system (47) and the system screened for enhanced release of inorganic phosphate.

As a general method for identifying actin in crude extracts the procedure has several limitations. Artifactual inhibition or stimulation induced by other nuclear proteins cannot readily be ruled out especially since purified tubulin has been shown to stimulate the Mg^{2+}-ATPase of myosin (49). In addition, if the regulatory proteins tropomyosin and troponin are also present in the chromatin extract these proteins may inhibit the ability of actin to stimulate enzyme activity (see reviews 24–27 for following statements). One mole of tropomyosin inhibits the myosin binding sites of seven actin monomers in the F-actin filament. Unless Ca^{2+} and troponin are also present in the proper stoichiometric relationships (actin : tropomyosin : troponin, 7 : 1 : 1) a significant stimulation of ATPase activity will not be observed. Ca^{2+} together with troponin removes the inhibitory effect of tropomyosin by unmasking the binding sites on the actin molecules. The Ca^{2+} activation of the ATPase of crude actomyosin is diagnostic however for the presence of tropomyosin and troponin.

III. Identification of Myosin

Myosin from skeletal muscle is composed of two identical polypeptides with molecular weights very near 200,000 (50–53). These polypeptides interact to form a dimer characterized by two terminal globular head portions and a long rodlike tail structure (1400 Å) formed by the helical coiling of a sizable portion of each chain. The terminal region of each chain folds to form the two globular heads (100 Å diameter) (50). In addition, native myosin (isolated under nondenaturing conditions) also contains in the globular head portion of the molecule four smaller polypeptides (light chains) each of about 20,000 daltons. In muscle the tail portions of the myosin molecule self-associate and form the "thick filament" in such a way that the globular head portions are positioned radially around the surface of the filament. The globular head portions of the molecule contain the ATPase activity of myosin. The remaining tail portions will self-associate to form smooth thick filaments. As pointed out earlier brief trypsin treatment will cleave myosin near the middle portion of the tail yielding HMM and light meromyosin LMM. HMM contains the globular heads and thus the ATPase and actin binding sites.

An interesting aspect of contractile protein biochemistry is the fact that methods for identifying one component are frequently useful in identifying other molecules of the contractile machine. This of course is consistent with our knowledge that all components of contractile systems (myosin, actin, tropomyosin, troponin) in one capacity or another specifically bind or interact with each other. In addition, either directly or indirectly the function of one component is dependent on the function of the other molecules in the contractile complex. As a relevant case in point force generation is dependent on the globular head portions of the self-associated myosin molecules binding G-actin monomers in the F-actin filament. However, this binding cannot take place in the absence of Ca^{2+} due to tropomyosin (bound to seven actin monomers) masking the myosin binding sites on the F-actin filament. Only when troponin is saturated with Ca^{2+} is tropomyosin roled off the myosin binding sites on the thin filament.

With regard to myosin identification the most important aspects of the above scheme are the tail-to-tail associations of myosin, the highly specific interaction of actin with myosin, and the stimulation of ATP hydrolysis which follows this interaction. Just as the many unique properties of actin are useful in actin identification as are the unique properties of myosin useful in myosin identification. However, while actins are highly similar among eukaryotes myosins differ slightly in molecular weight, light chain composition, and ATPase activity. Most of the significant differences occur between mammals and lower eukaryote forms, but in all cases the myosins and actins of higher and lower eukaryotes cross-interact.

Experimentally, the strongest criterion for confirmation of a myosinlike protein is its ability to interact with actin. This interaction should be reflected in physical binding and an actin stimulation of the myosin ATPase. However, as in attempts to confirm actin in chromatin fractions, the first procedure should be to determine whether in SDS–gels a particular protein migrates together with or closely to a myosin standard. Muscle myosin migrates in SDS–polyacrylamide gels with a molecular weight of 200,000, but this precision is primarily due to the fact that myosin is the best molecular-weight marker for this region. If chromatin from mammalian cells is examined, the putative protein should perfectly comigrate with skeletal muscle myosin. Myosins from lower eukaryotes such as *Dictyostelium* and *Physarum* migrate with apparently higher molecular weights (210,000 and 240,000, respectively) (*8,21,54*) than muscle myosin, while myosin from *Acanthamoeba* is lower at 180,0000 (*19*).

If a protein is found to comigrate with a myosin standard, then the most rapid procedure for confirmation is to add purified actin, dialyze to low salt (0.05 M KCl–5 mM MgCl$_2$), and determine if the putative protein is present as the high-molecular-weight component of an actomyosin precipitate (for details consult the procedures described previously for actin confirmation). If sufficient material is present ATP and Mg^{2+} should be added to test for superprecipitation. In nonmuscle cells and lower eukaryotes myosinlike proteins are present at significantly lower levels than actin, and in most cases an initial purification and concentration step is required. This can usually be achieved by taking advantage of the very high molecular weight of myosin. The first protein fractions from a Sephadex G-200 column [eluted with 1.0 M KCl–5 mM ATP–5 mM Mg^{2+}–phosphate buffer (pH 6.8) to aid dissociation of myosin from endogenous actin] should be enriched in myosin. This fraction should be examined on SDS–polyacrylamide gels. If the putative high-molecular-weight component is present the fractions can be concentrated and equilibrated in 0.6 M KCl–5 mM MgCl$_2$ (pH 6.8). Again, purified actin can be added and attempts to form an actomyosin precipitate on dialysis into low salt can be repeated. Other procedures for myosin enrichment can also be followed (*19*).

Small amounts of purified or partially purified myosinlike proteins can be tested for Mg^{2+}-dependent ATPase activity following microscale procedures (*47*). If an ATPase is detected, further confirmation of myosin can be obtained by demonstrating an actin stimulation of this activity. Usually some combination of actin binding or actin-stimulated ATPase is required for myosin confirmation. However, in addition to these procedures ultrastructural studies can provide evidence for myosin. When fractions enriched in myosin are diluted or dialyzed to 0.05 M KCl in the presence of Mg^{2+} or Ca^{2+} the tail portions of the myosin molecules will frequently associate terminally leading to the formation of bipolar dumbell-shaped structures which can be observed in negatively stained preparations. For details and examples of these structures from mammalian and lower eukaryote forms see Lazarides and Lindberg (*54*).

The procedures outlined above for myosin confirmation exist only as an overview of the most rapid and widely used procedures. There are countless methods for myosin purification and identification, and the reader is referred to the various relevant references for details. It is generally true that if actin is present in a particular subcellular compartment myosin will also be present since these molecules must interact in force generation. Other than differences in genome structure, contractile proteins exist as one of the most fundamental differences between eukaryotes and prokaryotes. Early evolutionary roles for actin such as in nuclease inhibition (*55*) or in cross-linking chromatin fibers (*8,10*) cannot as yet be eliminated as retained functions. Clearly these proteins were present before sarcomere evolution.

Emphasis in this chapter has been placed on methods for the rapid identification of actin and myosin in nuclear protein fractions. In addition to these two major components of contractile systems, the regulatory proteins tropomyosin and troponin are likely to be present. Evidence for tropomyosin in rat liver chromatin has been obtained (*12*). However, it is illogical to attempt identification of these proteins before actin confirmation. Methods for purifying these components are somewhat more complex than the methods for actin and myosin, as are the procedures for confirmation and characterization. Those interested in these components may consult the various reviews referred to previously.

REFERENCES

1. Ohnishi, T., Kawamura, H., and Yamamoto, T. *J. Biochem.* (*Tokyo*) **54**, 298 (1963).
2. Ohnishi, U. T., Kawamura, H., and Tanaka, Y. *J. Biochem.* (*Tokyo*) **56**, 6 (1964).
3. Forer, A. *in* "Cell Cycle Controls" (G. M. Padilla, I. L. Cameron, and A. M. Zimmerman, eds.), p. 319. Academic Press, New York, 1974.
4. Jockusch, B. M., Brown, D. F., and Rusch, H. P. *Biochem. Biophys. Res. Commun.* **38**, 279 (1970).
5. Jockusch, B. M., Brown, D. F., and Rusch, H. P. *J. Bacteriol.* **108**, 705 (1971).

6. Jockusch, B. M., Ryser, U., and Behnke, O. *Exp. Cell Res.* **76**, 464 (1973).
7. Jockusch, B. M., Becker, M., Hindennach, I., and Jockusch, H. *Exp. Cell Res.* **89**, 241 (1974).
8. LeStourgeon, W. M., Forer, A., Yang, Yea-Zu, Bertram, J. S., and Rusch, H. P. *Biochim. Biophys. Acta* **379**, 529 (1975).
9. Nations, C., LeStourgeon, W. M., Magun, B. E., and Rusch, H. P. *Exp Cell Res.* **88**, 207 (1974).
10. LeStourgeon, W. M., Totten, R., and Forer, A. *in* "Acidic Proteins of the Nucleus" (I. L. Cameron and J. R. Jeter, eds.), p. 59. Academic Press, New York, 1974.
11. Sanger, J. W. *Proc. Natl. Acad. Sci. U.S.A.* **72**, 2451 (1975).
12. Douvas, A. S., Harrington, C. A., and Bonner, J. *Proc. Natl. Acad. Sci. U.S.A.* **72**, 3902 (1975).
13. Rubin, R. W., Hill, M. C., Hepworth, P., and Boehmer, J. *J. Cell Biol.* **68**, 740 (1976).
14. Peterson, J. L., and McConkey, E. H. *J. Biol. Chem.* **251**, 548 (1976).
15. Blessing, J. *Cytobiologie* **6**, 342 (1972).
16. Elzinga, N., Collins, J. H., Kuehl, W. M., and Adelstein, R. S. *Proc. Natl. Acad. Sci. U.S.A.* **70**, 2687 (1971).
17. Hatano, S., and Oosawa, F. *Biochim. Biophys. Acta* **127**, 488 (1966).
18. Adelman, M. R., and Taylor, E. W. *Biochemistry* **8**, 4964 (1969).
19. Pollard, T. D., and Korn, E. D. *Cold Spring Harbor Symp. Quant. Biol.* **37**, 573 (1973).
20. Yang, Y., and Perdue, J. F. *J. Biol. Chem.* **247**, 4503 (1972).
21. Clarke, M., and Spudich, J. A. *J. Mol. Biol.* **86**, 209 (1974).
22. Nachmias, V. T., Huxley, H. E., and Kessler, D. *J. Mol. Biol.* **50**, 83 (1970).
23. Williamson, R. E. *Nature (London)* **248**, 801 (1974).
24. Ebashi, S., Endo, M., and Ohtsuki, I. *Quart. Rev. Biophys.* **2**, 351 (1969).
25. Weber, A., and Murray, J. M. *Physiol. Rev.* **53**, 612 (1973).
26. Cohen, C. *Sci. Am.* **233**, 36 (1975).
27. Huxley, H. E. *Cold Spring Harbor Symp. Quant. Biol.* **37**, 361 (1972).
28. Murray, J. M., and Weber, A. *Sci. Am.* **230**, 58 (1974).
29. Elzinga, M., Maron, B. J., and Adelstein, R. S. *Science* **191**, 94 (1976).
30. Huxley, H. E. *J. Mol. Biol.* **7**, 281 (1963).
31. Ikemoto, N., Kitagawa, S., Nakamura, A., and Gergely, J. *J. Cell Biol.* **39**, 620 (1968).
32. Asatoor, A. M., and Armstrong, M. D. *Biochem. Biophys. Res. Commun.* **26**, 168 (1967).
33. Johnson, P., Harris, C. I., and Perry, S. U. *Biochem. J.* **105**, 361 (1967).
34. Elzinga, M. *Biochemistry* **10**, 224 (1971).
35. Maruyama, K., and Gergely, J. *J. Biol. Chem.* **237**, 1095 (1962).
36. Adelstein, R. S., Conti, M. A., Johnson, G. S., Pastan, I., and Pollard, T. D. *Proc. Natl. Acad. Sci. U.S.A.* **69**, 3693 (1972).
37. Moos, C. *Cold Spring Harbor Symp. Quant. Biol.* **37**, 137 (1972).
38. Mommaerts, W. H. M. *Methods Med. Res.* **7**, 1 (1958).
39. Ebashi, S., and Ebashi, F. *J. Biol. Chem.* **55**, 604 (1964).
40. Spackman, D. H., Stein, W. H., and Moore, S. *Anal. Chem.* **30**, 1190 (1958).
41. Moore, S., and Stein, W. H., *in* "Methods in Enzymology" (S. P. Colowick and N. O. Kaplan, eds.), Vol. 6, p. 819. Academic Press, New York, 1963.
42. Moore, P. B., Huxley, H. E., and De Rosier, D. J. *J. Mol. Biol.* **50**, 279 (1970).
43. Huxley, H. E., *Nature (London), New Biol.* **243**, 445 (1973).
44. Pollard, T. D., Shelton, E., Wiehing, R. R., and Korn, E. D. *J. Mol. Biol.* **50**, 91 (1970).
45. Forer, A., Emmersen, J., and Benke, O. *Science* **175**, 774 (1972).
46. Finlayson, B., Lyman, R. W., and Taylor, E. W. *Biochemistry* **8**, 811 (1969).
47. Yang, Y., and Perdue, J. F. *J. Biol. Chem.* **247**, 4503 (1972).

48. Martin, J. B., and Doty, D. M. *Anal. Chem.* **21**, 965 (1949).
49. Puzkin, S., and Berl, S. *Nature (London)* **225**, 558 (1970).
50. Lowey, S., Slayter, H. S., Weeds, A. G., and Baker, H. *J. Mol. Biol.* **42**, 1 (1969).
51. Gershman, L. C., Stracher, A., and Dreizen, P. *J. Biol. Chem.* **244**, 2726 (1969).
52. Gazith, J., Himmelfarb, S., and Harrington, W. F. *J. Biol. Chem.* **245**, 15 (1970).
53. Lowey, S., and Holt, J. C. *Cold Spring Harbor Symp. Quant. Biol.* **37**, 19 (1972).
54. Nachmias, V. T. *Cold Spring Harbor Symp. Quant. Biol.* **37**, 607 (1972).
55. Lazarides, E., and Lindberg, V. *Proc. Natl. Acad. Sci. U.S.A.* **71**, 4742 (1974).

Chapter 17

Methods for Selective Extraction of Chromosomal Nonhistone Proteins

JEN-FU CHIU, HIDEO FUJITANI, AND
LUBOMIR S. HNILICA

Department of Biochemistry,
Vanderbilt University School of Medicine,
Nashville, Tennessee

I. Introduction

During rapid cell proliferation, chromosome puffing, and chromosomal stimulation, there is an increase in the content, synthesis, and metabolic activity of chromosomal nonhistone proteins (1–5). Several studies have implicated chromosomal nonhistone proteins as likely candidates for gene regulatory functions. For example, phosphoproteins, a significant component of the chromosomal nonhistone proteins, are heterogeneous and tissue-specific (6–9), can alter the rate of RNA synthesis *in vitro* (10–14), and bind specifically to homologous DNA (10,15,16). Changes in the phosphorylation of chromosomal nonhistone proteins were correlated with activation of gene activity in a variety of systems (17–20).

Since chromatin nonhistone proteins are very heterogeneous (perhaps over hundreds of individual species), they may have many biological functions. Presently, there are no methods available for a complete separation of all biologically active proteins by a single standard procedure. Techniques taking advantage of particular biological properties of chromosomal nonhistone proteins, such as enzymic activity, transcriptional regulations, immunological specificity, etc., are most useful and probably indicate the future direction of this rapidly moving field.

The extent of the tissue specificity of chromosomal nonhistone proteins

283

was shown immunologically by Chytil and Spelsberg (*21*). These authors used dehistonized chromatin to elicit tissue-specific antibodies in rabbits. Wakabayashi and Hnilica (*22*) confirmed these observations and reported that antibodies can be obtained which are specific for various tissues as well as for normal or transformed cells. They also found that the immunospecificity of dehistonized chromatin depended on the presence of tissue-specific complexes between DNA and some nonhistone proteins in chromatin. Only intact chromatin or samples obtained by the reconstitution of nonhistone proteins with the DNA isolated from the same species (homologous) were immunoreactive. Free rat DNA or nonhistone proteins were inactive. Conversely, Wakabayashi *et al.* (*23*) digested dehistonized chromatin with DNase I. The immunospecificity of dehistonized chromatin was lost after this treatment.

We have recently reported (*24*) that neoplastic growth changes the immunological specificity of chromatin nonhistone protein complexes with DNA to a new type which is characteristic for the malignant tumor. Complement fixation has shown that the tissue specificity of a fraction of chromatin nonhistone proteins changes gradually with the development of hepatomas in rats fed a carcinogenic diet (*24*). Nonhistone protein–DNA complexes from fast-growing Morris hepatomas were more immunoreactive than similar proteins from better differentiated, slow-growing tumors (*25*). We describe here a simple technique for separation of chromosomal proteins into three principal fractions: the bulk of chromosomal nonhistone proteins (UP); histones (HP); and a small, biologically active fraction of nonhistone proteins (NP) with affinity for DNA. The immunological tissue specificity of this latter fraction was used as a marker in developing the described fractionation scheme. In addition to its immunological specificity, the DNA-binding fraction NP was found to contain components active in the *in vitro* transcriptional regulation of specific genes and in mediating the binding of androgens by target tissue chromatin.

II.   Methods

A.   Preparation of Nuclei

Unless specified otherwise, all work is performed at $2°$–$4°C$. Novikoff ascites tumor cells are routinely carried in male Sprague–Dawley rats. After collection of the ascites fluid, the cells are diluted with 2.5 volumes of $0.25\ M$ sucrose–$5\ mM$ $MgCl_2$–$10\ mM$ Tris-acetate (pH 7.2) and washed with this buffer several times. Each wash is followed by low-speed centrifugation (100 g, 15 minutes) to separate the cells. About 20–30 ml of the washed, packed cells are homogenized vigorously in 80 ml of 10 mM Tris-acetate (pH 7.4) in a tight-fitting (Potter–Elvehjem type, 0.005 in. clearance) Teflon–glass homogenizer. The cells are pelleted (700 g, 5 minutes), resuspended by vigorous homogenization in the same buffer, and passed through a Chaikoff press using a 27-μm pestle. The crude nuclei are pelleted (700 g, 5 minutes) and washed twice by homogenization in 0.25 M sucrose–5 mM MgCl$_2$–10 mM Tris-HCl (pH 7.4). The nuclear pellet is resuspended in 60 ml of 2.2 M sucrose–5 mM MgCl$_2$–10 mM Tris-HCl (pH 7.4) and centrifuged at 50,000 g for 50 minutes. The pellet contains nuclei free of cytoplasmic tags, while whole or partially broken cells float. The purified nuclei are gently washed once in 40 ml of 0.25 M sucrose–1.5 mM MgCl$_2$–10 mM Tris-HCl (ph 7.4) and finally used for chromatin preparation.

Rat livers (15 gm) are homogenized in 150 ml of 0.32 M sucrose containing 5 mM MgCl$_2$ in a Teflon–glass (Potter–Elvehjem type) homogenizer using eight up and down strokes at 2000 rpm. After filtration through four layers of gauze, the homogenate is centrifuged at 1000 g for 10 minutes. The crude nuclear pellet is then suspended in 150 ml of 2.2 M sucrose containing 5 mM MgCl$_2$ and centrifuged at 75,000 g for 1 hour. The purified pelleted nuclei are suspended in 0.32 M sucrose–5 mM MgCl$_2$ and collected by centrifugation at 1000 g for 10 minutes.

B. Preparation of Chromatin

Several procedures for the isolation and purification of interphase chromatin have been tried in our laboratory. Best results were achieved by the procedure based on previous studies of several investigators (26–28) with some modifications (29). Nuclei are gently homogenized in 50 volumes of 0.08 M NaCl–0.02 M ethylenediaminetetracetate (EDTA) (pH 6.3) using a Potter–Elvehjem type Teflon–glass homogenizer driven by hand. The suspension is centrifuged at 5000 g for 10 minutes, and the pelleted chromatin is resuspended in the same buffer. The homogenization and centrifugation are repeated twice. The chromatin pellet is then extracted once with 5 volumes of 0.35 M NaCl. Finally, the cromatin is rehydrated by homogenization in 5 volumes of 1.5 mM NaCl–0.15 mM sodium citrate (pH 7.0) and centrifugation at 20,000 g for 10 minutes. This step is repeated once more. The purified chromatin can be stored frozen in 1.5 mM NaCl–0.15 mM sodium citrate (pH 7.0) at −20°C. When needed, the chromatin solution is thawed and rehomogenized in a hand-operated Teflon–glass homogenizer to disaggregate the chromatin.

C. Fractionation Procedure

The fractionation scheme developed in our laboratory (*24*) is based on the solubility properties of chromosomal proteins in 5 *M* urea at three different pH values and salt concentrations. The first extraction is performed at pH 7.6 and a relatively low ionic strength. It removes the bulk of proteins that are not firmly associated with DNA in chromatin. The second extraction at high ionic strength and relatively low pH (5.0) takes advantage of the solubility of histones under these conditions. The DNA-binding non-histone proteins are poorly soluble at this pH. They can be solubilized by increasing the pH to 8.0 in the final extraction step.

Isolated rat liver or Novikoff hepatoma chromatin is gently homogenized in 5.0 *M* urea containing 50 m*M* sodium phosphate buffer (pH 7.6). After adjusting the DNA concentration to approximately 6 OD units/ml at 260 nm, and stirring for 2 to 3 hours, the mixture is centrifuged at 20,000 *g* for 30 minutes. The combined supernatants contain 90–95% of chromosomal non-histone proteins (UP fraction). Histones (HP fraction) are removed from the remaining chromatin pellets (UC) by resuspending them gently in 5.0 *M* urea –2.5 *M* NaCl–50 m*M* sodium succinate (pH 5.0) (final concentration of chromatin is 6 OD/ml at 260 nm) and centrifuging the viscous solution at 110,000 *g* for 36 hours. DNA and the associated nonhistone proteins form a pellet (HC), while histones with small amounts of other proteins remain in the supernatant. Finally, the DNA-binding protein fraction (designated as NP fraction) is recovered by dissociation in 5.0 *M* urea–2.5 *M* NaCl–50 m*M* Tris-HCl (pH 8.0) and centrifugation at 110,000 *g* for 48 hours. The NP fraction in the supernatant represents about 3–5% of the total chromatin protein content and the pellet (NC) is DNA with a small amount of associated protein. This procedure is summarized in Fig. 1.

Because of the time-consuming centrifugations and relatively large volumes, the above fractionation procedure was modified by using hydroxylapatite column chromatography instead of centrifugation to separate histones (HP) and nonhistone proteins (NP) from DNA (*30*). The isolation of UP proteins follows the same procedure as described above. The residual chromatin pellets are rehomogenized in 1.5 m*M* NaCl–0.15 m*M* sodium citrate. After adjusting the DNA concentration of chromatin to approximately 1 mg/ml, the UC suspension is dialyzed against the same buffer overnight with three changes. The dialyzed suspension is then sonicated with a Branson sonifier cell disruptor in 15-second intervals for a total of 2 minutes and adjusted to 2 *M* NaCl–5 *M* urea–1 m*M* sodium phosphate (pH 6.8). It is important to remove urea from the chromatin suspension before its exposure to ultrasound if immunologically active NP proteins are desired. If the sonication is performed in the presence of urea, the immunological activity

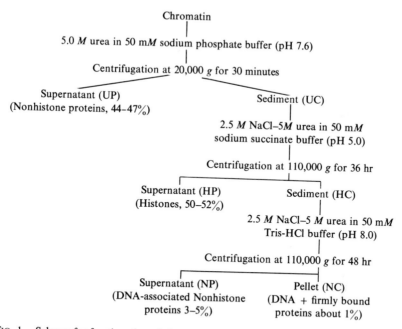

FIG. 1.   Scheme for fractionation of chromatin proteins by sequential extraction method.

is destroyed. After centrifugation at 10,000 g for 10 minutes the supernatant is applied to a hydroxylapatite column (8, 31). After elution of the unretained histone fraction (HP) with 2 M NaCl–5 M urea–1 mM sodium phosphate (pH 6.8), the nonhistone protein fraction NP is partially fractionated by stepwise elution using 50 mM and 200 mM sodium phosphate (pH 6.8) containing 2 M NaCl and 5 M urea. Finally, DNA can be eluted with 0.5 M sodium phosphate (pH 6.8) containing 2 M NaCl and 5 M urea. While the initial steps are performed at 4 °C, the temperature is increased to 25 °C when the phosphate concentration is raised to 200 mM and higher. The elution profile of hydroxylapatite column chromatography is shown in Fig. 2.

III.   Properties of the Chromatin Fractions

Selectivity of the described chromatin fractionation procedure is shown in Fig. 3. The polyacrylamide gel electrophoresis performed in the presence of sodium dodecylsulfate (SDS) shows that the first step of this fractionation schedule removes most of the nonhistone proteins together with small

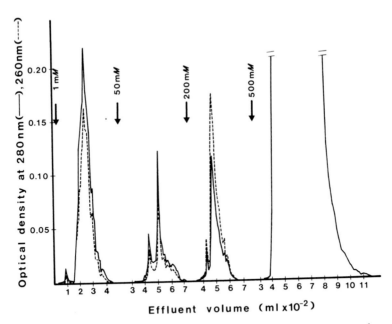

FIG. 2. Hydroxylapatite column chromotography of Novikoff ascites hepatoma chromatin (UC fraction). Isolated chromatin (200 mg DNA) was first extracted with 5 *M* urea–50 m*M* sodium phosphate–0.1 m*M* phenylmethylsulfonyl fluoride (pH 7.6) to remove the majority of loosely bound nonhistone proteins (UP). The resulting pellet (UC) was hydrated by dialysis against 1.5 m*M* NaCl–0.15 m*M* sodium citrate (pH 7.0), sonified, and the solution was brought to 5 *M* urea–2 *M* NaCl–1 m*M* sodium phosphate–0.1 m*M* phenylmethyl sulfonyl fluoride. After removal of insoluble debris by centrifugation (10,000 *g*, 10 minutes) the solution was applied to a 5.1 cm × 17 cm (bed volume = 350 ml) hydroxylapatite column and eluted with the same buffer at a flow rate of 1.5 ml/minute. The phosphate concentration was then raised stepwise to 50 m*M*, 200 m*M*, and finally to 500 m*M*.

amounts of histones. The second fractionation step removes essentially all the histones. The NP protein fraction consists of two to three low-molecular-weight and several high-molecular-weight components.

The efficiencies of the residual chromatin pellets from the individual fractionation steps in serving as a template for the *in vitro* RNA synthesis are shown in Table I. The initial extraction of chromatin with buffered 5.0 *M* urea which removed most of the nonhistone proteins approximately doubled the templating efficiency of the residual chromatin. As can be expected, the removal of histones during the second extraction step derepressed most of the DNA. Finally, the templating efficiency of the NC pellets was only slightly lower than that of the control DNA.

The change in templating efficiency of chromatin fractions can result from changes in the total number of available initiation sites or in the rate

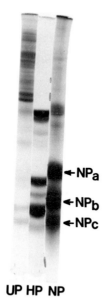

UP HP NP

FIG. 3.   Polyacrylamide gel electrophoresis of the UP, HP, and NP protein fractions from rat liver chromatin. The electrophoresis was performed in the presence of sodium dodecyl sulfate. The origin of migration is at the top of the gels. From Chiu *et al.* *(33)*.

of chain elongation. The number of initiation sites available for RNA polymerase on DNA and chromatin can be assayed under conditions that allow only one RNA molecule to be made at each available initiation site *(32,33)*. The experiment is performed by allowing a large excess of enzymes to initiate transcriptions in low salt with only three kinds of nucleotides present. The absence of the fourth nucleotide prevents extensive chain elongation and thus inhibits the formation of multiple initiations at the same site. After 15 minutes, the mixture is brought to $0.4 M$ $(NH_4)_2SO_4$. This prevents further initiations. The fourth nucleotide is then added to permit the elongation of the already initiated nucleotide chains. The number of growing chains is then calculated from sucrose gradient centrifugation of the transcripts. The number (average size) for the RNAs transcribed from pure DNA, NC, HC, and UC pellets, and intact chromatin is between 400 and 600 nucleotides, and the number of RNA molecules is equal to the number of initiation sites *(33)*. From the amount of nucleotides incorporated and the average chain length, the number of initiation sites can be calculated (Table II). Using 1.5 μg of template DNA (2.2 nmol of base pairs) we observed 1.72 pmol of initiation sites on rat DNA and 0.31, 0.57, 1.51, and 1.61 pmol on rat liver

TABLE I

TEMPLATING ACTIVITY OF RESIDUAL PELLETS
RESULTING FROM THE SCHEME IN FIG. 1[a,b]

Fraction	[³H]UTP (pmol/μg of DNA)	Percent free DNA activity
Chromatin	22.1	11.3
UC	50.0	25.6
HC	167.5	85.7
NC	176.8	90.5
DNA	195.4	100.0

[a] From Chiu *et al.* (*33*).

[b] The results are averages of several preparations of rat liver chromatin. Free DNA was isolated from rat spleen. The reaction mixture (0.25 ml final volume) consisted of 40 mM Tris-HCl buffer (pH 8.0)–120 mM KCl–0.1 mM EDTA–2.5 mM MnCl$_2$–1.0 mM dithiothreitol–0.08 mM each ATP, GTP, CTP–0.02 mM ³H (sp act 1 Ci/mol). The concentration of chromatin in each assay was 10–15 μg in respect to DNA, together with 20 units of *E. coli* RNA polymerase (sp act 600 units/mg of protein). The assay mixtures were incubated at 37°C for 15 minutes, and the reaction was terminated by adding 2 ml of 10% trichloroacetic acid–1% sodium pyrophosphate solution.

TABLE II

DETERMINATION OF THE NUMBER OF GROWING CHAINS BY
SUCROSE GRADIENT ANALYSIS[a,b]

Template	Nucleotides incorporated (pmol)	Chain length (nucleotides)	Initiations (pmol)
DNA	1015	590	1.72
NC	918	570	1.61
HC	906	600	1.51
UC	308	540	0.57
Chromatin	155	500	0.31

[a] From Chiu *et al.* (*33*).

[b] Assay conditions were as described in Table I. The average chain length was determined from 0.2 ml of each assay tube by sucrose gradient analysis. Each tube contained 1.5 μg of chromatin as indicated.

chromatin, UC, HC, and NC fractions, respectively. The removal of the nonhistone proteins UP only slightly increased the number of initiation sites, while the subsequent removal of histones HP increased the initiation considerably. Final removal of the NP proteins did not further increase the initiation. This indicates that histones may function as general repressors and the nonhistone proteins NP may serve as specific regulators of either negative or positive transcriptional control.

It was shown by Wakabayashi *et al.* (*16*) that the nonhistone protein fraction of chromatin which could be identified by its immunological tissue specificity also contained proteins with affinity for homologous native DNA. To show that the immunospecificity of chromatin nonhistone proteins was not lost during the fractionation schedule, the immunoreactivity of the intact chromatins and of the UC, HC, and NC residual chromatin pellets was assayed. The dehistonized Novikoff hepatoma chromatin was used as antigen for the immunization of rabbits. The removal of urea-soluble proteins UP and histones HP did not change much the extent of complement fixation (Fig. 4). However, the complement-fixing ability of the complex decreased considerably when the DNA-binding proteins NP were removed during the last fractionation step (NC pellet). The residual immunoactivity of the NC pellet was probably caused by traces of NP proteins still present in the NC pellet.

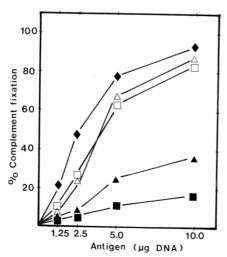

FIG. 4. Complement fixation of chromatin preparations performed in the presence of antiserum against dehistonized Novikoff hepatoma chromatin. All data were corrected for anticomplementarity. (◆) Novikoff hepatoma native chromatin; (△) Novikoff hepatoma UC pellet, i.e., chromatin devoid of UP proteins; (□) Novikoff hepatoma HC pellet, i.e., chromatin devoid of UP and HP proteins; (▲) Novikoff hepatoma NC pellet, i.e., chromatin devoid of UP, HP, and NP proteins; (■) normal rat liver chromatin. From Wang *et al.* (*34*).

The identity of immunologically tissue-specific proteins in the NP fraction can be further ascertained by reconstitution experiments. The nonhistone protein fraction NP can be isolated from one tissue type and reconstituted with the DNA isolated from another tissue, e.g., Novikoff hepatoma NP and normal rat liver DNA. As can be seen in Fig. 5, the immunospecificity of the resulting complexes is determined by the tissue donating the non-histone proteins NP.

To learn if the specific antibodies are directed against the nuclear material and not against some cytoplasmic components that could have associated with chromatin during its isolation, intracellular localization of these antigens can be performed. We use the horseradish peroxidase method to localize antigens of normal rat liver cells. As shown in Fig. 6, the antibodies localize in the nuclei and not in the cytoplasm.

Proteins comprising the immunologically active chromosomal nonhistone fraction NP also bind to the DNA. In 10 mM NaCl–10 mM Tris-HCl (pH 8.0) the binding sites available on homologous DNA (rat spleen) are saturated at the NP protein:DNA ratio of approximately 1.5:100 (w/w). The reciprocal plot of the saturation curve forms a straight line intercepting the abscissa at the K_m value of about 6.7×10^{-9} (34). The binding of NP proteins is species specific. As can be seen in Table III, both rat spleen and liver DNA bind the liver NP proteins equally well. Calf thymus DNA exhibits a small

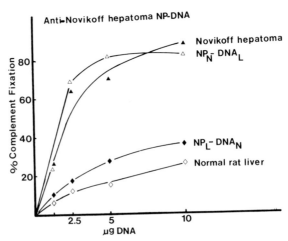

FIG. 5. Complement fixation of native and reconstituted NP–DNA complexes from rat liver and Novikoff hepatoma in the presence of antiserum against Novikoff hepatoma NP–DNA. All experimental points were corrected for anticomplementarity. (▲) Novikoff hepatoma chromatin (native). (△) reconstituted complex of Novikoff hepatoma NP and normal rat liver DNA (NP$_N$–DNA$_L$); (◇) normal rat liver chromatin (native); (◆) reconstituted complex of rat liver NP and Novikoff hepatoma DNA (NP$_L$–DNA$_N$). From Chiu et al. (25).

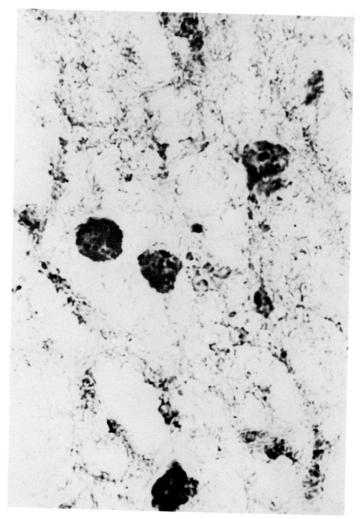

FIG. 6.   Localization of antigens in rat liver by the horseradish peroxidase bridge technique in the presence of rabbit serum against dehistonized rat liver chromatin.

but significant binding while the affinity of rat liver NP proteins to chicken erythrocyte or *Escherichia coli* DNA is negligible.

To study the association of NP proteins with DNA fractionated according to its renaturation kinetics, sheared rat spleen DNA can be separated by hydroxylapatite chromatography into repetitive, middle repetitive, and slowly renaturing fractions. The $\frac{1}{2}$ *Cot* values of these fractions in our experi-

TABLE III

INTERACTIONS OF RAT LIVER NP FRACTION WITH HOMOLOGOUS
AND HETEROLOGOUS DNA[a,b]

Source of DNA	DNA (μg)	Protein applied (μg)	Protein bound (μg)	Protein/DNA binding ratio
Rat spleen	400	40	5.8	0.0145
Rat liver	400	40	5.6	0.0141
Calf thymus	400	40	1.0	0.0025
Chicken erythrocyte	400	40	0.3	0.0008
E. coli	400	40	0.1	0.0003

[a] From Wang *et al.* (*34*).

[b] The formation of DNA–protein complexes was assayed by sucrose density gradient centrifugation using ^{125}I-labeled NP protein. The binding ratios represent weight percentages of protein retained by the DNA.

ment were 7.1, 79.4, and about 1000, respectively. The results in Table IV show about 2-fold preference of the NP proteins for the unique sequence DNA. The binding of NP proteins to single- and double-stranded unique sequence DNA is also shown in Table IV. The rat liver NP proteins exhibit a significant preference for native DNA.

IV. Conclusions

The fractionation scheme described here is based principally on solubility of the three protein groups in salt and urea at various pH values. Although fraction NP binds selectively to homologous DNA, its separation from histones in sodium succinate buffer (pH 5.0) is principally facilitated by its insolubility at relatively low pH. The NP fraction is heterogeneous. Recent attempts in our laboratory to fractionate the NP proteins resulted in separation of the three principal low-molecular-weight bands from the remaining proteins of higher molecular weight (over 30,000). As determined by complement fixation, the immunologically tissue-specific proteins are in the high-molecular-weight fraction of the NP proteins.

Isolated chromatin can be fractionated by a variety of techniques into transcriptionally active and inactive fractions. If the fractionation of chromatin is accomplished by shearing and subsequent sucrose density gradient centrifugation or divalent cation precipitation, the immunologically tissue-specific proteins are selectively accumulated in the extended, transcriptionally active chromatin (*34,35*). As was mentioned in the introduction,

TABLE IV

RAT LIVER NP PROTEIN BINDING TO FRACTIONATED DNA[a]

DNA fraction	DNA (μg)	Protein applied (μg)	Protein bound (μg)	Protein/DNA binding ratio
Double-stranded repetitive	200	20	2.2	0.0112
Double-stranded middle repetitive	200	20	2.9	0.0147
Double-stranded unique sequence	200	20	3.7	0.0188
Single-stranded unique sequence	200	20	1.8	0.0089

[a] Unless single-stranded DNA is used, fractionated DNAs are renatured to the same percentage of renaturation (80% reassociated).

the DNA-binding fraction NP contains macromolecules which can influence the *in vitro* transcription of chromatin. Using the chicken reticulocyte chromatin system, cDNA probes were prepared by reverse transcription of purified globin mRNA. It was found that the *in vitro* transcription of globin genes by reticulocyte chromatin depends on the presence of proteins contained in the reticulocyte chromatin fraction NP. Under normal circumstances, chicken liver or brain chromatins do not transcribe *in vivo* or *in vitro* RNA sequences complementary to chicken globin cDNA probes. However, when the reticulocyte fraction NP was reconstituted to liver or brain chromatin devoid of its own NP proteins, the final product transcribed the globin genes at a frequency comparable to that of isolated native reticulocyte chromatin (*30*).

Although the detailed mechanism of steroid hormone action is not known, it is anticipated that after its initial association with cytoplasmic receptor, the steroid hormone is transferred into the nucleus where it associates with acceptor sites specific for the target chromatin. It has been shown in several laboratories that chromosomal nonhistone proteins of the target tissue play an important role in the final and selective binding of the steroid–receptor complex. We have found (*36*) that the NP fraction proteins are principally responsible for the target tissue specific binding of steroid hormone complexes with cytoplasmic receptors.

It can be concluded from this brief discussion of the biological properties of proteins comprising the NP fraction that, in addition to its immunological tissue specificity, the NP fraction may be an important source and starting material for the purification and characterization of macromolecules active in the transcriptional regulation of specific genes.

ACKNOWLEDGMENT

This work was supported by National Cancer Institute Contract N01-CP-65730 and USPHS Grant CA-18389.

REFERENCES

1. Spelsberg, T. C., Wilhelm, J. A., and Hnilica, L. S., *Sub-Cell. Biochem.* **1**, 107 (1972).
2. Baserga, R., and Stein, G. S., *Fed. Proc. Fed. Am. Soc. Exp. Biol.* **30**, 1752 (1971).
3. Hnilica, L. S., "The Structure and Biological Functions of Histones." Chemical Rubber Publ. Co., Cleveland, Ohio, 1972.
4. Stein, G. S., Spelsberg, T. C., and Kleinsmith, L. J., *Science* **183**, 817 (1974).
5. Cameron, I. L., and Jeter, J. R., Jr., "Acidic Proteins of the Nucleus." Academic Press, New York, 1974.
6. Elgin, S. C. R., and Bonner, J., *Biochemistry* **9**, 4440 (1970).
7. Shaw, L. M. J., and Huang, R. C. C., *Biochemistry* **9**, 4530 (1970).
8. MacGillivray, A. J., Carroll, D., and Paul, J. *FEBS Lett.* **13**, 204 (1971).
9. Wang, T. Y., *Exp. Cell Res.* **69**, 217 (1971).
10. Teng, C. S., Teng, C. T., and Allfrey, V. G., *J. Biol. Chem.* **246**, 3597 (1971).
11. Kostraba, N. C., and Wang, T. Y., *Biochim. Biophys. Acta* **262**, 169 (1972).
12. Spelsberg, T. C., and Hnilica, L. S., *Biochim. Biophys. Acta* **195**, 63 (1969).
13. Kamiyama, M., Dastugue, B., and Kruh, J., *Biochem. Biophys. Res. Coomun.* **44**, 1345 (1971).
14. Shea, M., and Kleinsmith, L. J., *Biochem. Biophys. Res. Commun.* **50**, 473 (1973).
15. Kleinsmith, L. J., Heidema, J., and Carroll, A., *Nature (London)* **226**, 1025 (1970).
16. Wakabayashi, K., Wang, S., Hord, G., and Hnilica, L. S., *FEBS Lett.* **32**, 46 (1973).
17. Kleinsmith, L. J., *J. Cell. Physiol.* **85**, 459 (1975).
18. Kleinsmith, L. J., *in* "Acidic Proteins of the Nucleus" (I. L. Cameron and J. R. Jeter, Jr., eds.), p. 103. Academic Press, New York, 1974.
19. Allfrey, V. G., Johnson, E. M., Karn, J., and Vidali, G., *in* "Protein Phosphorylation in Control Mechanisms" (F. Huiging and E. Y. C. Lee, eds.), p. 217. Academic Press, New York, 1973.
20. Langan, T. A., *in* "Regulation of Nucleic Acid and Protein Biosynthesis" (V. V. Koningsberger and L. Bosch, eds.), p. 233. Elsevier, Amsterdam, 1967.
21. Chytil, F., and Spelsberg, T. C., *Nature (London), New Biol.* **233**, 215 (1971).
22. Wakabayashi, K., and Hnilica, L. S., *Nature (London), New Biol.* **242**, 153 (1973).
23. Wakabayashi, K., Wang, S., and Hnilica, L. S., *Biochemistry* **13**, 1027 (1974).
24. Chiu, J.-F., Hunt, M., and Hnilica, L. S., *Cancer Res.* **35**, 913 (1975).
25. Chiu, J.-F., Craddock, C., Morris, H. P., and Hnilica, L. S., *FEBS Lett.* **42**, 94 (1974).
26. Paul, J., and Gilmour, R. S. *J. Mol. Biol.* **34**, 305 (1968).
27. Dingman, W., and Sporn, M. B., *J. Biol. Chem.* **239**, 3483 (1964).
28. Commerford, S. L., Hunter, M. J., Oncley, J. L., *J. Biol. Chem.* **238**, 2123 (1963).
29. Spelsberg, T. C., and Hnilica, L. S., *Biochim. Biophys. Acta* **228**, 202 (1971).
30. Chiu, J.-F., Tsai, Y. H., Sakuma, K., and Hnilica, L. S., *J. Biol. Chem.* **250**, 9431 (1975).
31. MacGillivray, A. J., Cameron, A., Krauze, R. J., Rickwood, D., and Paul, J., *Biochim. Biophys. Acta* **277**, 384 (1972).
32. Cedar, H., and Felsenfeld, G., *J. Mol. Biol.* **77**, 237 (1973).
33. Chiu, J.-F., Wang, S., Fujitani, H., and Hnilica, L. S., *Biochemistry* **14**, 4552 (1975).
34. Wang, S., Chiu, J.-F., Klyzsejko-Stefanowicz, L., Fujitani, H., and Hnilica, L. S., *J. Biol. Chem.* **251**, 1471 (1976).
35. Hardy, K., Chiu, J.-F., Beyer, A., and Hnilica, L. S., in preparation.
36. Klyzsejko-Stefanowicz, L., Chiu, J.-F., Tsai, Y. H., and Hnilica, L. S. *Proc. Natl. Acad. Sci. U.S.A.* **73**, 1954 (1976).

Chapter 18

Low-Molecular-Weight Basic Proteins in Spermatids

ROBERT D. PLATZ[1] AND MARVIN L. MEISTRICH

Section of Experimental Radiotherapy,
The University of Texas System Cancer Center,
M. D. Anderson Hospital and Tumor Institute, Houston, Texas

SIDNEY R. GRIMES, JR.[2]

Department of Biochemistry,
Vanderbilt University School of Medicine
Nashville, Tennessee

I. Introduction

Low-molecular-weight basic proteins replace the histones and many of the nonhistone proteins during maturation of spermatids in the rat testis (*1–3*). This maturation process serves as a model of cellular differentiation in which the major morphological event is the condensation of chromatin and the major functional event is the total cessation of transcription. Both events occur over a short period of time and correlate with the replacement of histone and nonhistone proteins by a limited number of low-molecular-weight proteins. Several of these proteins have been identified in rat testis spermatids and include the testis proteins TP (*1,4*), TP2 (*5,6*), TP3 (*2,6*), and TP4 (*2*) and the sperm protein S1 (*4*). Besides these proteins which are predominant in late spermatids, there are at least three additional histonelike proteins found in early spermatids and spermatocytes (*7–9*), and it seems likely that additional spermatidal proteins will be identified in the near future.

This chapter focuses on the low-molecular-weight basic proteins in late

[1]*Present address*: Frederick Cancer Research Center, Biological Markers Laboratory, Frederick, Maryland.

[2]*Present address*: Veterans Administration Hospital, Shreveport, Louisiana.

spermatids. However, the techniques described here for the localization and identification of proteins in spermatid nuclei are applicable to other nuclear proteins and to a variety of cell types within the testis (*1, 2, 6*). A more complete description and evaluation of the techniques for separating spermatogenic cells and nuclei has been published elsewhere (*10*) and will be helpful in applying these procedures to other systems. The protocols presented here are those most commonly used in our laboratory for preparing and fractionating spermatids from rat testis. Methods are described for analyzing the proteins from spermatid nuclei, and procedures are given for purifying three of the low-molecular-weight basic proteins which have been isolated from rat testis spermatids.

II. Localization of Nuclear Proteins in Rat Testis Spermatids

A. Preparation and Separation of Spermatid Cells

1. PREPARATION OF CELLS USING EDTA–TRYPSIN

The following procedure is used to prepare a suspension of rat testis cells which can be separated using the Staput or Elutriator (*1, 11*). Volumes are based on 11 testes; one testis is processed by sucrose homogenization (Section II,B,1,b) as a control.

a. Solutions.

—Z: Calcium, magnesium-free phosphate-buffered saline (8 gm NaCl, 0.2 gm KCl, 1.15 gm $Na_2HPO_4 \cdot 2H_2O$, 0.2 gm KH_2PO_4 per liter)–5 mM ethylenediaminetetraacetic acid (EDTA, disodium salt)–0.1% glucose at pH 7.4. Mix 200 ml and warm to 31°C in four Erlenmeyer flasks containing magnetic stirring bars.

—Y: PBS [phosphate-buffered saline (8 gm NaCl, 0.2 gm KCl, 1.15 gm $Na_2HPO_4 \cdot 2H_2O$, 0.2 gm KH_2PO_4, 0.1 gm $MgCl_2 \cdot 6H_2O$, 0.1 gm $CaCl_2$ per liter) Gibco K-13]–0.02% STI (soybean trypsin inhibitor, Worthington, SI)–2 μg/ml purified DNase (Worthington, D)–5 mM NDA (2-naphthol-6,8-disulfonic acid, dipotassium salt, Eastman Chemicals). Mix 120 ml and bring to room temperature.

—Purified trypsin (Worthington, TRL): 2.5% solution in saline (0.9% NaCl).

—$MgCl_2$: 100 mM.

—Crude pancreatic DNase (Sigma, DN-25): 2 mg/ml in saline.

—Fetal calf serum (Gibco).

b. Procedure.   All steps are performed at room temperature unless stated otherwise. Remove the tunica from 11 testes and cut the testes into small pieces with a razor blade. Transfer $2\frac{2}{3}$ testes ($\frac{1}{4}$ of total) to 50 ml of solution Z in an Erlenmeyer flask at 31 °C and swirl to mix. Repeat the procedure with the remaining testes and then treat each flask of 50 ml as follows: Incubate at 31 °C with stirring for 10 minutes. Add 2.0 ml of 2.5% purified trypsin and incubate at 31 °C without stirring for 10 minutes. Then add 5.5 ml of 100 mM MgCl$_2$ and 0.5 ml of crude DNase, and incubate at 31 °C with stirring for 10 minutes. Add 5 ml of fetal calf serum and pool the four samples. Filter through a coarse screen and then through an 80-μm stainless-steel screen. Centrifuge the filtrate in 50-ml conical tubes at 500 g (g_{av}) for 15 minutes. Aspirate and discard the supernatant and suspend the pellet in 120 ml of solution Y at room temperature. After 5 minutes cool the sample to 4 °C. Dilute an aliquot of the suspension 1:10 and count the cells using a hemacytometer. The yield is about 2.5×10^8 cells per rat testis.

2. SEPARATION OF CELLS

Rat testis cells prepared by the EDTA–trypsin procedure may be separated by velocity sedimentation at unit gravity (Staput method) (*1, 12, 13*) or by centrifugal elutriation (*14*). While the latter method is considerably faster and accommodates more cells, it does require a sizable outlay for equipment. As an alternative, the Staput method yields similar results, and the Staput chamber can be simply constructed out of lucite or purchased as glass chambers from Johns Scientific Co., Toronto, Canada. A detailed discussion of these two cell-separation techniques has been presented elsewhere (*10*).

a. *Staput.*

(i) Solutions:

—PBS (phosphate-buffered saline, Gibco K-13).

—Y: (Section IIa,1,a)

—BSA (bovine serum albumin): 1%, 2%, and 4% w/v dissolved in PBS (pH 7.4) and containing 5 mM NDA to reduce cell clumping. These solutions are prepared from a sterile stock of 10% BSA dissolved in PBS and adjusted to pH 7.4. Penicillin (100,000 units/liter) and streptomycin (100 mg/liter) are added to the 10% BSA before filter-sterilization and storing at 4 °C.

(ii) Procedure: All steps are performed at 4 °C. For preparative scale separation of cells, we use a Lucite chamber 28 cm in diameter. After introducing 100 ml of PBS into the chamber, up to 1.5×10^9 cells prepared by the EDTA–trypsin procedure are introduced in a volume of 100 ml of solution Y, followed by a nonlinear 1–4% gradient of BSA generated as described elsewhere (*10*). The cells are allowed to sediment until the visible band of late pachytene spermatocytes has migrated 3.7 cm (about 3 hours) (*1*). The chamber is unloaded into a fraction collector based on volume collection (Gilson Escargot), and the appropriate fractions, identified by microscopic counts or Coulter volume spectrometry, are pooled and concentrated by centrifugation at 600 *g* for 10 minutes. The composition of pooled fractions after Staput separation of rat testis cells is presented in Table I. The protein composition of nuclei prepared from these separated cell fractions is shown in Figs. 1 and 2. Differences in the electrophoretic profiles of both the acid-soluble proteins (Fig. 1) and acid-insoluble proteins (Fig. 2) are apparent.

b. Elutriator.

(i) Solutions:

—Y: (Section II,A,1,a)

—Separation medium: PBS containing 0.5% BSA to preserve cell integrity and 5 m*M* NDA to reduce cell adhesion.

(ii) Procedure: All steps are performed at 4°C. Cell suspensions prepared by the EDTA–trypsin procedure are taken up in solution Y and filtered through a 25-μm nylon screen (Nitex, TET/Kressilk Products, Inc., Elmsford, New York) just prior to loading into the Elutriator. Cells are separated by centrifugal elutriation in 20–40 minutes using a Beckman JE-6 Elutriator rotor in a J21–B centrifuge and the medium described above (*10*). Following separation, the cells are concentrated by centrifugation at 600 *g*

TABLE I

PERCENTAGE CELLULAR COMPOSITION OF STAPUT FRACTIONS

Cell type	Staput fractions							
	B	C	D	E	F	G	H	I
Spermatogonia[a]	2	1	2	2	1	8	1	0
Spermatocytes[b]	65	49	30	7	10	6	1	0
Spermatids								
Steps 1–8	13	24	43	71	67	9	0	0
Steps 9–10	2	2	2	5	3	6	3	0
Steps 11–15	7	10	7	5	7	15	26	21
Steps 16–19	0	1	0	1	0	2	4	27
Nongerminal[c] and unknown	6	8	7	4	6	16[d]	6	1
Residual bodies[e]	5	6	8	5	6	38	59	50

[a] Includes preleptotene spermatocytes which are indistinguishable from type B spermatogonia.

[b] Includes leptotene, zygotene, pachytene, and secondary spermatocytes.

[c] Includes Sertoli cells, Leydig cells, and macrophages. Erythrocytes were not counted.

[d] Some of these cells appear to be lymphocytes.

[e] Residual bodies are defined as cytoplasmic fragments which contain ribonucleoprotein aggregates. Other cytoplasmic fragments were not scored.

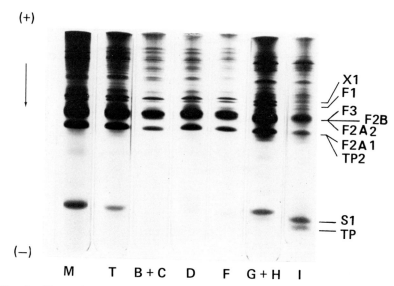

FIG. 1. Electrophoretic patterns of acid-soluble nuclear proteins from controls and from different Staput fractions (1). The basic proteins were extracted with 0.25 N HCl (Section II,D,1,a) and separated on acid-urea polyacrylamide gels (Section II,D,2,a) for 6 hours at 1 mA per gel. M: Basic proteins from rat testis nuclei prepared directly by sucrose homogenization (Section II,B,1,b). T: Basic nuclear proteins isolated from cells prepared using EDTA–trypsin (Section II,A,1) but not separated by the Staput. Nuclei were isolated from pooled Staput fractions as described in Section II,B,1,a. The cellular composition of the pooled fractions is given as a percent of total. B + C: 61% late pachytene primary spermatocytes and 14% round spermatids (steps 1–8). D: 19% primary spermatocytes, 51% round spermatids, and 9% late spermatids (steps 9–19). F: 56% early round spermatids and 27% late spermatids. G + H: 33% late spermatids and 48% residual bodies. I: 82% late spermatids and 16% small residual bodies. The position of migration of somatic histones F1, F3, F2B, F2A2, and F2A1 are identified together with the lysine-rich histonelike protein X1 (7, 9), the testis proteins TP (1,4) and TP2 (2,6), and the sperm protein S1 (4).

for 10 minutes. The separation obtained and the purity of the various fractions are essentially identical to those obtained with the Staput method.

B. Preparation and Separation of Spermatid Nuclei

1. PREPARATION OF NUCLEI

a. Hypotonic Lysis—Triton X-100. The following procedure is used to isolate nuclei from pooled fractions (after separation of cells by means of the Staput or Elutriator) when no further separation of nuclei is planned (1).

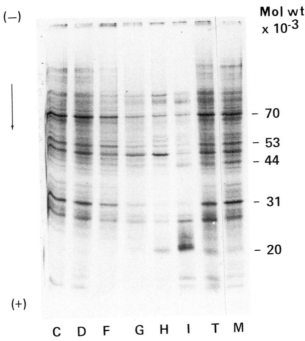

FIG. 2. Electrophoretic patterns of acid-insoluble nuclear proteins from controls and from different Staput fractions (*1*). After extraction of nuclei with 0.25 *N* HCl, the insoluble residue was separated on an SDS–polyacrylamide slab gel (Section II,D,2,b). The cellular composition of Staput fractions C, D, F, G, H, and I is given in Table I. T: Acid-insoluble nuclear proteins isolated from cells prepared using EDTA–trypsin (Section II,A,1) but not separated by the Staput. M: Acid-insoluble proteins from nuclei prepared directly by sucrose homogenization (Section II,B,1,b). The 20,000 mol wt protein which appears in late spermatids (fractions H and I) is TP4.

(i) Solutions:

—PBS (phosphate-buffered saline, Gibco K-13).

—L: 5 m*M* MgCl$_2$–5 m*M* sodium phosphate–0.25% Triton X-100–0.025% STI–5 m*M* NDA (pH 6.5). NDA is included to promote lysis, remove residual cytoplasm, and reduce aggregation of nuclei. Solution L is mixed fresh from stock solutions of individual components. The stock solutions are:

MgCl$_2$: 100 m*M*.

Sodium phosphate: 0.2 *M* (pH 6.5).

Triton X-100: 10% (v/v) solution in distilled water.

STI: 2.5% in distilled water (store at −20°C).

NDA: 100 mM.

—PMSF (phenylmethylsulfonyl fluoride): 10 mM in 30% isopropanol. Warm to room temperature to bring crystals into solution. Immediately before use, add 0.1 mM PMSF to solution L.

—MP: 5 mM MgCl$_2$–5 mM sodium phosphate (pH 6.5).

—Sucrose-MP: 2.4 M sucrose in solution MP.

(ii) Procedure: All steps are performed at 4°C. After concentration of cells in desired fractions by centrifugation at 600 g for 10 minutes, suspend the pellets in PBS by gentle pipetting and centrifuge again at 600 g for 10 minutes. To lyse the cells, suspend the pellets in solution L (containing 0.1 mM PMSF) at a concentration not exceeding 2 × 10^7 cells/ml. After centrifuging at 600 g for 10 minutes, suspend each pellet in 6.0 ml of solution MP using a Pasteur pipette. Use a syringe to force the suspension through a 25-gauge needle twice. Add 30 ml of sucrose–MP and mix well with a glass rod. Centrifuge at 24,000 rpm (76,000 g) for 60 minutes in a Beckman SW 27 rotor. After centrifugation, discard the sucrose supernatant and suspend each pellet in 10 ml of solution MP to remove the excess sucrose. Aliquot 0.1 ml for microscopic examination and counting, and centrifuge the remainder at 600 g for 10 minutes to recover the nuclei as a pellet.

When nuclei are to be isolated from pooled fractions (after cell separation on the Staput or Elutriator) and prepared for further separation on the Staput (section II, B, 2, a), centrifugation is avoided to minimize clumping of the nuclei. After lysing the cells in solution L (containing PMSF) and passage through the 25-gauge needle, the suspension is diluted in MP buffer to 5 × 10^6 nuclei/ml and loaded directly on the Staput without centrifugation (*1*).

b. Sucrose Homogenization. The following procedure is used to prepare nuclei from whole testes under conditions designed to minimize possible protein degradation (*1*). Proteins extracted from these nuclei and from nuclei of cells prepared by EDTA–trypsin and stored during the separation are compared by gel electrophoresis. Such controls indicate that no detectable nuclear protein degradation occurs during preparation of the cell suspensions or during cell separation (Fig. 3A and B).

(i) Solutions:

—H: 0.31 M sucrose–3 mM MgCl$_2$–10 mM potassium phosphate (pH 6.0)– 0.05% Triton X-100–0.1 mM PMSF. Add PMSF immediately before homogenization. (See Section II, B, 1, a.)

—L: (See Section II, B, 1, a.) No PMSF is added.

—MP: 5 mM MgCl$_2$–5 mM sodium phosphate (pH 6.5).

(ii) Procedure: Testes are cooled on ice immediately and all steps are performed at 4°C. Remove tunica from testes and homogenize gently in

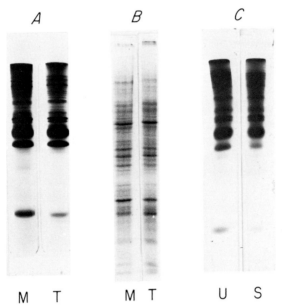

FIG. 3. Electrophoretic patterns of rat testis nuclear proteins showing the absence of degradation after preparation of cell suspensions or after sonication. (A) Acid-soluble nuclear proteins. (B) Acid-insoluble nuclear proteins. (C) Acid-soluble proteins from rat testis nuclei prepared by sucrose homogenization. M: Nuclei were isolated directly by sucrose homogenization (Section II,B,1,b) under conditions designed to minimize possible protein degradation. T: Nuclei were isolated from cells which were prepared using EDTA–trypsin and allowed to stand at 4°C for 3–4 hours to simulate conditions used for Staput fractionation. U: Nuclei prepared by sucrose homogenization were extracted with 0.25 N HCl without sonication. S: Nuclei were sonicated for a total of 3 minutes (Section II,B,1c) before the suspension was extracted with 0.25 N HCl. Basic proteins from both extracts were precipitated with 25% TCA and separated on acid-urea polyacrylamide gels (Section II,D,2,a). Gels shown were overloaded to detect any minor degradation products. No differences could be detected in the protein patterns.

5 volumes of solution H using a motor-driven Teflon pestle. Filter homogenate through four layers of gauze and centrifuge at 600 g for 10 minutes. Suspend pellet in 8–10 volumes of solution L using a Teflon pestle. Centrifuge at 600 g for 10 minutes and resuspend each pellet in 6.0 ml of solution MP. From this point, the procedure is identical with that described for preparation of nuclei by hypotonic lysis and centrifugation through dense sucrose (Section II,B,1,a).

   c. Sonication.   The following procedure employs sonication to prepare late spermatid nuclei (steps 13–19) in quantities ranging from 1×10^8 to 7×10^9 nuclei (*2,6*).

Electrophoretic profiles of the acid-soluble proteins extracted from testicular nuclei show no evidence of degradation as a result of sonication (Fig. 3). Late spermatid nuclei prepared from the testis by this method are about 99% pure, with about 1% contamination by ribonucleoprotein aggregates from residual bodies and less than 0.1% contamination by other nuclei (2). The absence of histones in acid extracts of these preparations confirms the absence of contamination by nuclear material from other cells (6) (Fig. 4). Based on the intensity of staining with hematoxylin, 33% of the sonication-resistant spermatid nuclei are in steps 13–15 of development and 67% are in steps 16–19. Steps 13–15 nuclei may be separated from steps 16–19 by equilibrium density centrifugation in metrizamide gradients (Section II,B,2,b).

(i) Solutions:

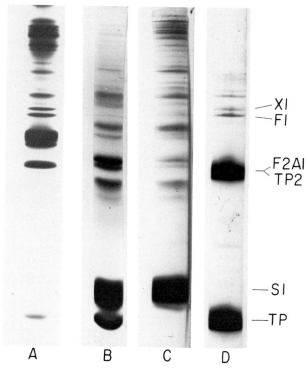

FIG. 4. Electrophoretic profiles of acid-soluble proteins from rat testis nuclei. (A) A 0.25 N HCl extract of whole nuclei prepared by sucrose homogenization. (B) A 0.25 N HCl extract of sonication-resistant nuclei (steps 13–19). (C) A 0.25 N HCl extract of whole nuclei showing the proteins precipitated with 5% TCA (D) A 0.25 N HCl extract of whole nuclei showing the proteins precipitated between 5 and 20% TCA.

—H: (Section II,B,1,b)
—PMSF: 10mM in 30% isopropanol (Section II,B,1,a).
—Sucrose: 1.5 M in distilled water.

(ii) Procedure: All steps are performed at 4°C. Cool testes on ice, remove tunica, and suspend tubules in solution H (12 ml/testis). Add PMSF to 0.1 mM immediately before homogenization. Homogenize for 10 seconds at 90% maximum power using a Polytron homogenizer (PT-10, Brinkmann Instruments) or a motor-driven Teflon homogenizer. Filter through four layers of gauze and centrifuge the filtrate at 600 g for 10 minutes. Suspend the pellet in cold distilled water (3–7 ml/testis) using a Teflon pestle. Add PMSF to a concentration of 0.1 mM and sonicate 12 times with 15-second bursts of ultrasound. Keep the sample on ice during sonication and allow sufficient time between bursts (45 seconds) to keep the temperature below 6°C. We use either a Branson S-185 sonifier fitted with a microprobe and set at 6 (25 W output), or a Bronwill Biosonik BP III Ultrasonic system with the large probe at 90% maximal output. The objective is to disrupt all cells and nuclei except those from elongated spermatids and to break the tails into short fragments roughly the length of a sperm head.

Filter the sonicate through an 80-μm screen to remove debris and layer 20 ml of the suspension over 10 ml of 1.5 M sucrose. Centrifuge in a round-bottom tube in a swinging-bucket rotor at 1000 g for 30 minutes. The sonication-resistant nuclei form a pellet while the tail fragments collect just below the sucrose–water interface. Resuspend the pellets in solution MP using a Teflon pestle. If large tail fragments or clumps of debris remain, sonicate again. Layer over 1.5 M sucrose and centrifuge again. Repeat (usually once more) until heads are free of tails and debris.

Sperm nuclei from the epididymis are prepared by the same procedure except that distilled water is used as the homogenization medium (in place of solution H) and as the resuspension medium (in place of solution MP). In addition, the pH of the suspension is raised to about 8.5 by adding 1–2 drops of 1 N NaOH before sonication and again before layering over sucrose. Water is used because isolated epididymal sperm nuclei tend to form clumps in the presence of Mg^{2+}. In contrast, testicular sperm nuclei tend to aggregate more readily in the absence of Mg^{2+}.

2. SEPARATION OF NUCLEI

a. Staput. The following procedure is used to prepare highly purified nuclei from round spermatids (steps 1–8) and elongated spermatids (steps 12–19) (*1*). The procedure for separating nuclei using the Staput is, with a few exceptions, the same as that used for cells (Section II,A,2,a) (*10*).

(i) Solutions:
—MP: 5 mM $MgCl_2$–5 mM sodium phosphate (pH 6.5).

—Sucrose: 1.3%, 4.0%, and 10% (w/w) dissolved in solution MP.

—Triton X-100: 10% (v/v) solution in distilled water.

(ii) Procedure: Nuclei are isolated and prepared for separation on the Staput as described in Section II,B,1,a. Nonlinear gradients are generated from sucrose dissolved in solution MP (1, 10). We allow 8 hours for the separation of round spermatid nuclei and up to 16 hours for elongated spermatids. Fractions are collected, nuclei are identified and counted using a hemacytometer, and desired fractions are pooled. After adding 0.2% Triton X-100 to facilitate recovery of the nuclei, the pooled fractions are centrifuged at 600 g for 10 minutes.

b. Metrizamide. Equilibrium density gradient centrifugation is used to separate sonication-resistant spermatids (steps 13–19) into at least two discrete developmental stages, steps 13–15 and steps 16–19 (2, 10). Gradients are formed from metrizamide, 2-(3-acetamido-5-N-methyl-acetamido-2,4,6-tri-iodobenzamido)-2-deoxy-D-glucose (Gallard-Schlesinger Corp.), a nonionic iodinated benzamide derivative which has no effect on either the morphological properties or basic protein composition of the elongated spermatid nuclei (2).

(i) Solutions:

—G: 5 mM NDA–1 mM MgCl$_2$–1 mM sodium phosphate (pH 6.5)–0.25% Triton X-100.

—Metrizamide light: 44.5% w/w (59% w/v) dissolved in solution G.

—Metrizamide dense: 50.0% w/w (69% w/v) dissolved in solution G.

(ii) Procedure: Prepare spermatid nuclei (steps 13–19) by sonication and centrifugation through sucrose (Section II,B,1,c). Suspend nuclei in a small volume of solution G and add 0.11-ml aliquots to 3.2 ml of each metrizamide solution. Using the light and dense metrizamide solutions, generate a linear 5-ml gradient containing between 10^7 and 2×10^8 nuclei uniformly dispersed throughout (15). Centrifuge gradients at 20,000 rpm (37,400 g) for 20 minutes in a Beckman SW 50.1 rotor. Collect fractions from the top of the gradient or remove bands by puncturing the side of the tube. Measure the refractive index (n) of each fraction at room temperature and calculate the density using the formula: $\rho = 3.496n - 3.661$.

Bands of nuclei are formed at $\rho = 1.31$ gm/cm^3 and at $\rho = 1.34$ gm/cm^3. The lighter band contains steps 13–15 nuclei with up to 93% purity, while the denser band is more concentrated and contains steps 16–19 nuclei with as high as 99% purity.

C. Identification of Cells and Nuclei

Separated testis cells are identified on air-dried smears fixed in Bouin's and stained with periodic-acid Schiff (PAS)–hematoxylin. Criteria for the identification of cells have been described previously (*1,16*). Round and elongated spermatid nuclei are identified either by phase-contrast microscopy or by staining with acetic–orcein as described elsewhere (*17*). The stages of development of elongated spermatid nuclei were determined on air-dried smears fixed in methanol and stained with hematoxylin (*2*).

D.   Separation and Quantitation of Nuclear Proteins

The acid-soluble proteins may be extracted with 0.25 N HCl from nuclei prepared by hypotonic lysis (Section II,B,1,a) or by sonication (Section II,B,1,c). These proteins may be fractionated by differential precipitation with trichloroacetic acid (TCA) (*2, 4*) and separated on acid-urea polyacrylamide gels.

Proteins in the acid-insoluble residue from these nuclei, or in whole nuclei not subjected to acid extraction, may be analyzed by gel electrophoresis in the presence of SDS (sodium dodecyl sulfate). Major proteins migrating as single bands after gel electrophoresis may be cut out and analyzed for their amino acid composition according to the procedure of Houston (*18*).

1.   CHEMICAL FRACTIONATION

a. Extraction with HCl.   Suspend nuclei in 1.0 ml of cold distilled water using a Teflon homogenizer. Add 0.33 ml of 1 N HCl to make the solution 0.25 N HCl and then add 1.0 ml of 0.25 N HCl and allow to stand on ice for 20 minutes with occasional homogenization. Centrifuge at 12,000 g for 10 minutes and retain the supernatant. Reextract the pellet with 1.0 ml of 0.25 N HCl, centrifuge, and combine supernatants. Precipitate proteins from the pooled extract by adding an equal volume of 50% TCA. Recover the precipitate by centrifugation (12,000 g for 10 minutes), wash with acidified acetone (200 ml of acetone + 0.1 ml of 12 N HCl), wash with acetone, and then dry under vacuum.

b. Extraction with HCl after Guanidine–HCl–Mercaptoethanol Treatment. The extractability of S1 and other proteins from late spermatids may be incomplete using dilute acids because of the formation of disulfide crosslinks. Sonication appears to enhance the extractability of S1. To obtain complete extraction, nuclei are first treated with guanidine–HCl and β-mercaptoethanol followed by acid extraction. The nuclei are dissolved initially in the smallest possible volume of 5 M guanidine–HCl–β-mercaptoethanol solution in order to facilitate the final precipitation of the acid-soluble proteins and their separation from the excess guanidine–HCl with acidified acetone.

Dissolve nuclei in a small volume (0.5–1.0 ml) of 5 M guanidine–HCl– 0.5 M Tris (pH 8.5)–0.28 M β-mercaptoethanol (*19*) (roughly 10^8 nuclei/ml).

To each ml of this solution add 0.33 ml of 1 N HCl and 1.0 ml of 0.25 N HCl. Allow the solution to stand on ice for 15 minutes with occasional homogenization and centrifuge at 12,000 g for 10 minutes Reextract any insoluble material in the pellet with 1.0 ml of 0.25 N HCl and measure the volume of the combined supernatants. When this procedure is used with small amounts of protein for separation on acrylamide gels, 200 μg of BSA is often added as carrier at this point. Then add 3.0 ml of 33% TCA per milliliter of supernatant and allow the solution to stand on ice for 30 minutes. Centrifuge at 12,000 g for 10 minutes. Slowly add cold, acidified acetone to the pellet (which contains guanidine–HCl as well as protein) until the pellet is dissolved. Continue adding the cold acetone until the solution becomes slightly turbid, presumably indicating reprecipitation of the basic proteins. Centrifuge at 12,000 g for 10 minutes and discard the supernatant. Suspend the pellet in cold acetone and centrifuge again. Discard the supernatant and allow the pellet to dry in the vacuum desiccator.

c. Differential Precipitation with TCA. Basic proteins may be chemically fractionated by differential precipitation with TCA (*4*). This procedure is based on the fact that most of the histones (and sperm protein, S1) are precipitated by 5% TCA, while testis proteins TP, TP2, and lysine-rich histones F1 and X1 remain in solution. These testis proteins and histones are subsequently precipitated by increasing the TCA concentration to 20% (Fig. 4).

Mix an equal volume of cold 10% TCA with the 0.25 N HCl extract of nuclei (final concentration 5% TCA), and allow the mixture to stand on ice for 15 minutes. Centrifuge at 12,000 g for 10 minutes. The pellet contains most of the histones and S1. Add to the supernatant enough 80% TCA to give a final concentration of 20% TCA, and allow the mixture to stand on ice for 15 minutes. Centrifuge at 12,000 g for 10 minutes and carefully discard the supernatant. The pellet contains TP, TP2, X1, and F1. Wash the pellets once with acidified acetone, once with acetone, and dry in a vacuum desiccator.

2. ELECTROPHORETIC SEPARATION

a. Acid-Urea Polyacrylamide Gels. Acid-soluble proteins are separated at acid pH in 15% polyacrylamide gels containing 2.5 M urea as described by Panyim and Chalkley (*20*). Prepare 50 ml of gel solution (enough for 10 tubes, 0.6 × 16 cm) as follows: Dissolve 7.5 gm of acrylamide (Bio-Rad), 7.5 gm urea (Schwartz-Mann, Ultrapure), and 50 mg N,N'-methylenebisacrylamide in water. Add 2.58 ml of glacial acetic acid (0.9 N final concentration), gently warm the solution to room temperature, and bring the volume up to 50 ml with water. Add 50 mg ammonium persulfate and 25 μl of N,N,N',N'-tetramethylethylenediamine (TEMED). Filter the solution through a 0.45-μm filter using a gentle vacuum. Degas the solution using a good vacuum pump until no further bubbles form (1–2 minutes). Fill the glass tubes to within 1 cm of the top (about 4 ml/tube) and overlayer with 2.5 M urea using a microliter syringe. Allow gels to polymerize for about 2 hours. Subject the gels to preelectrophoresis overnight at 0.5 mA/gel, using 0.9 N acetic acid in both upper and lower chambers.

Fill buffer chambers with fresh 0.9 N acetic acid before loading the proteins. Dissolve proteins in loading solution (0.9 N acetic acid–7.0 M urea–2% β-mercaptoethanol) at a concentration of 1–2 mg/ml and allow samples to stand at room temperature for several hours before loading on the gels. Load 20–50 μg of protein on each gel and add 10 μl of 0.1% pyronine-Y in loading solution to one gel as a tracking dye. Run gels for 15 hours at 0.5 mA per gel or for 8 hours at 1.0 mA per gel. Stop the run when the tracking dye reaches the end of the tube. Remove the gels from the glass tubes using a 3-in., 22-gauge needle and a syringe filled with 10% Triton X-100. If the gel tubes are siliconized before using, the Triton X-100 is unnecessary. After each use, clean the tubes with detergent and soak overnight in sulfuric acid–dichromate cleaning solution. Stain gels overnight in 0.1% Amido black in 40% ethanol–5% acetic acid, and destain by transverse electrophoresis in the solvent. Separation of the basic nuclear proteins from rat testis using this gel system is illustrated in Figs. 1, 3 and 4.

b. SDS–Polyacrylamide Gels. Total nuclear proteins or the acid-insoluble proteins are separated in 12% polyacrylamide slab gels containing SDS using the buffer system of Laemmli (pH 8.3) (*21*). The procedure and apparatus we use for slab gel electrophoresis is essentially as described by Ames (*22*). Proteins are dissolved in a loading solution (2% SDS–5% β-mercaptoethanol–10% glycerol–60 mM Tris (pH 6.8)–0.0005% Bromphenol blue) and run at 25 mA for about 5 hours. Gels are stained overnight with 0.25% Coomassie brilliant blue in 25% isopropanol–10% acetic acid and destained as recommended by Fairbanks *et al.* (*23*). Under these conditions rat sperm protein S1 does not migrate, presumably due to its low solubility in SDS. Separation of the nuclear proteins from rat testis using the SDS gel system is illustrated in Figs. 2 and 3.

3. Quantitation of Protein

a. TCA Precipitation. The concentration of proteins in solutions containing mercaptoethanol (which interferes with the Lowry procedure) were estimated by measuring turbidity after precipitation with 25% TCA (*24*). Small aliquots of a protein solution (10–100 μg) are mixed with an equal volume of 50% TCA. After 20 minutes on ice, the optical densities of samples and BSA standards are read at 400 nm, and the protein concentration is determined from the standard curve.

b. Densitometry of Polyacrylamide Gels. Cylindrical gels stained with

Amido Black are scanned at 600 nm, and slab gels stained with Coomassie brilliant blue are scanned at 550 nm using a Gilford model 240 spectrophotometer equipped with a linear transport attachment. Slab gels are prepared for scanning by drying the gel onto cellophane under a vacuum with heat using a commercial gel drier (Bio-Rad). After drying, each sample channel is cut as a strip about 10–12 mm wide using a sharp razor. The gel should be cut immediately before scanning because the narrow strip tends to curl with time. The gel strip is mounted in a scanning cuvette (6 mm wide, 100 mm long, 9 mm deep, e.g., Gilford 2412) and held in place by a glass block (5.5 mm wide) which nearly fills the cuvette. The remaining air space is filled with glycerol or immersion oil. If glycerol is used, the strip must be scanned immediately because the gel will begin to soften and curl after about an hour in glycerol. Immersion oil can be used in place of glycerol, but it is more expensive and more difficult to clean off.

The concentration of basic proteins in stained bands after electrophoresis can be estimated from densitometric scans of the gels using a standard curve prepared from electrophoretic profiles of purified histone fractions. The areas under peaks from the scans are quantitated either by planimetry or by cutting out and weighing the peaks.

Similar methods are used for quantitation of the nonhistone proteins. However, quantitation of individual proteins is possible only when each peak on the density scan represents a single protein, a condition which is difficult to achieve with such a complex mixture of proteins.

4. MEASUREMENT OF RADIOACTIVITY

a. Liquid Scintillation Spectrometry. Radioactivity in cylindrical gels is measured by liquid scintillation spectrometry. Gels are sliced at 1.0-mm intervals in a Gilson Aliquogel fractionator. This device uses a calibrated piston to force 1 mm of the gel out of a glass tube and through a wire screen followed by a jet of water which washes the gel fragments into a scintillation vial. The gel fragments are dried in an oven and then dissolved in 0.2 ml of 30% hydrogen peroxide by incubating the capped vials at 80 °C for 3 hours. After cooling the vials to room temperature, 7 ml of a scintillation cocktail [1 liter of toluene, 1 liter of ethanol, and 1 liter of dioxane, containing 240 gm of naphthalene, 15 gm of 2,5-diphenyloxazole (PPO), and 150 mg of 1,4-bis-2-(5-phenyloxazoyl)-benzene (POPOP)] are added. Radioactivity is measured using a liquid scintillation spectrometer.

b. Autoradiography and Fluorography The distribution of radioactivity in slab gels is most easily detected by autoradiography when ^{32}P is used as a label or by fluorography when proteins are labeled with ^{3}H, ^{14}C, or ^{35}S. These techniques are more convenient than conventional slicing methods and provide superior resolution of closely spaced bands.

For autoradiography of ^{32}P, a slab gel is dried onto either filter paper or cellophane as described in Section II,D,3,b and loaded together with an appropriate film (e.g., Cronex 20C, Dupont) into an 8 × 10 x-ray cassette.

Fluorography of slab gels containing ^{3}H-, ^{14}C-, or ^{35}S-labeled proteins has been described in detail by Bonner and Laskey (25). The gel is dehydrated in dimethyl sulfoxide, impregnated with PPO, dried, and exposed to x-ray film at -70 °C. Preexposure of the film to a brief flash of light increases the sensitivity of the process and permits accurate quantitation by microdensitometry of the image obtained (26).

III.   Isolation of Specific Low-Molecular-Weight Basic Proteins from Rat Testis Spermatids

A.   Isolation of S1

S1 is the major protein in rat sperm nuclei (4,27). Since S1 is only present after the onset of sonication resistance (late step 12) (2), purifying this protein from rat testis is facilitated by starting with a preparation of sonication-resistant spermatid nuclei (Section II,B,1,c). With this procedure, S1 may be purified from the testis without blocking the sulfhydryl groups.

The acid-soluble proteins are extracted from sonication-resistant heads using guanidine–HCl (Section II,D,1,b), and the acetone-dried protein is dissolved in 2 ml of 7 M urea–0.01 N HCl–4% β-mercaptoethanol. After centrifugation of this solution at 12,000 g for 10 minutes, the clarified supernatant is applied to a 3 × 85 cm column of Sephadex G-100 equilibrated at room temperature with a solution of 2.5 M urea–0.01 N HCl–1% β-mercaptoethanol–0.02% sodium azide (Fig. 5). Fractions of 3 ml are collected at a flow rate of 12 ml/hour. The quantity of protein in each fraction is estimated by measuring the turbidity after TCA precipitation (Section II,D,3,a). Proteins eluting in the last peak (between 300 and 360 ml) are recovered by adding enough cold 100% TCA to the pooled fractions to bring the final concentration to 20%. After centrifugation at 12,000 g for 15 minutes, the pellet containing S1 and TP is washed with acidified acetone and then with acetone and dried under vacuum. The protein is then dissolved in 2.0 ml of 12% guanidine–HCl–4% β-mercaptoethanol–0.1 M sodium phosphate (pH 6.8). After centrifuging the sample at 12,000 g for 10 minutes to clarify the solution, the supernatant is applied to an Amberlite IRC-50 column (2.5 × 20 cm) equilibrated with 12% guanidine–HCl–0.2 M sodium phosphate (pH 6.8) at room temperature. The column is eluted with 300 ml of a 12–40% linear gradient of guanidine–HCl in the same buffer.

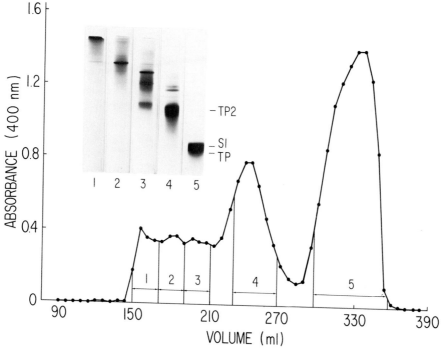

FIG. 5. Gel filtration chromatography of basic proteins from rat testis sonication-resistant nuclei. Approximately 30 mg of protein were loaded on a 3 × 85 cm column of Sephadex G-100. Fractions were pooled as indicated on the elution profile and separated by electrophoresis on acid-urea polyacrylamide gels (inset). The positions of migration of TP2, S1, and TP are indicated on the right.

S1 elutes as a broad peak between 29% and 35% guanidine–HCl (peak fraction at 30.5% guanidine–HCl) which appears free of TP by gel electrophoresis. Further purification of S1 is possible by precipitation with 3% TCA.

B. Isolation of TP

A procedure for isolating TP has been described in detail by Kistler *et al* (*4*). We employ two modifications to reduce the level of contaminating proteins early in the procedure. Since most of the synthesis of TP occurs after the acquisition of sonication resistance (*2*), the purification of this protein is facilitated by first isolating testicular sperm heads (Section II,B,1,c). The proteins extracted with 0.25 *N* HCl are then subjected to differential precipitation with TCA (Section II,D,1,c). We routinely precipitate contaminating proteins from the acid extract with 5% TCA instead of 3% TCA as originally described by Kistler *et al.* (*4*). Raising the TCA concentration to 20% precipitates TP and additional proteins. Separation of these proteins on acid-urea polyacrylamide gels shows more than 50% of the protein migrating as TP. The other major protein precipitated at 20% TCA under these conditions has been identified as TP2 (*2, 6*) (Fig. 4). Procedures of Kistler *et al.* (*4*) may then be used to obtain TP in pure form.

C.   Isolation of TP2

The acid-soluble proteins are extracted from sonication-resistant nuclei using 0.25 N HCl (Section II,D,1,a or b), and the acetone-dried protein is dissolved in 2 ml of 7 M urea–0.01 N HCl–4% β-mercaptoethanol. The solution is clarified by centrifugation and applied to a Sephadex G-100 column at room temperature as described above in Section III,A and shown in Fig. 5. This time, however, proteins eluting in an earlier peak (between 230 ml and 265 ml) are pooled and precipitated at 20% TCA. The precipitate is recovered by centrifugation and dissolved in 0.4 N H_2SO_4. TCA is added to 5% and precipitating proteins are removed by centrifugation. TP2 is then precipitated by raising the TCA concentration to 20%. When run on acid-urea gels, more than 95% of the protein migrates as TP2 (*2*). The absence of histone F2A1 contamination has been verified by showing that less than 1% of the ^{125}I-labeled F2A1 added before the Sephadex column is recovered in the pellet after differential TCA precipitation.

IV.   Concluding Remarks

The rat testis system offers the possibility of analyzing the role of nuclear proteins in terms of their effect on nuclear structure and gene transcription. During its maturation, the spermatid nucleus undergoes a transition from a round body of diffuse, "beaded-fiber" (*28*) chromatin, which is transcriptionally active, to an elongated body of condensed, smooth-fiber chromatin, which is transcriptionally inactive. During this transitional period, the three low-molecular-weight basic proteins TP, TP2, and S1 are synthesized and along with other proteins replace the somatic and testis-specific histones and most of the nonhistone chromatin proteins. Thus, the transition in nuclear protein composition during spermatid maturation correlates closely with the change in nuclear morphology and chromatin template activity.

The potential of the rat testis system to contribute to our understanding of nuclear protein function is enhanced by the availability of techniques for the isolation of cells at different stages of the transition. We have described methods for isolating spermatids in steps 1–8, 13–15, and 16–19 in 90 to 99% purity. Procedures for isolating step 9–12 spermatids are currently being developed. These procedures should enable us to localize the stages during which each of the spermatidal proteins is synthesized. The stage at which these proteins become bound to the chromatin, are enzymically modified, and are subsequently displaced from the chromatin can also be determined.

The availability of spermatidal proteins in purified form permits the design of *in vitro* experiments to test mechanisms of nuclear protein replacement. For example, one might investigate the *in vitro* conditions necessary for protein replacement to occur and then study the effects of this replacement process on nuclear condensation and transcriptional inactivation. Thus, the rat testis system permits both *in vitro* and *in vivo* approaches to the analysis of nuclear protein function.

ACKNOWLEDGMENTS

We are grateful to Dr. L. S. Hnilica for helpful insights and guidance during the course of these investigations. We also wish to thank Gwen Hord, Patricia Trostle, and Betty Reid for their assistance in the development of these procedures. This work was supported by grant CA-17364 from the NIH, grant ACS IN-430 from the American Cancer Society, and a grant from the National Science Foundation.

REFERENCES

1. Platz, R. D., Grimes, S. R., Meistrich, M. L., and Hnilica, L. S., *J. Biol. Chem.* **250**, 5791 (1975).
2. Grimes, S. R., Platz, R. D., Meistrich, M. L., and Hnilica, L. S., Submitted.
3. Kumaroo, K. K., Jahnke, G., and Irvin, J. L., *Arch. Biochem. Biophys.* **168**, 413 (1975).
4. Kistler, W. S., Geroch, M. E., and Williams-Ashman, H. G., *J. Biol. Chem.* **248**, 4532 (1973).
5. Platz, R. D., Grimes, S. R., Hord, G., Meistrich, M. L., and Hnilica, L. S., *in* "Chromosomal Proteins and Their Role in the Regulation of Gene Expression" (G. S. Stein and L. J. Kleinsmith, eds.), pp. 67–92. Academic Press, New York, 1975.
6. Grimes, S. R., Jr., Platz, R. D., Meistrich, M. L., and Hnilica, L. S., *Biochem. Biophys. Res. Commun.* **67**, 182 (1975).
7. Branson, R. E., Grimes, S. R., Yonuschot, G., and Irvin, J. L., *Arch. Biochem. Biophys.* **168**, 403 (1975).
8. Grimes, S. R., Chae, C. B., and Irvin, J. L., *Biochem. Biophys. Res. Commun.* **64**, 911 (1975).
9. Shires, A., Carpenter, M. P., and Chalkley, R., *Proc. Natl. Acad. Sci. U.S.A.* **72**, 2714 (1975).
10. Meistrich, M. L., *Methods Cell Biol.* **15**, 15 (1976).
11. Meistrich, M. L., and Trostle, P. K., *Exp. Cell Res.* **92**, 231 (1975).
12. Lam, D. K. M., Furrer, R., and Bruce, W. R., *Proc. Natl. Acad. Sci. U.S.A.* **65**, 192 (1970).
13. Go, V. L. W., Vernon, R. G., and Fritz, I. B., *Can. J. Biochem.* **49**, 753 (1971).
14. Grabske, R. J., Lake, S., Gledhill, B. L., and Meistrich, M. L., *J. Cell. Physiol.* **86**, 177 (1975).

15. Grdina, D. J., *Methods Cell Biol.* **14**, 213 (1976).

16. Meistrich, M. L., Bruce, W. R., and Clermont, Y., *Exp. Cell Res.* **79**, 213 (1973).

17. Meistrich, M. L., and Eng, V. W. S., *Exp. Cell Res.* **70**, 237 (1972).

18. Houston, L. L., *Anal. Biochem.* **44**, 81 (1971).

19. Coelingh, J. P., Rozijn, T. H., and Monfoort, C. H., *Biochim. Biophys. Acta.* **188**, 353 (1969).

20. Panyim, S., and Chalkley, R., *Arch. Biochem. Biophys.* **130**, 337 (1969).

21. Laemmli, U. K., *Nature (London)* **227**, 680 (1970).

22. Ames, G. F., *J. Biol. Chem.* **249**, 634 (1974).

23. Fairbanks, G., Steck, T. L., and Wallach, D. F. H., *Biochemistry* **10**, 2606 (1971).

24. Layne, E., *in* "Methods in Enzymology," Vol. 3: Preparation and Assay of Substrates (S. P. Colowick and N. O. Kaplan, eds.), pp. 447–454. Academic Press, New York, 1957.

25. Bonner, W. M., and Laskey, R. A., *Eur. J. Biochem.* **46**, 83 (1974).

26. Laskey, R. A., and Mills, A. D., *Eur. J. Biochem.* **56**, 335 (1975).

27. Marushige, Y., and Marushige, K., *J. Biol. Chem.* **250**, 39 (1975).

28. Kierszenbaum, A. L., and Tres, L. L., *J. Cell Biol.* **65**, 258 (1975).

Chapter 19

Isolation and Characterization of Nonhistone Phosphoproteins

TUNG YUE WANG AND NINA C. KOSTRABA

Division of Cell and Molecular Biology,
State University of New York at Buffalo,
Buffalo, New York

I. Introduction

The nonhistone chromosomal proteins are believed to play a regulatory role in the control of gene expression. One evidence supporting this regulatory role for the nonhistone proteins is their ability to alter transcription *in vitro*. In this chapter, we limit ourselves to the preparative procedures and characteristic descriptions of only those nonhistone phosphoproteins that directly stimulate or inhibit transcription *in vitro* from DNA or chromatin. At present, three nonhistone phosphoprotein fractions have been shown to exert such direct effects on transcription. For convenience, these three nonhistone protein fractions are referred to here as NHP-I, NHP-II, and NHP-III. One fraction (NHP-I) enhances the transcription from chromatin, as shown in liver, spleen, and kidney of rat (*1–5*), in Walker 256 carcinosarcoma (*2,3*), in frog liver (*6*), in calf endometrium (*7*), and in cultured carrot cells (*8*). The second fraction (NHP-II) binds selectively to, and stimulates transcription from, homologous DNA (*9–12*). This nonhistone protein fraction has been demonstrated in rat liver (*11,13*), in KB cells (*14*), and in Ehrlich ascites tumor cells (*12*). The third fraction (NHP-III), isolated from Ehrlich ascites tumor (*15*) and calf thymus inhibits DNA-directed RNA synthesis *in vitro*.

NHP-I, the nonhistone protein fraction that stimulates transcription from chromatin *in vitro*, is a highly heterogeneous group of proteins containing RNA. It stimulates transcription from chromatin isolated from the same tissue of origin as the nonhistone protein fraction as well as that from other tissues. However, NHP-I stimulates the chromatin template of homologous tissue to a greater extent than heterologous chromatin (2,3,6).

Like NHP-I, NHP-II is also heterogeneous but, as prepared, contains no nucleic acid. NHP-II selectively binds to, and enhances transcription of, homologous DNA (9–13). The specificity of NHP-II is such that it interacts with and activates transcription of only unique sequences in DNA (16). The mechanism of activation by NHP-II is stimulation of the initiation of RNA synthesis (16,17).

NHP-III has been isolated as a near-homogeneous nonhistone protein. This nonhistone protein binds to, and inhibits transcription of, reiterated sequences in DNA, but not unique DNA sequences, nor chromatin (17). It inhibits the transcription of repetitive DNA sequences by acting at the initiation step of RNA synthesis (15,17).

In brief, two of the nonhistone protein fractions (I and II) can be shown to stimulate transcription *in vitro*, one (NHP-I) by acting on chromatin and the other (NHP-II), on DNA. The third fraction (NHP-III) inhibits transcription from DNA. The varied but specific effects of these nonhistone proteins indicate the involvement of nonhistone chromosomal proteins in both positive and negative controls in the differential regulation of gene expression.

II.   NHP-I. The Nonhistone Proteins That Stimulate
Transcription of Chromatin

A.   Preparative Procedure

Buffers:
   Buffer A. 0.02 M Tris-HCl (pH 7.6)–0.15 M NaCl.
   Buffer B. 0.02 M Tris-HCl (pH 7.6).
   Buffer C. 0.02 M Tris-HCl (pH 8.0)–2.0 M NaCl.
   Buffer D. 0.02 M Tris-HCl (pH 7.2)–0.4 M NaCl.
   Buffer E. 0.015 M Tris-HCl (pH 8.0)–1 mM EDTA–1 mM β-mercaptoethanol.
   Buffer F. 0.015 M Tris-HCl (pH 8.0)–0.05 M NaCl–1 mM EDTA–1 mM β-mercaptoethanol.
   Buffer G. 0.015 M Tris-HCl (pH 8.0)–0.15 M NaCl–1 mM EDTA–1 mM β-mercaptoethanol.
   Buffer H. 0.015 M Tris-HCl (pH 8.0)–0.30 M NaCl–1 mM EDTA–1 mM β-mercaptoethanol.
   Buffer I. 0.02 M Tris-HCl (pH 8.0).

Step 1. NHP-I is prepared from crude chromatin (saline-extracted nuclei). The method of Chauveau *et al.* (*18*) is applicable to the isolation of nuclei from rat and most other solid tissues, but other methods may also be used. For the isolation of nuclei, the tissue, trimmed and cleansed of fat and connective tissue, is forced through a Harvard press into a pulp. The macerated tissue is weighed and then homogenized with a Dounce homogenizer in 9 volumes of 2.4 M sucrose–10^{-4} M CaCl$_2$ to a 10% (w/w) homogenate. The suspension is centrifuged at 75,000 g for 1 hour. At the end of the centrifugation, the pellets are collected, resuspended in 10 volumes of 0.33 M sucrose, and centrifuged at 1000 g for 5 minutes. The isolated nuclei are recovered in the pellets.

Step 2. The isolated nuclei are extracted with 100 volumes of Buffer A by first homogenizing them in a loose-fitting Dounce homogenizer and then stirring for 1 hour. The nuclear suspension is centrifuged at 25,000 g for 15 minutes, discarding the supernatant. This step of extraction with Buffer A is repeated twice.

Step 3. The saline-extracted nuclei are suspended in 20 volumes of Buffer B and stirred. During the stirring, an equal volume of Buffer C is added slowly into it, and the stirring is continued for 4–6 hours. The mixture is then centrifuged at 75,000 g for 90 minutes. The supernatant is collected and its volume measured. To this supernatant solution 1.5 volumes of Buffer B are added drop by drop with stirring to a final concentration of 0.4 M with respect to NaCl. The resulting mixture with its formed precipitate, which consists mostly of DNA and histones, is centrifuged at 75,000 g for 90 minutes. The supernatant solution from this centrifugation is recovered and dialyzed against Buffer D with three changes of the buffer.

Step 4. Bio-Rex 70 (Na$^+$ form, minus 400 mesh) is suspended in Buffer D, and the pH of this mixture is adjusted to 7.2 with HCl. The mixture is centrifuged at 15,000 g for 5 minutes, and the sedimented resin is equilibrated with Buffer D for 30 minutes and again centrifuged to remove the buffer. The resin is then mixed with the dialyzed solution obtained from Step 3 in a protein-to-resin ratio of approximately 1/100 (w/w). This mixture is stirred for 10 minutes and is allowed to settle for another 30 minutes. The unadsorbed acidic proteins are collected by centrifugation. The resin is washed with Buffer D, centrifuged, and the wash is combined with the first supernatant solution. The combined solution is dialyzed against 3 volumes of Buffer B for 2 hours and lyophilized. The lyophilized powder is taken up in a small volume of distilled water and dialyzed against several changes of Buffer E.

Step 5. A O-(diethylaminoethyl) cellulose (DEAE-cellulose) column is preequilibrated with Buffer E onto which the nonhistone protein sample obtained in Step 4 is charged. The column with its adsorbed proteins is then washed with two bed volumes of Buffer F and eluted with Buffer G until the absorbance at 280 nm of the eluate is less than 0.05. Both the wash and the eluate are discarded. The column is finally eluted with Buffer H. The eluate, representing NHP-I, is collected and the NaCl removed by dialysis against Buffer I. The NHP-I is then concentrated by negative pressure or by Sephadex G-200 and finally clarified at 20,000 g before use.

B. Assay

The activity of NHP-I is measured by its stimulation of template activity of chromatin in an RNA polymerase reaction. The source of chromatin can be either the same tissue from which NHP-I is prepared (homologous) or a different tissue (heterologous). The homologous chromatin is preferred because it gives a greater activation than heterologous chromatin (2,3,6).

The stimulatory activity of NHP-I varies among different preparations as well as different tissues. The activation is dose-related, increasing with the amount of NHP-I added to the chromatin-templated RNA polymerase system. Both bacterial RNA polymerase and eukaryotic RNA polymerase II (α-amanitin sensitive) can be used to provide the RNA synthesizing system in vitro. Two assay methods, one employing Micrococcus luteus RNA polymerase (19) and the other using rat liver nucleoplasmic RNA polymerase (20), are given here as examples.

1. Assay in M. luteus RNA Polymerase Reaction

The purification of RNA polymerase from M. luteus follows the procedure of Nakamoto et al. (19), and the assay is essentially that of the same authors. The reaction mixture, in a final volume of 0.50 ml, contains: Tris-HCl (pH 8.0), 50 μmol; KCl, 30 μmol; MgCl$_2$, 2.5 μmol; MnCl$_2$, 1.25 μmol; ATP, CTP, GTP, and UTP, one of which is radioactively labeled, 400 nmol each; chromatin, 20 μg equivalent of DNA; RNA polymerase, 1 unit; NHP-I, 20–100 μg. The chromatin template is mixed with NHP-I, and the other components, except MnCl$_2$ and the enzyme, are then added. The reaction is initiated by addition of the enzyme and MnCl$_2$, and the reaction mixture is incubated for 10 minutes at 30 °C. At the end of incubation, the reaction is stopped by chilling on ice and addition of 0.10 ml of cold 50% trichloroacetic acid. Processing of acid-insoluble radioactive precipitate for counting is carried out as routinely performed in laboratories.

2. Assay in Rat Liver RNA Polymerase II Reaction

Form II RNA polymerase from rat liver is partially purified by the procedure of Chesterton and Butterworth (20). The assay is conducted at 37 °C for 30 minutes in the following reaction mixture in a final volume of 0.25 ml: Tris-HCl (pH 8.0), 14 μmol; KCl, 2 μmol; MnCl$_2$, 0.4 μmol; NaF,

1.5 μmol; β-mercaptoethanol, 0.4 μmol; phosphoenol pyruvate, 1.0 μmol; pyruvic kinase, 5 μg; ATP, CTP, and GTP, 150 nmol each; [^{3}H]UTP, 15 nmol; chromatin, 20 μg equivalent of DNA; RNA polymerase, 0.5 units; NHP-I, 20–100 μg.

C. Properties

The heterogeneity of NHP-I, some of the characteristics of the RNA product synthesized from chromatin activated by NHP-I, and the ability of NHP-I to activate transcription from homologous as well as heterologous chromatins all have been described previously. Regarding the latter, it also has been reported (8) that nonhistone proteins from carrot cells of induced embryonic stages have a greater stimulatory activity than those from non-induced cells in promoting the transcription of chromatin.

Nonhistone protein fractions also have been shown to restore histone-inhibited transcription from DNA (8,21–23). In the assay procedure for NHP-I, a prior interaction between the nonhistone proteins and DNA template before the addition of histone results in a more effective counter-action toward histone inhibition (8,22). If NHP-I is added after DNA and histone have already been interacted, the restoration of histone-inhibited template activity of DNA for RNA synthesis is still effective but requires higher concentrations of the nonhistone proteins than those required if the nonhistone proteins are added after histone and DNA have reacted first (4, 8). This suggests that NHP-I stimulates transcription from chromatin by acting on chromatin proteins, perhaps by dissociating or displacing histones.

The stimulatory action of NHP-I in promoting transcription from chromatin is tissue-specific. When chromatin-templated RNA synthesis is activated by NHP-I prepared from a different tissue, the synthesized product contains RNA species resembling the RNA transcribed from chromatin of the heterologous tissue, as determined by DNA–RNA hybridization (2,3,6). The RNA product transcribed from the activated chromatin is capable of stimulating amino acid incorporation into protein in a cell-free ribosomal system (1).

III. NHP-II. The Nonhistone Proteins That Stimulate Transcription of DNA

The activation of transcription from DNA by NHP-II was originally demonstrated by Teng et al. (13) and by Shea and Kleinsmith (11) in rat liver nuclei. The former workers used phenol extraction and the latter used

1.0 M NaCl extraction, Bio-Rex 70 treatment, and calcium phosphate gel adsorption to isolate the active nonhistone proteins. In our laboratory, we have prepared NHP-II either by the phenol extraction method or by selective binding to homologous DNA (12).

The starting material for the isolation of NHP-II is purified chromatin. There are a number of procedures for the isolation of chromatin. The advantages and disadvantages of several methods of preparation, in at least one aspect of the characteristics of chromatin (i.e., its template properties in RNA synthesis), have been discussed by DePomerai et al. (24). We have routinely used the procedure of Seligy and Miyagi (25), which is modified after Marushige and Bonner (26), to prepare chromatin from isolated nuclei. The procedure uses minimal shearing and is detailed elsewhere (12).

A.   Preparative Procedure

Buffers and solvents (all contain 1.0 mM phenylmethylsulfonyl fluoride):

Buffer A. 0.02 M Tris-HCl (pH 7.5)–0.35 M NaCl.

Buffer B. 0.01 M Tris-HCl (pH 7.0)–0.4 M NaCl.

Buffer C. 0.1 M Tris-HCl (pH 8.4)–0.01 M EDTA–0.14 M β-mercaptoethanol.

Buffer D. 0.1 M acetic acid–0.14 M β-mercaptoethanol.

Buffer E. 0.05 M acetic acid–9.0 M urea–0.14 M β-mercaptoethanol.

Buffer F. 0.01 M Tris-HCl (pH 8.4)–8.6 M urea–0.01 M ethylenediaminetetraacetic acid (EDTA)–0.14 M β-mercaptoethanol.

Buffer G. 0.01 M Tris-HCl (pH 7.4)–0.001 M EDTA.

Buffer H. 0.01 M Tris-HCl (pH 7.4)–0.05 M NaCl–0.001 M EDTA.

Buffer I. 0.01 M Tris-HCl (pH 7.4)–0.6 M NaCl–0.001 M EDTA.

Buffer J. 0.01 M Tris-HCl (pH 7.4)–0.05 M NaCl.

Step 1.   The isolated chromatin is suspended in 1000 volumes of Buffer A with the aid of a loose-fitting Dounce homogenizer, and the mixture is stirred for 20 minutes. The chromatin suspension is centrifuged at 105,000 g for 2 hours. The pellets obtained from this centrifugation are cut into small pieces with scissors and reextracted as in the first extraction. The two extracts are combined and dialyzed against Buffer B, with several changes of the buffer, and then centrifuged at 20,000 g for 15 minutes to remove a small amount of insoluble material. The supernatant solution is then mixed with Bio-Rex 70 (Na$^+$) which has been previously equilibrated with Buffer B as described in Section II,A, Step 4. The Bio-Rex-treated nonhistone proteins are recovered by centrifugation at 12,000 g for 5 minutes and the supernatant solution is collected.

Step 2(a) (13).   The nonhistone proteins obtained from Step 1 are concentrated by lyophilization, dissolved in a minimal volume of distilled water, and dialyzed against Buffer C. An equal volume of freshly redistilled phenol saturated with Buffer C is added to the protein solution. The mixture is kept at 2–4 °C. for 10–12 hours with occasional gentle swirling. This suspension is centrifuged at 12,000 g for 10 minutes to separate phenol from the aqueous layer. The phenol phase is saved, and the aqueous phase and interphase are collected and reextracted with phenol as above. The two phenol extracts are combined and dialyzed against 100 volumes of Buffer D until the phenol phase is reduced to one-fifth its original volume. The reduction of the phenol phase to no less than one-fifth the original volume is critical because the protein will irreversibly precipitate if the volume is further reduced. The resulting phenol phase is dialyzed for 24 hours against Buffer E, followed by dialysis against Buffer F for 4 hours. The urea is removed from the sample by exhaustive dialysis against 0.02 M Tris-HCl (pH 8.0), and the sample is clarified at 20,000 g for 10 minutes prior to use.

Step 2(b). An alternative procedure for preparing NHP-II is by selective binding to homologous DNA. DNA–cellulose is prepared essentially according to Alberts and Herrick (27); the method is as follows. The DNA, purified according to a modified procedure of Marmur (28) as described elsewhere (12), is dissolved in 24 ml of Buffer G to a concentration of 5 mg DNA/ml. Eight grams of Munktell 410 cellulose, which have been washed three times with boiling ethanol and then washed successively with 0.1 N NaOH, 0.001 M EDTA, 0.01 N HCl, and thoroughly with distilled water, are mixed with the DNA solution. The resulting thick paste is spread on a watch glass, which is then covered with cheesecloth, and air-dried for 24 hours. The nearly dry mixture is ground to a fine powder and lyophilized to complete dryness. The powder is again mixed with 20 ml of DNA solution, and the above process is repeated. The dried DNA–cellulose powder is suspended in 200 volumes of Buffer G and left to stand at 2–4 °C for 24 hours with occasional swirling. The buffer is then decanted and the DNA–cellulose is washed with Buffer H until no absorbance at 260 nm is detected in the wash.

The amount of DNA bound to the cellulose is determined on an aliquot of the DNA–cellulose suspension in Buffer G. The suspension is heated in a boiling-water bath for 20 minutes, centrifuged, and the supernatant's absorbance at 260 nm is read against the buffer. An absorbance at 260 nm of 32 is taken as a DNA concentration of 1 mg/ml.

To prepare the DNA–cellulose column, the DNA–cellulose suspension is poured into a short column 3 cm in diameter and packed to a height of 2–2.5 cm. Two such columns, each containing approximately 40 mg of DNA, are prepared: one contains *E. coli* DNA–cellulose, and the other, DNA–cellulose prepared with DNA purified from homologous tissue. Prior to use, the DNA–cellulose columns are once again washed with Buffer H to insure that no free DNA is present. Approximately 10–15 mg of the nonhistone proteins, isolated as described in Step 1, in 10 ml are dialyzed against Buffer H and are initially passed through a DNA-free cellulose column which has previously been equilibrated with Buffer H to remove nonspecific adsorbing material. The nonhistone proteins are then applied to the *E. coli* DNA–cellulose column. The column is washed with Buffer H until the $A_{280\,nm}$ of the wash is less than 0.02. The nonadsorbed proteins in the wash are collected and applied to the homologous DNA–cellulose column. The column is washed with Buffer H. The flow-through and wash fractions are discarded. The column is then eluted with Buffer I at the rate of approximately 1 ml/3 minutes. Fractions of 1 ml of the eluate are collected until the $A_{280\,nm}$ of the eluate is less than 0.02. The eluate fractions are pooled, dialyzed against Buffer J, and concentrated by Sephadex G-200. Prior to use, the sample is dialyzed against 0.01 M Tris-HCl (pH 8.0) to remove the NaCl.

B. Assay

The activity of NHP-II isolated from Ehrlich ascites tumor has been assayed with RNA polymerase II purified from homologous cells (*29*). Form II RNA polymerase from other eukaryotic sources presumably can be used with similar effectiveness, but this has not been tested. NHP-II shows no activity if assayed in *M. luteus* RNA polymerase reaction (*12*). Whether this applies to any other prokaryotic RNA polymerase is not known.

The assay system of Ehrlich ascites tumor RNA polymerase II used in our laboratory is essentially that of Natori *et al.* (*30*) which varies somewhat from the rat liver enzyme assay. The reaction mixture, in a total volume of 0.25 ml, contains the following: Tris-HCl (pH 8.0), 10 μmol; $(NH_4)_2SO_4$, 12.5 μmol; $MnCl_2$, 0.75 μmol; $MgCl_2$, 1.15 μmol; β-mercaptoethanol, 1.0 μmol; ATP, CTP, and UTP, 62.5 nmol each; [^{3}H]GTP, 6.25 nmol; DNA, 5.0 μg; RNA polymerase II, 0.2–0.5 unit. The reaction is incubated at 37 °C for 30 minutes or 1 hour.

C. Properties

When analyzed by dodecylsulfate polyacrylamide gel electrophoresis, NHP-II, as prepared from Ehrlich ascites tumor chromatin either by phenol extraction or by specific DNA binding, is heterogeneous. Its molecular complexity, however, is relatively simplified as compared with the total nonhistone chromosomal proteins or the loosely bound nonhistone proteins obtained by extraction with 0.35 M NaCl. It consists of subunits mostly of molecular weights 36,000 and less and contains 0.90% (w/w) alkali-labile phosphorus (*12*). It has a higher acidic amino acids content than basic amino acid residues.

Whether NHP-II prepared from the loosely bound nonhistone chromosomal proteins is identical to the activating fractions isolated from rat liver is unknown (9,10,11,13). However, the essential characteristics of these fractions prepared from various tissues and by different methods are similar. They all contain approximately 1% phosphorus and bind selectively to, and stimulate transcription of, only homologous DNA. As mentioned earlier, only unique homologous DNA sequences are involved in the interaction with NHP-II and subsequent stimulation of RNA synthesis. The mechanism of the activation of transcription is at the initiation step of RNA chain growth (16,31). The activation apparently requires an eukaryotic RNA polymerase (12) and depends on the phosphoprotein components (11). The gross amino acid compositions of the activating fractions isolated from rat liver, calf thymus, and Ehrlich ascites tumor cells are different. This is probably due to the heterogeneous nature of the fractions or to the different methods of preparation. If it is assumed that the activator fractions from different tissues are all derived from the loosely bound nonhistone chromosomal proteins, the chemical differences in various preparations may also suggest tissues variations for NHP-II (32).

IV. NHP-III. The Nonhistone Protein That Inhibits Transcription of DNA

A. Preparative Procedure

Buffers (all contain 1.0 mM phenylmethylsulfonyl fluoride):
  Buffer A. 0.05 M Tris-HCl (pH 8.0).
  Buffer B. 0.05 M Tris-HCl (pH 8.0)–4.0 M NaCl.
  Buffer C. 0.01 M Tris-HCl (pH 8.0).
  Buffer D. 0.1 M Tris-HCl (pH 8.4).
  Buffer E. 0.02 M Tris-HCl (pH 7.0) –0.4 M NaCl.

NHP-III was isolated originally from the DNA–protein complex of Ehrlich ascites tumor chromatin (15) and, employing the same procedure, recently from calf thymus (17). The method for preparation seems to be applicable to other tissues. The starting material is chromatin.

Step 1. The isolated chromatin is suspended in 500 volumes of Buffer A with the aid of a Dounce homogenizer. To this suspension an equal volume of Buffer B is added slowly with stirring. The stirring is continued for 6–12 hours, and the mixture is centrifuged at 76,000 g for 90 minutes. The viscous supernatant solution is collected and dialyzed against 13 volumes of Buffer C. During dialysis, the content of the dialysis tubing is occasionally mixed. Dialysis is carried out for 6–12 hours until heavy precipitation of the DNA–protein complex occurs. The DNA–protein precipitate is recovered after centrifugation of the dialyzed mixture at 76,000 g for 1 hour.

Step 2.　The DNA–protein pellets are cut into small pieces, suspended in 1000 volumes of 0.4 N H_2SO_4, blended at 4400 rpm in an Omni-mixer, and stirred in a beaker for 1 hour. The acid-soluble proteins, mostly histones, are removed by centrifugation at 20,000 g for 10 minutes. The pellet is reextracted with H_2SO_4 twice in a similar manner, except that the blending is omitted.

Step 3.　The DNA–protein complex extracted 0.4 N H_2SO_4 is washed twice with 10 volumes of Buffer D and extracted with phenol as described in Section III, A, Step 2(a), except that the final removal of urea is accomplished by dialysis against Buffer E, with three changes of the buffer. The dialyzed sample is then treated with Bio-Rex 70, as described in Section II, A, Step 4. The NHP-III thus obtained is dialyzed against Buffer C and concentrated by Sephadex G-200. NHP-III may be further purified by DNA-cellulose chromatography according to Section III, A, Step 2(b), but this is generally not necessary.

In the case of Ehrlich ascites tumor cells, large amounts of starting ascites fluid from 1000 mice are required to give a yield of 50–100 μg of NHP-III. Yield from calf thymus is approximately 100 μg/100 gm of tissue.

B.　Assay

NHP-III has thus far been assayed with only form II RNA polymerase purified from homologous tissues. RNA polymerase from other eukaryotic sources has not been investigated, but it is presumed to be equally operative. The assay system using Ehrlich ascites tumor RNA polymerase II is described in Section III, B.

C.　Properties

The NHP-III as prepared above is nearly homogeneous when subjected to polyacrylamide gel electrophoresis. The NHP-III isolated from Ehrlich ascites tumor cells contains 2.7% alkali-labile phosphorus and is isoelectrically focused at pH 5.3, having an acidic-to-basic amino acid residues ratio of 1.42. It consists of one polypeptide of molecular weight 10,273, which is determined by dodecylsulfate polyacrylamide gel electrophoresis and calculated from its amino acid composition (15). The NHP-III prepared from calf thymus has a sedimentation coefficient of $s_{20,w} = 3.0$. Equilibrium ultracentrifugation of the nonhistone protein yielded a molecular weight of 30,800 $\pm$ 2,400. It consists of two subunits, with approximate molecular weights of 16,000 and 13,000 (*17*). The amino acid composition of the calf thymus complete protein is quite different from that of the Ehrlich ascites tumor cells, notably in the relative molar contents of proline, glycine, isoleucine, leucine, tyrosin, and phenylalanine. The NHP-III from both tissues lacks cystine and methionine which, if present, are only in trace amounts.

As mentioned previously, NHP-III inhibits the *in vitro* transcription of DNA at the initiation step; only reiterated sequences in DNA interact with NHP-III and are inhibited for transcription (*15,17*).

ACKNOWLEDGMENTS

This work was supported in part by research grants from the National Foundation–March of Dimes and the National Institute of Health (HD-09443).

REFERENCES

1. Kamiyama, M., and Wang, T. Y., *Biochim. Biophys. Acta* **228**, 563 (1971).
2. Kostraba, N. C., and Wang, T. Y., *Biochim. Biophys. Acta* **262**, 169 (1972).
3. Kostraba, N. C., and Wang, T. Y., *Cancer Res.* **32**, 2348 (1972).
4. Wang, T. Y., *Exp. Cell Res.* **61**, 455 (1970).
5. Wang, T. Y., *Exp. Cell Res.* **69**, 217 (1971).
6. Wang, T. Y., and Kostraba, N. C., *in* "The Role of RNA in Reproduction and Development" (M. C. Niu and S. Segal, eds.), p. 324. North-Holland Publ., Amsterdam, 1973.
7. Teng, C. S., and Hamilton, T. H., *Proc. Natl. Acad. Sci. U.S.A.* **63**, 465 (1969).
8. Matsumoto, H., Gregor, D., and Reinert, H., *Phytochemistry* **14**, 41 (1975).
9. Teng, C. T., Teng, C. S., and Allfrey, V. G., *Biochem. Biophys. Res. Commun.* **41**, 690 (1970).
10. Kleinsmith, L. J., Heidema, J., and Carroll, A., (*London*) **266**, 1025 (1970).
11. Shea, M., and Kleinsmith, L. J., *Biochem. Biophys. Res. Commun.* **50**, 473 (1973).
12. Kostraba, N. C., Montagna, R. A., and Wang, T. Y., *J. Biol. Chem.* **250** 1548 (1975).
13. Teng, C. S., Teng, C. T., and Allfrey, V. G., *J. Biol. Chem.* **246**, 3597 (1971).
14. Conrad, D., and Wang, T. Y., Unpublished results (1976).
15. Kostraba, N. C., and Wang, T. Y., *J. Biol. Chem.* **250**, 8938 (1975).
16. Kostraba, N. C., Montagna, R. A., and Wang, T. Y., *Biochem. Biophys. Res. Commun.*, **72**, 334 (1976).
17. Wang, T. Y., Kostraba, N. C., and Newman, R. S., *Progr. Nucleic Acid Res.* **19**, 447 (1977).
18. Chauveau, J., Moulé, Y., and Roueller, C., *Exp. Cell Res.* **11**, 317 (1956).
19. Nakamoto, T., Fox, F. C., and Weiss, S., *J. Biol. Chem.* **239**, 169 (1964).
20. Chesterton, C. J., and Butterworth, P. H. W., *FEBS Lett.* **15**, 181 (1971).
21. Langan, T., *in* "Regulation of Nucleic Acid and Protein Synthesis" (V. V. Koningsberger and L. Bosch, eds.), p. 233. Elsevier, Amsterdam, 1967.
22. Spelsberg, T. C., and Hnilica, L. S., *Biochim. Biophys. Acta* **195**, 63 (1969).
23. Wang, T. Y., *Exp. Cell Res.* **53**, 288 (1968).
24. DePomerai, D. I., Chesterton, C. J., and Butterworth, P. H. W., *Eur. J. Biochem.* **46**, 461 (1974).
25. Seligy, V., and Miyagi, M., *Exp. Cell Res.* **58**, 27 (1969).
26. Marushige, K., and Bonner, J., *J. Mol. Biol.* **15**, 160 (1966).

27. Alberts, B. M., and Herrick, G., *in* "Methods in Enzymology," Vol. 21: Nucleic Acids, Part D (L. Grossman and K. Moldave, eds.) p. 198. Academic Press, New York, 1971.
28. Marmur, J., *J. Mol. Biol.* **3**, 208 (1961).
29. Chan, J. Y. H., Loor, R. M., and Wang, T. Y., *Arch. Biochem. Biophys.* **173**, 564 (1976).
30. Natori, S., Takeuchi, K., and Mizuno, D., *J. Biochem. (Tokyo)* **73**, 345 (1973).
31. Kostraba, N. C., and Wang, T. Y., unpublished results.
32. Burckard, J., Mazen, A., and Champagne, M., *Biochim. Biophys. Acta* **405**, 434 (1975).

Chapter 20

Fractionation of Nonhistone Chromosomal Proteins Utilizing Hydroxyapatite Chromatography

A. J. MacGILLIVRAY*

*Beatson Institute for Cancer Research,
Wolfson Laboratory for Molecular Pathology,
Bearsden, Glasgow, Scotland*

I. Introduction

Much interest has centred recently on the nonhistone proteins of chromatin since they have been found to contain many of the biological activities of the nucleus, e.g., enzymes of nuclear metabolism such as the nucleic acid polymerases. However, perhaps more biological significance has been attached to the findings that these proteins have activities which appear to control the synthesis of RNA. These properties include the ability to affect the template activity of both DNA and chromatin *in vitro* and, in particular, the control of the expression of specific genes (*1,2*). Work in this laboratory has been directed toward the study of eukaryotic gene-regulatory proteins, and one of our first steps involved the development of procedures to isolate and fractionate the nonhistone proteins of chromatin.

Perhaps the most efficient of the methods currently available for the preparation of nonhistone proteins involves their isolation from chromatin which has been dissociated in solutions of high ionic strength containing denaturants such as urea (*3*). Our approach has been to conduct such a fractionation of chromatin using a column of hydroxyapatite, whereby at near–neutral pH the application of salt–urea dissociated chromatin results in the basic histones being unretained and the nonhistone proteins (and RNA) being adsorbed by the column. By using appropriate concentrations of phosphate buffer in the eluting medium, the nonhistone proteins can be

Present address: Biochemistry Laboratory, School of Biological Sciences, University of Sussex, Falmer, Brighton, Sussex BN1 9QG, England.

eluted while the DNA is still retained by the hydroxyapatite. Moreover, a preliminary fractionation of the nonhistone proteins can be obtained by using stepwise increase in phosphate concentrations (3–5).

II.   Preparation of Hydroxyapatite

Hydroxyapatite is prepared from $Na_2HPO_4 \cdot 2\ H_2O$ (Merck 6580) and $CaCl_2$ (Merck 2382) according to the method of Bernardi (6) and stored at 4° in 1 mM sodium phosphate (pH 6.8) containing a few drops of chloroform. It has been our experience that hydroxyapatite prepared by other procedures or obtained from some commercial sources either possesses slow flow rates or gives unsatisfactory separations. However, another laboratory (7) has reported satisfactory results with hydroxyapatite from a supplier. Prior to use the hydroxyapatite is washed and defined by sedimentation after gentle suspension in 2 M NaCl–5 M urea–1 mM sodium phosphate (pH 6.8) containing 2 mM Tris.[1] After repeating the procedure several times columns of hydroxyapatite are packed and equilibrated at room temperature with the same solution.

III.   Preparation of Nuclei and Chromatin

For most purposes nuclei are prepared from tissues and cells by sucrose and Triton X-100 procedures described by Birnie (7a). Except where stated all procedures are carried out at 4°. Chromatin is prepared by extraction of nuclei by means of homogenization in 0.14 M NaCl–0.05 M Tris-HCl (pH 7.5)–5 mM ethylenediaminetetraacetic acid (EDTA)–0.1 Mm phenylmethylsulforylfluoride (PMSF).[2] After stirring for 20 minutes the suspension is centrifuged at 15,000 g for 15 minutes. This procedure is repeated twice.

IV.   Chromatography on Hydroxyapatite

The final pellet of chromatin is homogenized in sufficient 2 M NaCl–5 M urea–1 mM sodium phosphate (pH 6.8)–2 mM Tris[3]–0.1 mM PMSF to give approximately 0.5 mg of DNA/ml. After centrifugation at 15,000 g for 15

---

[1] The pH of the final solution is adjusted to 6.8 using concentrated HCl.
[2] PMSF acts as a protease inhibitor (7b). A 10 mM stock solution is prepared in isopropanol.
[3] The pH of the final solution is adjusted to 6.8 using concentrated HCl.

minutes the pellet is treated in the same way and the two extracts combined. The solution is sonicated[4] for two periods of 15 seconds each and then centrifuged at 15,000 g for 15 minutes to remove traces of residual material.

The supernatant (50–100 ml; $OD_{260} = 5$–10) is then applied to the equilibrated column of hydroxyapatite.[5] Solutions containing up to 15 mg of DNA are normally applied to a 25 × 1.6 cm column, and those containing between 15 and 50 mg are normally applied to a column of dimensions 25 × 2.2 cm. The flow rate is set between 5 and 10 ml/hour, and with concentrated chromatin solutions it is sometimes necessary to occasionally gently stir the top of the column with a glass rod.

After the sample has entered the column a small volume of the salt–urea solution is added to the top of the hydroxyapatite and allowed to sink in. This solution is then applied to the column until, as seen by monitoring fractions at 280 nm, the histone fraction, hydroxyapatite fraction 1 (HAP1), is eluted. The nonhistone proteins are then obtained by passing the following solutions through the column:

2 M NaCl–5 M urea–2 mM Tris–50 mM phosphate (pH 6.8)   (HAP2)
2 M NaCl–5 M urea–2 mM Tris–200 mM phosphate (pH 6.8)   (HAP3)
2 M guanidine hydrochloride–2 mM Tris–200 mM phosphate (pH 6.8) (HAP4)

If necessary a fraction containing DNA as the major constituent can then be eluted by increasing the phosphate concentration in the last solution to 500 mM (HAP5).

Analyses of the fractions prepared from mouse leukemic Friend cells are given in Table I. Fraction HAP1 represents the bulk of the chromatin protein which electrophoretic and amino acid analyses show to consist of histones (4, 5). However, labeling of nonhistone proteins with [³H]tryptophan shows that some 15% of the labeled proteins are eluted in this fraction. Two-dimensional gel electrophoresis confirms the presence of high-molecular-weight basic proteins in fraction HAP 1 (5). Fraction HAP 2 consists of the major part of the nonhistone proteins, accounting for some 12% of the total chromatin proteins, together with a small quantity of RNA. Both fractions HAP 3 and 4 contain protein and a considerable quantity of RNA, the former constituent representing 9 and 12% of the tryptophan-labeled nonhistone proteins, respectively. The DNA is strongly retained by the column, only traces being found in fractions HAP3 and 4 after thymidine labeling. Hence the bulk of the DNA is eluted along with RNA and a small quantity of protein in fraction HAP5.

This fractionation of chromatin proteins appears to depend largely on the degree of posttranslational modification of the proteins by, e.g., phospho-

[4] An M.S.E. Ultrasonic Power Unit set at 1.5 A is used.
[5] The column is normally run at room temperature. However, by substituting potassium for sodium salts we have achieved identical separations and recoveries at 4° (5).

TABLE I

Analyses of Hydroxyapatite Fractions of Mouse Leukemic Friend Cell Chromatin

Fraction	Phosphate concentration (mM)	Protein			$[^{32}P]$ADPR^d			RNA		DNA	
		Chemical^a	$[^3H]$tryptophan^{a,b}	Acidic/basic amino acids	$^{32}P^c$ (cpm/mg $\times 10^{-6}$)	(cpm/mg $\times 10^{-4}$)	Mean chain length	Chemical^e	$[^3H]$uridine^{e,f}	Chemical^e	$[^{14}C]$thymidine^{a,b}
HAP1	1	72	15	0.5	0.1	1.0	1.5	0	0	0	0
HAP2	50	12	41	1.3	3.3	21.6	5.8	0.03	0.02	0	0
HAP3	200	3	9	1.3	13.5	38.3	11.5	0.5	1.15	0	1.2
HAP4	200	4	12	1.2	5.2	81.2	8.0	0.75	2.9	0	1.2
HAP5	500	1	3	nd^g	4.5	nd	nd	1	nd	100	92

^{a} Percentage in each fraction.

^{b} Isotope labeling for one generation time.

^{c} Isotope labeling for nuclei with γ-$[^{32}P]$ATP *in vitro*.

^{d} Isotope labeling of nuclei with $[^{32}P]$nicotinamide adenine dinucleotide (NAD) *in vitro*.

^{e} Ratio using protein concentration as 1.

^{f} Isotope for 30-minute pulse.

^{g} nd = not determined.

rylation and adenosine diphosphoribosylation (5, 8). As indicated by specific activities and the adenosine diphosphoribose (ADPR) chain length, fractions HAP1–3 appear to elute in an order of increasing modification. Being intermediate between HAP2 and 3 fractions, HAP4 proteins are an exception to the rule, and it is possible that these are aggregated polypeptides requiring the denaturing powers of guanidine hydrochloride for their removal from the column. The small quantity of protein eluting with the DNA in fraction HAP5 also appears to be highly phosphorylated, but the large amount of nucleic acid has prevented further characterization of this material.

Yields of protein from the hydroxyapatite column are high. The total recovery from the type of experiment shown in Table I in which chromatin equivalent to 10 mg of DNA was applied to a 25 × 1.6 cm column was over 90%. Lower yields are experienced when the procedure is scaled up. On the whole this affects the recovery of the nonhistone proteins, a factor perhaps associated with losses due to adsorption–desorption effects on hydroxyapatite.

V.   Removal of Polynucleotides

As described above nucleic acids, particularly RNA, contaminate the nonhistone protein fractions obtained from hydroxyapatite columns. In addition, free poly-ADPR of chain length up to 20 units also coelutes with the nonhistone proteins (8). It is often necessary to remove these polynucleotides since they can interfere with certain separation techniques and also with nucleic acid hybridization procedures. The following procedure removes nucleic acids and allows some degree of concentration of the proteins.

Hydroxyapatite column fractions are concentrated to not more than 1 mg of protein/ml by dialysis against Carbowax[6] and then dialyzed at room temperature against 8 M urea–0.2 M Tris = HCl (pH 8)–0.2 mM PMSF. The samples are mixed with an equal volume of 60% (w/w) CsCl, and 5-ml volumes are centrifuged at 120,000 g_{avg} for 35–44 minutes at 4 °C. The proteins, free of nucleic acids, are found in the top 1.5 ml of the gradient. After this volume has been collected by unloading the gradient the CsCl is removed by dialysis against a suitable buffer, e.g., 8 M urea–0.3 M Tris-HCl (pH 8.3) which is the medium used for reduction of the proteins prior to isoelectric focusing (9).

[6]Carbowax (mol wt 15,000–20,000) is supplied by G. T. Gurr, High Wycombe, Bucks, England.

VI.   Characterization of Hydroxyapatite Fractions

A.   Electrophoresis

We have characterized the proteins in the hydroxyapatite fractions by two-dimensional gel electrophoresis employing a combination of isoelectric focusing and sodium dodecyl sulfate (SDS)–electrophoresis in polyacrylamide gels (2, 5, 9). Figure 1 shows the polypeptide pattern of HAP1–4 fractions of mouse liver chromatin after nucleic acids have been removed by centrifugation in CsCl–urea and the proteins subsequently reduced.

The HAP1 fraction is seen to consist of the very basic low-molecular-weight histones together with a few species of high-molecular-weight nonhistone proteins, one group of which has almost neutral isoelectric points (Fig. 1A). The HAP2 proteins, on the other hand, show a wide range of both molecular weights and isoelectric points, amounting to some 40 distinct polypeptide species (Fig. 1B). The HAP3 proteins consist of a more homogeneous group, all with molecular weights greater than 30,000. The major polypeptide species have isoelectric points near pH 6 (Fig. 1C). A similar situation exists in relation to the HAP4 fraction, except that the major protein species have isoelectric points near neutrality (Fig. 1D).

An interesting feature of the technique is that there appears to be little overlap of the major species from one fraction to another. Other experiments (2,5,9) show that the HAP2 proteins consist of a mixture of phosphorylated and nonphosphorylated species, the major phosphoproteins being acidic and basic components, both of which are of low molecular weight. On the other hand, in the HAP3 and 4 fractions the major protein species appear to be the most highly phosphorylated.

B.   Biological Assay

We have attempted to locate the activity responsible for the expression of the globin genes by reconstituting mouse embryo liver chromatin from DNA, histones, and individual hydroxyapatite fractions of nonhistone proteins (10). After transcription of the chromatin by E. coli RNA polymerase, the RNA produced was tested for the presence of globin gene transcripts using complementary DNA prepared from mouse globin mRNA as a probe. The most efficient expression of the globin genes was given by chromatin containing HAP2 proteins, followed by that containing HAP3 proteins. This result indicates that the proteins which regulate the expression of the globin genes are found among the bulk of the nonhistone proteins in the HAP2 fraction with a small amount also being eluted with the HAP3 proteins.

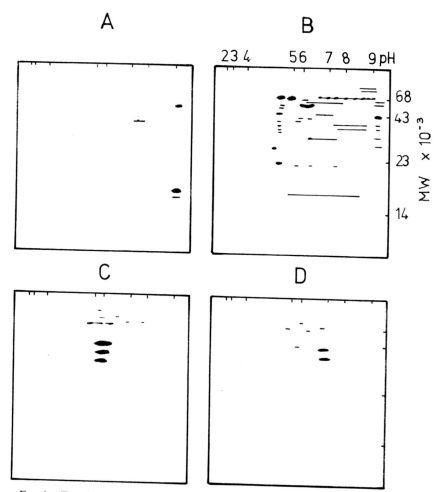

FIG. 1. Two-dimensional gel electrophoresis of protein fractions from mouse liver chromatin chromatographed on hydroxyapatite. The horizontal axis represents isoelectric focusing in the first dimension; the pH gradient is indicated as shown in (B). The vertical axis shows separation in the second dimension using gel electrophoresis in SDS; the position of marker proteins of known molecular weight is given in (B). Proteins stained with Coomassie blue are represented by lines or spots. (A) HAP1; (B) HAP2; (C) HAP3; (D) HAP4.

In collaboration with others we have also investigated certain immunological properties of chromatin nonhistone proteins (11). It is well established that the infection of human lymphoid cells by Epstein-Barr (E.B.) virus is associated with the presence of a specific Epstein–Barr nuclear antigen (EBNA) to which the patient raises antibodies. Using IgG isolated from anti-

EBNA human serum we have tested hydroxyapatite fractions of chromatin from an E.B. cell line for the presence of the antigen. In this case most of the immunological reactivity was detected in the HAP4 fraction, HAP3 and HAP2 proteins giving much-reduced reactions.

VII. Conclusion

The hydroxyapatite procedure is a relatively simple one-column procedure which not only separates the nonhistone proteins from the histones in high yields, but also provides a preliminary fractionation of these proteins. Electrophoretic analyses show that the three subfractions of the nonhistone proteins consist of different polypeptide species and, moreover, other analyses indicate that different biological activities can be separated by this procedure. It should also be noted that other laboratories have used variations of the hydroxyapatite procedure to isolate nonhistone proteins.[7]

REFERENCES

1. MacGillivray, A. J., and Rickwood, D., in "Biochemistry of Differentiation and Development" (J. Paul, ed.), Vol. 9, p. 301. Med. Tech. Publ. Co., Oxford, 1974.
2. MacGillivray, A. J., in "The Organization and Expression of the Eukaryotic Genome" (E. M. Bradbury and K. Javaherian, eds.), Academic Press, New York, in press.
3. MacGillivray, A. J., in "Subnuclear Components" (G. D. Birnie, ed.), p. 209. Butterworth, London, 1976.
4. MacGillivary, A. J., Cameron, A., Krauze, R. J., Rickwood, D., and Paul, J., Biochim. Biophys. Acta 277, 384 (1972).
5. Rickwood, D., and MacGillivray, A. J., Eur. J. Biochem. 51, 593 (1975).
6. Bernardi, G., in "Methods in Enzymology, Vol. 21: Nucleic Acids, Part D (L. Grossman and K. Moldave, eds.), p. 95. Academic Press, New York, 1971.
7. Appels, R., Bolund, L., and Ringertz, N. R., J. Mol. Biol. 87, 339 (1974).
7a. Birnie, G., Methods Cell Biol. 17 (in press).
7b. Nooden, L. D., van der Brock, H. W. J. and Sevall, J. S., FEBS Lett. 29, 326 (1973).
8. Rickwood, D., MacGillivray, A. J., and Whish, W. (in press).
9. MacGillivray, A. J., and Rickwood, D., Eur. J. Biochem. 41, 181 (1974).
10. Gilmour, R. S., and MacGillivray, A. J., Proc. Orono Meet. Soc. Dev. Biol. (in press).
11. Brown, T. D. K., Rickwood, D., MacGillivray, A. J. Klein, G., and Paul, J. (in press).
12. Bluthmann, H., Mrozek, S., and Gierer, A., Eur. J. Biochem. 58, 315 (1975).

[7]Appels et al. (7) elute the nonhistone proteins in two fractions. The first is obtained by washing the column with 200 mM phosphate in salt–urea and would be equivalent to our HAP2 and HAP3 fractions. The second fraction is then eluted using 0.5 M NaCl–5 M urea–200 mM phosphate–0.2% SDS and is probably similar to our HAP4 fraction. Bluthmann et al. (12) obtain two histone and four nonhistone protein fractions by eluting the hydroxyapatite column with 5 M urea containing different concentrations of both NaCl (0.45–2 M) and phosphate (1–175 mM).

Chapter 21

Isolation of Nuclear Proteins Associated with the Nuclear Pore Complex and the Nuclear Peripheral Lamina of Rat Liver

ROBERT PETER AARONSON

Department of Microbiology,
Mt. Sinai School of Medicine,
New York, New York

I. Introduction

Nuclear pore complexes are ubiquitous (*1*) morphologically well-characterized (*2,3*) organelles situated in pores (*4*) characteristic of the bileaflet membrane surrounding the cell nucleus. The nuclear peripheral lamina (*5*) has been described less often and occurs in a variety of forms (*5–9*) as a layer of relatively constant thickness subsuming the inner nuclear membrane. Although the presence of a nuclear peripheral lamina has not been reported in rat liver nuclei, a subnuclear fraction exhibiting laminas with characteristics consistent with a nuclear peripheral lamina can be isolated from rat liver (*10*). The laminas extend for several microns, do not have a unit membrane structure, and exhibit nuclear pore complexes attached in a specific orientation. This subnuclear fraction is at least 95% protein, and is made up of approximately equal amounts of three polypeptide species as shown upon analysis by sodium dodecyl sulfate (SDS)–polyacrylamide gel electrophoresis. The protein of the nuclear pore complex–peripheral lamina fraction represents approximately 5% of the total protein of rat liver nuclei.

Isolation of the subnuclear fraction containing proteins associated with the nuclear pore complexes and the peripheral lamina is achieved in three steps: isolation of pure nuclei by an aqueous procedure (*11*), preparation of intact nuclear envelopes by deoxyribonuclease digestion of chromatin under alkaline conditions (*12*), and solubilization of lipid (*13*) and contaminating chromatin. The original procedure (*10*) relied on the use of a high concentration of $MgCl_2$ to solubilize the remaining chromatin. High concentrations of NaCl are equally effective (G. Blobel, personal communication). An alternate procedure is also presented utilizing the polyanion polyvinylsulfate during DNase digestion. Inclusion of the polyvinyl sulfate allows complete hydrolysis of the DNA, thus permitting isolation of the nuclear pore complex–fibrous lamina subfraction without subjecting it to high salt concentrations.

II. Method

A.   Isolation of Rat Liver Nuclei (*11*)

Four 150-gm male rats are fasted overnight (approximately 18 hours) and sacrificed by decapitation using a Harvard guillotine. The livers are removed, rinsed briefly with ice-cold deionized water, and then minced in ice-cold 0.25 M sucrose–50 mM Tris-Cl (pH 7.5)–25 mM KCl–5 mM $MgCl_2$ (0.25S–TKM). All further steps in isolating nuclei are performed at 4° or in ice-cold buffer. The pieces of liver are rinsed twice and then homogenized with approximately 2 volumes of 0.25S–TKM in 40-ml batches with five up-and-down strokes of a loose-fitting Teflon-pestle Potter–Elvehjem homogenizer. The homogenate is filtered through a nylon monofilament screen with 80-μm openings (Nitex HC3-100, Tetko, Inc., Elmsford, New York). Then 80 ml of the homogenate are mixed well with 160 ml of 2.3 M sucrose–50 mM Tris-Cl (pH 7.5)–25 mM KCl–5 mM $MgCl_2$ (2.3S–TKM). The mixture is distributed equally to six cellulose nitrate tubes (35 ml per tube) which fit the Beckman SW27 rotor. Then 5 ml of 2.3S–TKM are added at the top of each tube. As the 2.3S–TKM settles to the bottom of the tube a small amount of mixing with homogenate occurs so that the resulting cushion does not have a sharp boundary. The tubes are then centrifuged for 1 hour at 25,000 rpm at 4°.

The nuclei form a white pellet. The supernatant is poured off along with the congealed layer of red cells, cell debris, and mitochondria at the top of the tubes. The tube walls are wiped clean and the pellet of nuclei is gently disrupted with a spatula and resuspended in a small volume of 2.3S–TKM using a Vortex mixer. The suspension of nuclei is diluted to 40 ml with 0.25S–TKM and centrifuged at 1000 rpm for 20 minutes at 4° in a Beckman JS-13 rotor ($g_{av} = 100$). The nuclei are resuspended in 40 ml of 0.25S–TKM, and a 20 μl aliquot is diluted to 1 ml with 0.25S–TKM. The absorbance of the dilute nuclear suspension at 260 nm is approximately 0.25 OD.

The nuclear suspension is poured slowly into another tube, leaving small clumps of aggregated nuclei behind, and then centrifuged at 1000 rpm for 20 minutes as above. The supernatant is removed by aspiration leaving a loose white pellet of approximately 500 A_{260} units of nuclei. There are 3.0 ± 0.3 × 10⁶ nuclei per A_{260} unit.

B.   Preparation of Nuclear Envelopes from Isolated Rat Liver Nuclei (12)

To the loose pellet of nuclei add, slowly initially with gentle mixing, 125 ml of 0.25 M sucrose–10 mM Tris-Cl (pH 8.5)–0.1 mM MgCl$_2$–1 μg/ml bovine pancreatic DNase I at 23°. Swirl gently at 5-minute intervals for 15 minutes at 23° before adding 125 ml of ice-cold deionized water. Centrifuge the partially digested suspension of nuclei at 3000 rpm for 10 minutes at 4° in the Beckman JS-13 rotor ($g_{av} = 900$). Remove the supernatant by aspiration. Gently resuspend the pellet with 125 ml of 0.25 M sucrose–10 mM Tris-Cl (pH 7.5)–0.1 mM MgCl$_2$–1 μg/ml DNase I at 23°. Swirl gently at 5-minute intervals for 20 minutes before adding 125 ml of ice-cold deionized water. Centrifuge at 3000 rpm for 15 minutes at 4° as above. The supernatant is removed by aspiration leaving a small white pellet of nuclear envelopes.

C.   Solubilization of Lipid and Chromatin (13, 14)

1. The pellet of nuclear envelopes is carefully suspended with 2 ml of ice-cold 0.25S–TKM. To this add slowly 0.5 ml of ice-cold 10% Triton X-100 and let stand at 0° for 10 minutes. Centrifuge at 3000 rpm for 15 minutes at 4° as above. Remove the supernatant by aspiration leaving the small pellet of phospholipid-depleted material.

2. The pellet is suspended with 5 ml of ice-cold 0.25S–TKM. Add 2.5 ml of cold 1 M MgCl$_2$ or 5 ml of cold 2 M NaCl. Centrifuge at 11,000 rpm for 20 minutes at 4° in the JS-13 rotor ($g_{av} = 12,000$). The pellet consisting of the nuclear pore complex–peripheral lamina contains up to 3 mg of protein.

D.   Alternate Procedure for the Preparation of Nuclear Envelopes from Isolated Rat Liver Nuclei and Detergent-Solubilization of Phospholipid

1. To the loose pellet of nuclei add, slowly initially with gentle mixing, 125 ml of 10 mM Tris-Cl (pH 8.5)–0.1 mM MgCl$_2$–0.1 mM CaCl$_2$–1 μg/ml bovine pancreatic DNase I at 23°. Immediately add 2.5 ml of 10 mg/ml poly-vinylsulfate. Swirl at 5-minute intervals for 30 minutes. Centrifuge at 3000 rpm for 30 minutes at 4° in the Beckman JS-13 rotor ($g_{av} = 900$). Suspend the pellet with 30 ml of 10 mM Tris-Cl (pH 8.5)–0.1 mM MgCl$_2$–0.1 mM CaCl$_2$ and centrifuge as above. Remove the supernatant by aspiration leaving a pellet of nuclear envelopes.

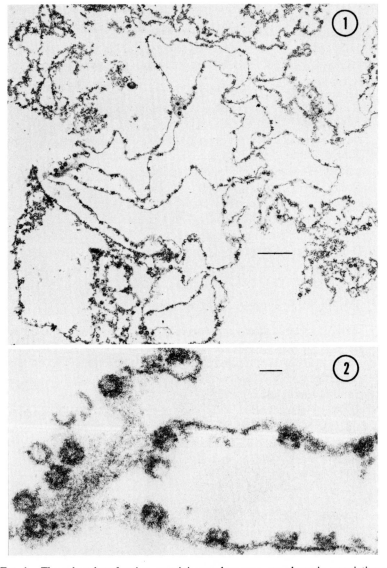

FIG. 1. The subnuclear fraction containing nuclear pore complexes in association with the nuclear peripheral lamina. This survey micrograph demonstrates the extensiveness and flexibility of the lamina. The profiles observed in these thin sections are consistent with the possibility that the nuclear peripheral lamina subsumes large areas, if not all, of the inner nuclear membrane. × 10,000. The bar indicates 1 μm.

FIG. 2. A higher magnification of a portion of Fig. 1 demonstrating in *en face* and lateral views the specific orientation and attachment of figures resembling the nuclear pore complex to a folded portion of the nuclear peripheral lamina. × 61,000. The bar indicates 100 nm.

2. Suspend the nuclear envelope pellet in 2 ml of ice-cold 10 mM Tris-Cl (pH 8.5)–0.1 mM MgCl$_2$–0.1 mM CaCl$_2$ and add 0.5 ml of ice-cold 10% Triton X-100. Incubate at 0° for 10 minutes and centrifuge at 11,000 rpm for 20 minutes at 4° in the Beckman JS-13 rotor (g_{av} = 12,000). Remove the supernatant by aspiration. The pellet contains the nuclear pore complex–peripheral lamina.

III. Comments

The nuclear pore complex–peripheral lamina is a delicate structure whose integrity depends on the lack of agitation during preparation and gentle sedimentation. Optimal morphological preservation necessitates fixation for electron microscopy in suspension prior to the final centrifugation.

Figures 1 and 2 demonstrate the appearance of a preparation of nuclear pore complex–fibrous lamina which was fixed in suspension overnight at 4° in 1% glutaraldehyde–0.1 M triethanolamine-Cl (pH 8.5) prior to centrifugation. Both circular figures and goblet-shaped figures resembling nuclear pore complexes are arrayed, with a specific orientation, on a rather extensive flexible 15-nm thick amorphous lamina.

Figure 3 demonstrates that the pore complex–peripheral lamina subfraction consists primarily of three polypeptides with similar molecular weights. These three polypeptides (or classes of polypeptides) appear to be present in nearly equivalent amounts. Furthermore, it can be seen that more common structural proteins [e.g., tubulin (52,000 MW), actin (44,000 MW), and myosin (190,000 MW)] are, if present, negligible components of the subfraction.

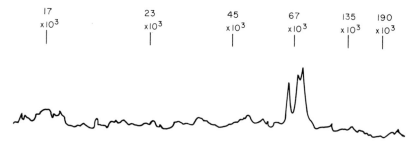

FIG. 3. Densitometer recording of SDS–PAGE analysis of the nuclear pore complex–peripheral lamina subfraction. The subfraction was solubilized in sodium dodecyl sulfate, heated under reducing conditions at 100° for 3 minutes, alkylated with iodoacetamide, and subjected to electrophoresis in the presence of sodium dodecyl sulfate on a 150 mm × 1 mm slab gradient (10–15% polyacrylamide) gel. The short vertical lines indicate the positions of migration of protein standards (molecular weights noted) applied in an adjacent slot.

ACKNOWLEDGMENT

This research was supported in part by grant GM 21950 from the National Institutes of Health.

REFERENCES

1. Franke, W. W., Z. Zellforsch. Mikrosk. Anat. **105**, 405, 1970.
2. Feldherr, C., Adv. Cell Molec. Biol. **2**, 273 (1972).
3. Kessel, R. G. Progr. Surface Membrane Sci. **6**, 243 (1973).
4. Stevens, B. J., and André, J., in "Handbook of Molecular Cytology" (A. Lima-de-Faria, ed.), p. 837. North-Holland Publ. Co., Amsterdam, 1969.
5. Fawcett, D. W., Am. J. Anat. **119**, 129 (1966).
6. Gall, J. G., in "Protoplasmatologia, Handbuch der Protoplasmaforschung." Vol. 5, p. 4. Springer-Verlag, Berlin and New York, 1964.
7. Patrizi, G., and Poger, M., J. Ultrastruct. Res. **17**, 127 (1967).
8. Kalifat, S. R., Bouteille, M., and Delarue, J., J. Microscopie **6**, 1019 (1967).
9. Mazanec, K., J. Microsc. (Paris) **6**, 1027 (1967).
10. Aaronson, R. P., and Blobel, G., Proc. Natl. Acad. Sci. U.S.A. **72**, 1007 (1975).
11. Blobel, G., and Potter, V. R., Science **154**, 1662 (1966).
12. Kay, R. R., Fraser, D., and Johnston, I. R., Eur. J. Biochem. **30**, 145 (1972).
13. Aaronson, R. P., and Blobel, G., J. Cell Biol. **62**, 746 (1973).
14. Monneron, A., Blobel, G., and Palade, G. E., J. Cell Biol. **55**, 104 (1972).

Chapter 22

Posttranscriptional Modifications of Nonhistone Proteins of Salivary Gland Cells of Sciara coprophila

REBA M. GOODMAN AND ELENA C. SCHMIDT

Division of Natural Sciences,
College of Physicians and Surgeons,
Columbia University,
New York, New York

WILLIAM B. BENJAMIN

Department of Physiology and Biophysics,
Health Sciences Center, School of Basic Sciences,
State University of New York at Stony Brook,
Stony Brook, New York

I. Introduction

The Dipteran salivary gland cells are especially suited to the study of the role of nucleoproteins in the control of differentiation. This model has been used extensively over the past 35 years both biochemically and cytochemically in analyses of chromosome morphology and physiology (*1–13*). *In vivo* and *in vitro* studies have demonstrated that, in salivary gland cells, chromosomal puffing is the cytological manifestation of genetic activity and that there is a correspondence between puff activity and biochemical events (*11,13–21*).

Salivary gland cells are in constant interphase, i.e., they are engaged in cyclical DNA replication and constant RNA synthesis. We have demonstrated, in *Sciara coprophila* salivary gland cells, that during puff formation there is both phosphorylation and methylation of nonhistone chromosomal proteins (*22–24*). It had been demonstrated previously that following puff induction by ecdysone (insect molting hormone) there is an increase in the amount of nonhistone proteins associated with the chromosomes and particularly with puff regions (*24,25*). Recently, we have shown that early in puff induction by ecdysterone, preexistent proteins are transported to many sites on the chromosomes, some of which develop into puffs (*24*). Furthermore, in larvae that are not responsive to ecdysterone ("noncompetent"), little transport of cytoplasmic protein was demonstrable. This indicates that specific proteins, destined to be transported to the nucleus, are synthesized during normal larval development *before* the chromosomes are competent to puff. Thus we demonstrated, autoradiographically, that cytoplasmic proteins can be induced by ecdysterone to move from the cytoplasm to regions on the chromosomes during the process of puff induction.

In this chapter we describe techniques we have found useful in studying nucleoprotein metabolism of polytene chromosomes.

II.   Methods[1]

A.   *Sciara coprophila* Stock

For these studies only female larvae of *Sciara coprophila* were used because their chromosomes are larger than those of the males as a result of one more endoduplication cycle during salivary gland development. In addition, they exhibit better staining properties (*27*). *Sciara coprophila* are

[1]The media, reagents, hormone, isotopes, and autoradiographic supplies used in our studies and the companies from which they were obtained are: brewer's yeast, the Schiff reagent for the Feulgen reaction, and Tris (THAM) from Fisher Scientific, Springfield, New Jersey; Cannon Insect Medium (complete) from Grand Island Biological Company, Grand Island, New York; ecdysterone (B grade) from Calbiochem, Los Angeles, California; Ilford K-5 nuclear research emulsion in gel form from Ilford, Ltd., Ilford, Essex, England; deoxyribonuclease (DNAse I, RNAse-free, grade DPFF) and ribonuclease (RNAse A) both from Worthington Biochemical Corporation, Freehold, New Jersey; the alkaline fast-green stain from FCF, Chroma-Gesellschaft, Schmid and Company, Stuutgart, Germany; cycloheximide from Sigma Chemical Company, St. Louis, Missouri; streptovitacin was a gift from Dr. Arline Deitch of Columbia University; [^{3}H]uridine, [^{3}H]thymidine, and ^{3}H-L-amino acid mixture (0.443 mg/ml) from Schwarz-Mann, Orangeburg, New York; *S*-adenosyl-^{3}H-methyl methionine from Nuclear Chicago, Chicago, Illinois; and L-(methyl-^{3}H)methionine (1 mCi/1.1 mg), [^{14}C]methionine carboxyl (0.25 mCi/8.9 mg), [^{32}P]orthophosphate (250–500 μCi/ml), [γ-^{32}P]ATP (100 μCi/ml), and [^{35}S]methionine (180 Ci/mM) all from New England Nuclear, Boston, Massachusetts.

raised on 2.5% agar in flint glass vials, stored at 20 °C in Precision-Scientific Incubators, and fed a mixture of Brewer's yeast and crushed straw [1:1] from day one after hatching until pupation. Larvae are fed three times a week and the incubators opened daily for airing.

Sciara coprophila LINTNER, a highly monogenic species, was used in these studies. The males of this species have a sex constitution of X0 in their somatic cells and the females are XX (*28*). The females are either male producers or female producers, the difference being determined by the genetic constitution of the sex chromosomes; females that are XX produce only male offspring, whereas females that are X'X (XX') produce only female progeny (*6–8,28*). In the stock used for the present work, the X' chromosome has a paracentric inversion (*29,30*) and carries the dominant Wavy (wavy wings), the latter allowing easy selection of the X'X adult females for mating. This permits the investigator to produce a clone of exclusively female offspring.

B.   Staging of the Larvae

In the *Sciaridae* the germ cells develop synchronously in both the ovary and the testis. Each egg is fertilized at ovoposition. Larvae obtained from eggs laid within 10 minutes of each other develop marked asynchrony, making it difficult to define an exact developmental stage, except at the molts.

In *Sciara coprophila* there are three larval molts and four instars. The third molt occurs (when larvae are raised at 20 °C) at about 8½ days after egg hatching. The duration of the fourth instar, following this last molt, is approximately 10–11 days. After about 6 or 7 days of fourth instar development the anlage of the adult eye, the so-called "eyespot," becomes visible. The eyespots are located just behind the head capsule and serve as an index of late fourth instar development. The eyespot consists typically of rows of pigmented granules, and its development occurs in an ordered fashion (*27*). With an increase in the number of pigmented granules per row, new rows of granules are laid down and the shape of the pigmented area approximates a triangle. Thus, a given stage of larval development can be expressed by an eyespot formula. At the onset of the prepupal stage, eyespots reach considerable size and undergo a characteristic migration laterally. Since the time of puff formation can be related to eyespot size it has been useful to designate the age of late fourth instar larvae by a numerical index of eyespot size. Stage e (20-day posthatching females; see chromosome maps, Fig. 1) is characterized by maximal development of large DNA puffs; the stage of eyespot development for stage e corresponds to a formula which

Larvae were decapitated by compression with two sterile 27G needles. The larvae eviscerate spontaneously. The salivary glands were separated from the larval body and transferred to fresh insect medium. Adhering fat bodies were carefully dissected away and the glands collected by centrifugation (1000 *g*) for 2 minutes. This step separates most of the remaining fat, and the resulting salivary gland pellet is virtually fat-free. Ecdysterone is dissolved in methanol and added to the culture medium so that the final concentration of ecdysterone is 0.5 μg/ml in a 2-ml volume. The culture is shaken gently 3 to 4 times during the incubation. In most experiments glands were removed and squashed at 30-minute intervals up to 240 minutes at which time induced puffs were usually observed. The stock ecdysterone solution is kept at $-20\,°C$ for a maximum of 3 weeks.

III. Methylation, Phosphorylation, Synthesis and Transport of Proteins to Cell Nucleus

A. Methylation

Salivary glands (16- to 20-day-old female larvae), dissected free from the larval bodies, were incubated in modified Cannon medium (Table I) with L-(methyl-^{3}H) methionine (20–60 μCi/ml; 1 mCi/1.1 mg) from 30 seconds to 2 minutes. At the conclusion of each incubation glands were transferred to albuminized slides, fixed briefly (2–3 seconds) in ice-cold (freshly prepared) 45% acetic acid containing 1% lactic acid, and squashed in 1% lactic-acetic–orcein stain (*33*) with a siliconized coverslip. The slides were immersed in liquid nitrogen and the coverslips popped off with a razor blade,

TABLE I

PREPARATION OF MODIFIED CANNON MEDIUM[a,b]

Solution 1: 608 mg MgCl$_2$ · 6 H$_2$O; 740 mg MgSO$_4$ · 7 H$_2$O; 596 mg KCl; 220 mg NaH$_2$PO$_4$.
  Dissolve in 60 ml of double-distilled water.

Solution II: 162 mg CaCl$_2$.
  Dissolve in 14.4 ml of double-distilled water.

Solution III: 74 mg α-ketoglutaric acid; 11 mg fumaric acid.
  Dissolve in 10 ml of double-distilled water.

Solution IV: 1000 mg trehalose; 140 mg glucose; 80 mg fructose; 80 mg sucrose.
  Dissolve in 10 ml of double-distilled water.

---

[a] From Cannon (*31*).
[b] Composition of the medium is in mg/200 ml.

leaving the squashed glands on the albuminized slide. The slides were then processed to selectively remove lipid, histone, and nucleic acids. Lipid was removed by ether–ethanol extraction (1:1) and hydrochloric acid–chloroform–methanol extraction (34), and the slides were hydrated to water or saline. Histone was removed by acid extraction (0.2 N HCl, 20 minutes at 0–4°C) *before* formalin fixation in order to assure completeness of histone extraction since formalin has been reported to prevent removal of histone if used prior to cold acid extraction (35). Following histone extraction, slides were fixed (5 minutes) in 10% neutral buffered formalin, rinsed in distilled water, and processed for light microscopic autoradiography as follows (36): slides were dipped individually in a 1:1 dilution of Ilford K-5 nuclear research emulsion, dried overnight in the dark (in a light-tight box), and stored in light-tight boxes in the refrigerator. For light microscopic examination slides were developed in a 1:1 dilution of Kodak D-19 (5 minutes), rinsed (30 seconds) in water, and fixed (10 minutes) in a 1:3 dilution of liquid GAF Rapafix. Slides were collected in gently running tap water and washed for 10 minutes. Photomicrographs were taken on Adox-KB 14 film, H and W VTE Ultrapan film, and Tri-X Pan. The Adox was developed in Rodinal and the H and W developed in D-19.

In experiments where salivary glands were first incubated in L-(methyl-^{3}H) methionine, nucleic acids were extracted by incubating the fixed squashes in solutions containing deoxyribonuclease (DNAse I; RNAse-free, grade DPFF) made up in insect medium containing 0.015 M MgSO$_4$ (pH 6.7) at a final concentration of 0.2 mg/ml. Digestions were at 37°C for 2 hours. The squash preparations were then incubated with ribonuclease (RNAse A; previously boiled for 10 minutes to remove DNAse-contaminating activity) in insect medium at a final concentration of 0.5 mg/ml (pH 6.7) at 37°C for 2 hours. Slides were rinsed in distilled water. Nucleic acids were further extracted by immersing slides in 10% trichloroacetic acid (TCA) at 90°C for 20 minutes with frequent agitation (37). Slides were immediately dipped in cold 70% ethanol (10 seconds) followed by cold 5% TCA (10 seconds) and rinsed gently with running tap water. The slides were collected in water or saline prior to coating with Ilford nuclear research emulsion.

Selective or complete nucleic acid extraction can be monitored by incubating salivary glands with [^{3}H]uridine or [^{3}H]thymidine (18.8 Ci/mM or 10 Ci/mM, respectively), diluted to a final concentration of 10 μCi/ml, for 10 minutes and demonstrating that all radiolabeled molecules could be removed by enzyme and hot TCA extractions. The completeness of the procedure can be assessed further by the modified Feulgen reaction for DNA content (38). All preparations treated to remove nucleic acids were Feulgen-negative, and the chromosomes could be visualized only with phase contrast microscopy. The completeness of the histone extraction procedure was assessed by the alkaline fast-green stain (*39*).

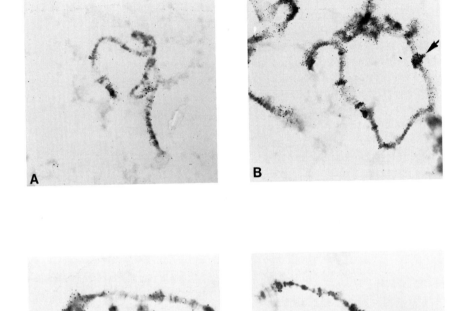

FIG. 2. (A) Autoradiogram of salivary gland chromosome from 20-day-old female larva of *Sciara coprophila*. Glands were incubated 30 seconds in L-(methyl-³H) methionine (20 μCi/ml) at 20 °C. Lipids, histone, and nucleic acids were extracted and slides processed for autoradiography as described in text. All squashes remained stained with orcein. Grains have extremely discrete distribution pattern. ×1020. (B) Autoradiogram of salivary gland chromosome following 4-minute incubation in L-(methyl-³H) methionine (40 μCi/ml) and processed as described for (A). Most of chromosome is labeled including heavy labeling in DNA puffs. ×650. (C) Autoradiogram of salivary gland chromosome (20-day-old female larva) from a split-gland experiment after 30-second incubation in L-(methyl-³H) methionine (60 μCi/ml) and processed as described for (A). Note discrete labeling of some bands and particularly in puffs. ×650. (D) Autoradiogram of chromosome from the other gland of the split-gland experiment described in (C) following 30-second incubation in [³H]thymidine (12.5 μCi/ml). Lipids and histone were removed as described in the text and the slide was processed as for (A). Nucleic acids were *not* removed. Note similarity of labeling pattern to chromosomes in (C). ×650.

To determine whether radioactivity associated with chromosomal proteins was derived predominantly from methylation or from *de novo* protein synthesis, salivary glands were incubated in inhibitors of protein synthesis (either 4 hours in cycloheximide or 20 minutes in streptovitacin; 10 μg/ml). Following these treatments, either L-(methyl-^{3}H) methionine (20–60 μCi/ml), [^{14}C]methionine carboxyl (125 μCi/ml), or ^{3}H-L-amino acid mixture (125 μCi/ml) were added for periods ranging from 30 seconds to 10 minutes. In some experiments S-adenosyl-^{3}H-methyl methionine (750 μCi/ml) was used as a methyl group donor. The glands were squashed and prepared for autoradiography as described above. The chromosomes from glands incubated with L-(methyl-^{3}H) methionine (30 seconds to 4 minutes) in the presence of inhibitors of protein synthesis were labeled (Fig. 2A and B). Glands incubated with [^{14}C]methionine carboxyl or ^{3}H-L-amino acid mixture (30 seconds to 10 minutes) were virtually unlabeled. Glands incubated *without* inhibitors of protein synthesis but with the ^{3}H-L-amino acid mixture were labeled. Therefore, methylation of the chromosomal proteins occurs in the absence of *de novo* protein synthesis. When S-adenosyl-^{3}H-methyl methionine was used as the methyl group donor, even more specificity was noted than when L-(methyl-^{3}H) methionine was used. After only 1 minute of incubation, chromosomes and specifically puff regions were discretely labeled.

A correlation was noted between cyclical patterns of DNA replication and protein methylation. To demonstrate this phenomenon gland pairs were dissected free and one gland of a pair was incubated up to 10 minutes in [^{3}H]thymidine (12.5 μCi/ml) and the other gland was incubated in L-(methyl-^{3}H) methionine (20–60 μCi/ml). We found that there were stages during which neither methylation nor DNA replication occurred, that when there was methylation in one gland of a pair there was also DNA replication in some nuclei of the other gland of the same pair (Fig. 2C and D), and that no such correlation was observed for RNA synthesis. Furthermore, methylation of chromosomal proteins appeared to be dependent on the developmental stage of the gland. Only minimal methylation could be demonstrated in younger larvae (16-day-old female larvae) and then only after more than 10 minutes of incubation with the isotope. These results imply that chromosomal nucleoprotein methylation may be involved in some metabolic aspects of control of DNA replication during development.

B. Phosphorylation of Nucleoproteins

Since nucleoprotein methylation has been established, studies on nucleoprotein phosphoprotein metabolism were undertaken in both untreated and puff-induced salivary gland chromosomes.

1. PHOSPHOPROTEIN PHOSPHORYLATION IN (UNTREATED)
LATE FOURTH INSTAR SALIVARY GLANDS

Salivary glands of late fourth instar larvae were incubated for 30 seconds, 1, 2, 4, and 10 minutes with either [^{32}P]orthophosphate (carrier-free; $NaH_2$$^{32}PO_4$, 10 mCi/ml in 0.02 M HCl) or [γ-^{32}P]ATP (100 μCi/ml). The isotopes were diluted in modified Cannon medium (Table I), pH 6.7, containing inorganic salts, organic acids, and sugars, but without amino acids. The various solutes were dissolved in 0.02 M Tris [tris-(hydroxy-methyl)-amino methane; THAM].

Following short incubations radioactivity was localized to specific segments of the chromosomes (Fig. 3A). With increasing incubation times radioactivity was spread more diffusely over the chromosomes. Experiments in which glands were incubated with [^{32}P]orthophosphate for 30 seconds and then with the unlabeled compound (0.02 M NaH_2PO_4; i.e., "chase experiments") for 9 minutes and 30 seconds demonstrated that the radioactivity appeared to come off the chromosomes. Turnover of phosphoryl label was thus indicated.

Nuclei from salivary gland cells that had been incubated in [^{32}P]ortho-phosphate were isolated, the phosphoprotein was extracted and hydrolyzed in acid, and the product was separated by high-voltage electrophoresis. Radioactivity was found to be associated predominantly with phosphoserine and, to a lesser extent, with phosphothreonine. This was taken as evidence of active phosphorylation of the chromosomal protein by the salivary gland cells. To assess further that the phosphorus was directly transferred to the chromosomal phosphoproteins, salivary glands were incubated in [γ-^{32}P] ATP. Autoradiographic preparations of chromosomes were extensively labeled when the source of phosphate was the specifically labeled ATP. One of the prerequisites for implicating a class of proteins in the control of gene action is that those proteins undergo active metabolism at the sites of RNA synthesis. The phosphorylation of nonhistone nucleoproteins demonstrated in this series of experiments paralleled the uptake of [^{3}H]-uridine in an accompanying series of studies; i.e., the autoradiographic localization of ^{32}P and ^{3}H was similar along the chromosomes. However, more precise localization of ^{32}P to a particular band(s) or puff(s) of specific chromosomes could not be demonstrated in this series of experiments.

2. PHOSPHORYLATION OF NUCLEOPROTEINS DURING PUFF
INDUCTION WITH ECDYSTERONE

Studies of nucleoprotein metabolism demonstrated specific hormone-induced radiolabeling of specific chromosomal regions. To examine phosphorylation during puff induction, isolated salivary glands from "com-

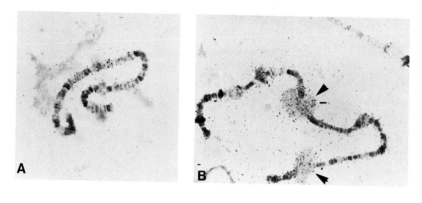

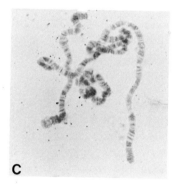

FIG. 3.   (A) Autoradiogram of salivary gland chromosome, midfourth instar female larva. Glands were incubated 2 minutes in [^{32}P]orthophosphate (1 mCi/ml) at 20°C in Cannon medium, fixed, and processed for light microscopic autoradiography. Grains are localized over bands in interbands. ×1020. (B) Autoradiogram of experimental salivary gland chromosome from 17- to 18-day-old female larvae. Glands were pretreated with ecdysterone (0.5 μg/ml) for 240 minutes at 20°C, postincubated with [^{32}P]orthophosphate for 2 minutes, squashed, and prepared for light microscopic autoradiography. Silver grains are localized predominantly over full-blown puffs and lightly scattered over bands and interbands. ×1020 (C) Autoradiogram of control salivary gland chromosome from 17- to 18-day-old female larva. Glands were pretreated with methanol carrier (0.5 μl/ml) (240 minutes) and then postincubated (2 minutes) in [^{32}P]orthophosphate. Chromosomes are virtually unlabeled as compared with (B). ×1020.

petent" larvae (17–18 days posthatching) were incubated for up to 240 minutes in ecdysterone. At "zero time" (5 minutes after initiation of induction with ecdysterone) and every 30 minutes thereafter until puffs were induced, salivary glands were removed and incubated with [^{32}P] orthophosphate for periods ranging from 30 seconds to 10 minutes. Figure 3B illustrates phosphorylation at two sites of induced puffs, while the control preparation was unpuffed and relatively unlabeled (Fig. 3C).

Ecdysterone induction of puff formation is associated with specific phosphoprotein labeling at sites of puff formation on the chromosomes. The kinetics of labeling, however, are not known. Whether protein is phosphorylated in the cytoplasm and then transported to specific sites on the chromosomes, or whether there is regulation of turnover of nucleoprotein phosphoryl groups at specific sites, are important questions amenable to investigation using techniques such as those described above.

C.   Synthesis and Transport of Proteins to Cell Nucleus in Ecdysterone-Treated Cells

Cytologic techniques to assess nucleoprotein side-chain metabolism have been described. However, questions as to the origin of puff proteins could not be answered by studying methylation or phosphorylation. Using modifications of these cytological techniques we examined the origin and time of appearance of nonhistone proteins at the puff sites.

Isolated glands from 17- to 18-day-old larvae were incubated for 60 minutes in modified Cannon medium (Table I) containing [^{35}S]methionine (10 μCi/ml; sp act 180 Ci/mM). At the conclusion of the incubations, an equal volume of *complete* medium, containing 0.8 mg/ml methionine, was added to both experimental and control preparations to halt further radioactive label incorporation into protein. Five minutes after the addition of complete medium, 0.5 μg/ml ecdysterone (25) was added to the experimental samples and an equal volume of methanol carrier was added to the controls. In most experiments glands were removed, fixed, and squashed at 30-minute intervals for up to 240 minutes, by which time induced puff formation was usually observed. Slides containing squashed salivary glands were treated for the removal of lipid, histone, and nucleic acids and prepared for autoradiography (as described in Section III,A).

In autoradiograms of glands incubated for 5 minutes with either methanol carrier or ecdysterone ("zero time") grains were localized primarily over the cytoplasm and the chromosomes were only lightly labeled (Fig. 4A). After 60 minutes of incubation in ecdysterone, however, silver grains were localized predominantly over the chromosomes (Fig. 4B), while in the control preparations the distribution of silver grains was similar to that in the zero-time preparations (Fig. 4C). With increasing incubation time with ecdysterone, the association of radioactivity with chromosomes was more pronounced (Fig. 4D) when compared to chromosomes in the control preparations (Fig. 4E). After 240 minutes puff formation was induced by ecdysterone and these puffs were labeled (Fig. 4F and G).

To demonstrate that ecdysterone treatment causes the transport of cytoplasmic proteins to the nucleus only in cells that are "competent" to puff, experiments similar to those described above were performed using salivary glands from 14- to 15-day-old larvae. Salivary glands incubated with or without ecdysterone over a period of 120–240 minutes of incubation showed no increase in label associated with chromosomes in ecdysterone-treated preparations as compared with controls and, of course, puffs were not induced.

To determine the effect of inhibiting protein synthesis on puff induction, salivary glands were incubated with cycloheximide (10 μg/ml) and ecdysterone, or with ecdysterone alone for up to 240 minutes. Figure 4H shows that in *Sciara coprophila* inhibiting *de novo* protein synthesis does *not* prevent puff induction by ecdysterone.

To determine whether ecdysterone induces an increase in synthesis of protein which then appears at puffed regions, isolated salivary glands from "competent" larva were incubated with and without ecdysterone for 240 minutes. Following hormone pretreatment, the glands were *post*incubated with [^{35}S]methionine for 5 and 60 minutes. Although puffs were induced by the molting hormone, chromosomal bands, interbands, and puffs were only lightly labeled indicating that most of the *newly* synthesized protein was transported to the nucleus during the 120 and 240 minutes of hormone pretreatment. These experiments suggest that there is a decrease in the radiolabeling of suitable tracer molecules with which to follow the movement of these proteins.

IV. Discussion

The steroid molting hormone, ecdysone, has been shown to induce puff formation only at specific stages of development. Either before or at the onset of puff formation there is a substantial accumulation of pre-existing protein in and around the developing puff. Since evidence suggests that the amount of histone protein remains unchanged as compared with chromosome regions *not* affected by puff formation, nonhistone proteins account for this accumulation. The first event in puff formation, accumulation of nonhistone proteins, is followed by despiralization of a DNA segment that will be transcribed. As RNA synthesis begins, protein accumulation diminishes. The gradual reduction of protein in the puffs has been explained by either turnover, transport, or loss of staining properties due to binding to newly synthesized RNA. It has been suggested that newly synthesized

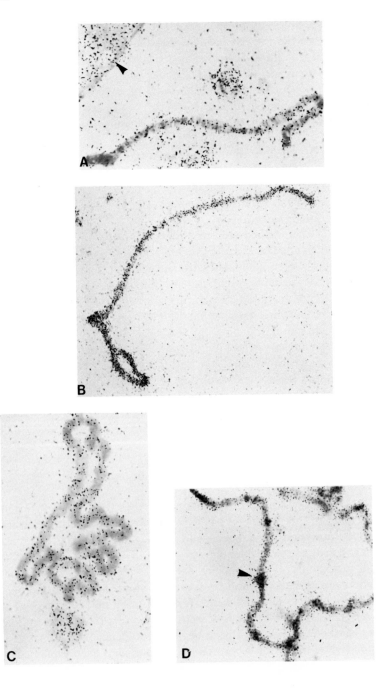

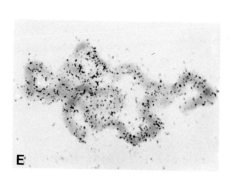

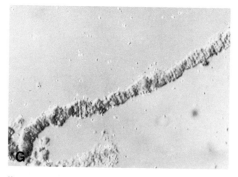

FIG. 4. Autoradiograms of *Sciara coprophila* chromosomes. Isolated salivary glands (17-day-old females) were incubated (60 minutes) at 20 °C with 10 μCi/ml [^{35}S]methionine in amino acid-free medium, followed by incubation with either ecdysterone, 0.5 μg/ml (experimental), or methanol carrier (control) in complete medium for 5 minutes (= zero time) to 240 minutes. Squashes stained with lactic–acetic–orcein were prepared for light microscopic autoradiography (lipid, histone, and nucleic acids were extracted as described in Section III,A). (A) Chromosome from experimental salivary gland. Incubation: zero time. Grains: distribution primarily over cytoplasm. Chromosomal bands and interbands lightly and nonspecifically labeled. ×650. (B) Chromosome from ecdysterone-treated salivary glands. Incubation: 60 minutes. Grains: clustered predominantly over chromosome bands and interbands. ×650. (C) Chromosomes from control salivary gland. Incubation: 60 minutes. Grains: similar to (A), randomly scattered over chromosomal bands and interbands. ×650. (D) Chromosome from experimental salivary gland. Incubation: 120 minutes. Grains: clustered over chromosome and associated with regions of incipient puff formation (arrow). ×650. (E) Chromosome from control salivary gland. Incubation: 120 minutes. Grains: few grains associated with chromosomes in contrast to (D). ×650. (F) Chromosome from experimental salivary gland. Incubation: 240 minutes. Grains: chromosomes are heavily labeled with grains concentrated along bands and interbands as well as with puffs (arrow). Nomarski optics. ×2040. (G) Chromosome from control salivary gland. Incubation: 240 minutes. Grains: along chromosome are scattered randomly and lightly. Differential interference technique (Nomarski). ×2040.

RNA is transformed into final messenger form within the puff. This product is combined with protein in the form of discrete ribonucleoprotein particles and transported through the nuclear envelope to the cytoplasm *(4,5,40)*.

The early appearance of nonhistone proteins in chromosome regions which will form puffs suggests that there may be a relationship between these proteins and the "activating" stimulus. These observations lead us to postulate that there is a pool of preexisting cytoplasmic proteins which, when acted upon either directly or indirectly by a specific hormone, allows recognition by this protein(s) of specific sites on the genome. Thus, the early appearance of nonhistone proteins may be part of the biochemical mechanism for the control of chromosomal loci designated to become activated by ecdysone. Since the effects of endogenous ecdysone in *Sciara coprophila* are permanent, the transfer of cytoplasmic proteins to the nucleus may represent a mechanism for stable differentiation.

ACKNOWLEDGMENTS

The authors wish to thank Dr. Donald West King for his generous support of this work and Joeline Spivack for her skillful assistance with the photographic processing. W. B. B. was supported by USPHS AM189050-1, in part by RCDA 5KD-4CA-2461, and the Alma Toorock Memorial Fund.

REFERENCES

1. Beermann, W., *Chromosoma* **5**, 139 (1952).
2. Beermann, W., *in* "Developmental Cytology" (D. Rudnick, ed.), p. 83. Ronald Press New York 1959.
3. Beermann, W., *Am. Zool.* **3**, 23 (1963).
4. Beermann, W., *J. Exp. Zool.* **157**, 49 (1964).
5. Beermann, W., and Bahr, G. F., *Exp. Cell Res.* **6**, 195 (1954).
6. Metz, C. W., and Schmuck, M. L., *Proc. Natl. Acad. Sci. U.S.A.* **15**, 863 (1929).
7. Metz, C. W., and Schmuck, M. L., *Proc. Natl. Acad. Sci. U.S.A.* **15**, 867 (1929).
8. Metz, C. W., and Smith, H. B., *Proc. Natl. Acad. Sci. U.S.A.* **17**, 195 (1931).
9. Pavan, C., *Proc. Int. Congr. Genet., 10th* **1**, 321 (1959).
10. Pavan, C., and Breuer, M. E., *J. Hered.* **48**, 151 (1952).
11. Pavan, C., and da Cunha, A. B., *Annu. Rev. Genet.* **3**, 425 (1969).
12. Pelc, S. R., and Howard, A., *Exp. Cell Res.* **10**, 549 (1956).
13. Pelling, C., *Cold Spring Harbor Symp. Quant. Biol.* **35**, 521 (1970).
14. Beermann, W., and Clever, U., *Sci. Am.*, April (1964).
15. Beermann, W., and Pelling, C., *Chromosoma* **16**, 1 (1965).
16. Berendes, H. D., *Chromosoma* **17**, 35 (1965).
17. Berendes, H. D., *Chromosoma* **22**, 274 (1967).
18. Berendes, H. D., *Chromosoma* **24**, 418 (1968).
19. Crouse, H. V., and Keyl, H.-G. *Chromosoma* **25**, 357 (1968).
20. Pelling, C. *Nature (London)* **184**, 655 (1959).
21. Pelling, C., *Chromosoma* **15**, 71 (1964).
22. Benjamin, W. B., and Goodman, R. M., *Science* **166**, 629 (1969).
23. Goodman, R. M., and Benjamin, W. B., *Exp. Cell Res.* **77**, 63 (1973).

24. Goodman, R. M., Schmidt, E. C., and Benjamin, W. B., *Cell Differ.*, in press.
25. Helmsing, P. J., and Berendes, H. D., *J. Cell Biol.* **50**, 893 (1971).
26. Holt, Th. K. H., *Chromosoma* **32**, 64 (1970).
27. Gabrusewycz-Garcia, N., *Chromosoma* **15**, 312 (1964).
28. Metz, C. W., *Am. Nat.* **72**, 485 (1938).
29. Crouse, H. V., *Genetics* **45**, 1429 (1960).
30. Crouse, H. V., *Chromosoma* **11**, 146 (1960).
31. Cannon, G. B., *Science* **146**, 1063 (1964).
32. Cannon, G. B., *J. Cell. Comp. Physiol.* **65**, 163 (1965).
33. Fullmer, H. M., and Lillie, R. D., *J. Histochem. Cytochem.* **4**, 64 (1952).
34. Allerton, S. E., and Perlman, G. E., *J. Biol. Chem.* **240**, 3892 (1965).
35. Cave, M. D., *Chromosoma* **25**, 392 (1968).
36. Rogers, A. W., "Techniques of Autoradiography." Elsevier New York, 1973.
37. Wolstenholme, D. R., *Chromosoma* **17**, 219 (1965).
38. Jordanov, J., *Acta Histochem.* **15**, 135 (1963).
39. Alfert, M., and Geschwind, I. I., *Proc. Natl. Acad. Sci. U.S.A.* **39**, 991 (1953).
40. Stevens, B. J., and Swift, H., *J. Cell Biol.* **31**, 55 (1966).

Chapter 23

The Mitotic Apparatus: Methods for Isolation

ARTHUR M. ZIMMERMAN

Department of Zoology, University of Toronto,
Toronto, Ontario, Canada

SELMA ZIMMERMAN

Division of Natural Sciences,
Glendon College, York University,
Toronto, Ontario, Canada

ARTHUR FORER

Biology Department, York University,
Downsview, Ontario, Canada

I. Introduction

The mitotic apparatus (spindle–aster–chromosome complex) functions during cell division to bring about an equal distribution of nuclear material to the daughter cells and is probably responsible for initiation of cytokinesis (see the review in *1–3*). Extensive investigation has been directed toward an understanding of the assembly and functional mechanisms of the mitotic apparatus (MA). These studies have been concerned with biochemical, physicochemical, and structural analyses of *in vitro* MA; the results of these analyses depend on the method of isolation used. Since the first report of mitotic isolates by Mazia and Dan (*4*), numerous isolation techniques have been developed, and one of the aims has been to obtain functional MA. Nevertheless, the mass isolation of a functional mitotic apparatus has not yet been achieved.

Since functional activity (chromosome movement) cannot be used to assess the merits of the MA isolation method in question, other criteria must be employed. These involve comparisons of optical and physicochemical properties of isolated MA with MA *in vivo*, and a review of these properties is in order before discussing isolation methods. Using a compound microscope with bright-field optics, one cannot see the MA *in vivo* and isolates are barely visible. With phase-contrast optics, *in vivo* MA cannot be easily seen (e.g., *in vivo* MA in *Arbacia punctulata* or *Strongylocentrotus purpuratus* zygotes are visible only if the cells are severely pressed between coverslip and slide), but in isolated MA, chromosomes, as well as clearly defined spindle fibers and asters, can be seen (Fig. 1). Polarization microscopy permits visualization of *in vivo* and *in vitro* spindles by virtue of the birefringence (retardation) in the spindle–aster regions (Fig. 2). The electron microscope reveals a similar structure for *in vivo* and isolated MA; both contain membranous material and microtubular components (Fig. 3). The membranous material is seen as vesicular profiles in isolated MA, or as endoplasmic reticulum *in vivo* (*4a*). These are distributed within the MA in the same regions in which microtubules are seen. The spindle microtubules have closely associated "ribosomelike" particles in many, but not all, organisms (*7*). MA microtubules in sea urchin spindles contain 13 protofilaments (*8*) similar to microtubules in other cells (*9*).

The MA *in vivo* are cold-labile (*10,11*), pressure-labile (*12–17*), and colchicine-sensitive (*10*). Depending upon the method of isolation, *in vitro* MA also display lability to cold (*5,6,18,19*), hydrostatic pressure (*5,6,20,21*), and sensitivity to colchicine (*19*).

II. General Isolation Methods

A. Isolation Procedures Reported up to 1974

Previous articles have evaluated the various isolation procedures reported in the literature up to 1974 (*22–25*), and thus these methods will be discussed only briefly in this paper.

FIG. 1. A photograph of an isolated mitotic apparatus taken through a phase-contrast microscope. The chromosomes appear dark, at the equator; some dark spindle fibers can be seen. ×950.

FIG. 2. A photograph of an isolated mitotic apparatus taken through a polarization microscope, with compensation [see Forer and Zimmerman (*5*) for details]. Spindle fibers appear bright, while astral fibers appear as a Maltese cross (with alternate light and dark quadrants). The chromosomes are not clearly seen and are identified as the equatorial termini of the birefringent spindle fibers. ×800.

FIG. 3. Electron micrographs of embedded and section-isolated mitotic apparatus. (A) Low-power (overview) electron micrograph from a somewhat oblique section. The chromosomes appear dark at the equator, while dark lines extend between the chromosomes and the leftward pole. ×6000. (B) At higher magnification, the microtubules and electron-dense particles lined up along the lengths of the microtubules are visible. ×59,000. From the work of Forer and Zimmerman (*6*).

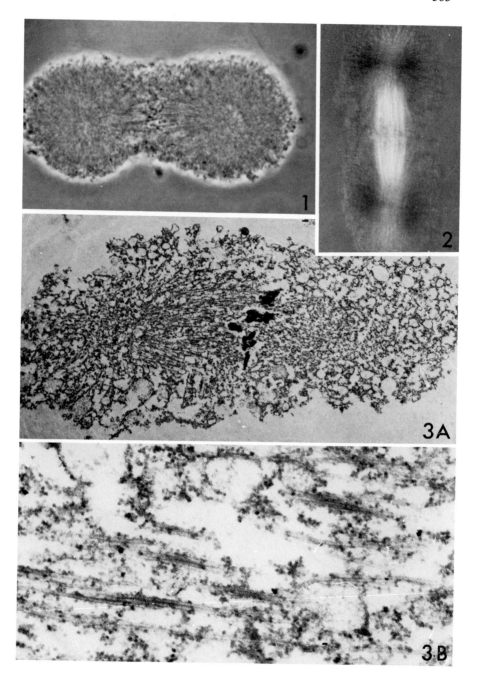

The first mass isolation of the mitotic apparatus was achieved by Mazia and Dan (*4*) who isolated MA from sea urchin zygotes. They treated *Strongylocentrotus purpuratus* zygotes with cold (−10°C) 35% ethanol, and then they selectively dispersed the cytoplasm with the detergent, Duponol; the isolated MA were separated from cell debris by differential centrifugation. This procedure was slightly modified in 1955 when digitonin was used in place of the Duponol (*26*). Isolation of mitotic apparatus directly from living zygotes of *S. purpuratus* was subsequently achieved by Mazia and co-workers (*27*). This technique was based on the premise that mitotic-apparatus stability was dependent upon the disulfide linkages within and between peptides. Dithiodiglycol was employed as the penetrating disulfide which, in addition to causing swelling, would stabilize the mitotic apparatus. The medium (at pH 5) was supplemented with sucrose which maintained an isotonic milieu and versene (EDTA) which helped disperse the cytoplasm. Dithiodipropanol was later substituted for dithiodiglycol (*28*). Subsequently, Mazia *et al.* (*29*) modified the original ethanol isolation procedure by replacing the Na$^+$ in sea water with Li$^+$. Isolation of MA was achieved following storage in cold ethanol by lysis in Triton X solution. The advantage of this technique is that large amounts of mitotic apparatus can be isolated because Li$^+$ reversibly stabilizes the MA *in situ* and allows the zygotes to develop to metaphase. Kane (*30*) isolated large quantities of mitotic apparatus directly from *Arbacia* sea urchin zygotes using hexanediol, a six-carbon glycol at pH 6.4. He also reported that mitotic apparatus could be isolated in the absence of organic solvents if the hydrogen ion concentration was sufficiently low (pH 5.5). However, this method was not suitable for mass isolation. Subsequently, hexylene glycol was substituted for hexanediol (*31*).

The sea urchin has been the most common organism from which the mitotic apparatus has been isolated. This is probably due to the fact that sea urchin zygotes divide synchronously and are easily obtained in large numbers for any given experiment. The methods summarized above were all developed using sea urchin zygotes. The glycol isolation method (*30,31*) has also been used to isolate mitotic apparatus from a number of different cell types, which include the first polar-body eggs and division zygotes of the surf clam *Spisula solidissima* (*32*), oocytes of the sea star *Pisaster* (*33, 34*), crane fly spermatocytes (*35*), cultured mammalian cells (*36,37*), and *Drosophilia* embryos (*38*).

B. Isolation Procedures Reported from 1974 to Present

1. Isolation Procedures with Tubulin-Polymerizing Medium

Rebhun *et al.* (*19*) isolated MA from surf clam eggs (and eggs from eight other species) using an isolation medium containing organic buffer [2-(*N*-morpholino) ethanesulfonic acid (MES) or piperazine-*N*-*N*'-bis (2-

ethane sulfonic acid) (PIPES), pH 6.85], $MgCl_2$, ethylene bis-(oxyethylen-enitrilo) tetraacetate (EGTA), p-tosyl arginine methylester HCl (TAME), and in some cases Triton X-100. Following cell lysis the MA are resuspended in this isolation medium to which up to 10 mg/ml of chick brain tubulin are added. Such MA are cold-labile for up to 1 hour after isolation, but this lability gradually decreases with time. At room temperature, the bire-fringence of these MA increases over a 15-minute interval with little observ-able change in shape when an appropriate amount of exogenous microtubule material by cold thermal treatment display partial restoration of birefring-ence when brain tubulin is added.

Inoué *et al.* (*18*) studied the MA from meiotic metaphase-arrested *Chaeto-pterus* oocytes. A limited quantity of meiotic MA were obtained in 0.1 M PIPES–1 mM EGTA–2.5 mM guanosine 5′-triphosphate (GTP) (pH 6.9) using a perfusion chamber constructed with a slide and coverslip. The bire-fringence of the isolated spindles was stabilized or augmented in the pre-sence of exogenous tubulin. This method is suitable for optical studies, but it was not designed for mass biochemical isolations.

The recent report (*39*) of lysis of rat kangaroo fibroblasts in a solution containing Carbowax, EGTA, GTP, and Triton-X at pH 6.9 does not con-stitute an MA isolation since the resulting cellular fragments contain an abundance of cytoplasm surrounding the mitotic mechanism. Nevertheless, this method is worthy of note since these cell fragments displayed chromo-some movement when anaphase cells were lysed in the presence of isolation medium supplemented with adenosine 5′-triphosphate (ATP) and tubulin.

2. GLYCEROL ISOLATIONS

Sakai and Kuriyama (*40*) isolated MA from zygotes of the sea urchin, *Hemicentrotus pulcherrimus*, using a solution containing 1 M glycerol–1 mM EGTA–5 mM MES (pH 6.5). The cells were disrupted by agitation, and then glycerol was added to a final concentration of 3 or 4 M. The suspension was centrifuged at 2000 g for 10 minutes, and the pellet was resuspended in a glycerol isolation medium containing 1 mM ATP. This material was used for subsequent experimentation. Recently, Sakai *et al.* (*41*) reported that they observed some chromosome movement when MA were isolated in a medium containing 1 M glycerol–10 mM MES–1 mM EGTA–1 mM GTP–20 μM $CaCl_2$–4 μM cyclic adenosine monophosphate (cAMP)–2.5 mM gluta-thione–3.5 mM ascorbic acid–1 mg trypsin inhibitor (pH 6.15). Movement occurred when the isolation medium was supplemented with ATP, Mg^{2+}, and 0.5 mg/ml of tubulin. The chromosome movement which occurred in early anaphase MA consisted of chromosomes moving $\frac{1}{3}$ to $\frac{1}{2}$ the way to the poles during 1 hour; the poles also moved apart by about 15%. This movement is not the same as *in vivo*, but the results seem a step toward attaining chromosome movement *in vitro*.

3. DMSO–Glycerol Isolation

A dimethyl sulfoxide–glycerol (DMSO–glycerol) isolation medium has been used to isolate MA from zygotes of the sea urchins *S. purpuratus* (*5*), *Lytechinus pictus* (*20*), and *S. droebrachiensus* (A. Forer, unpublished). This technique will be described in detail in the following section.

III. DMSO–Glycerol Isolation Method

The rationale for the DMSO–glycerol method is to isolate functional MA with stable microtubules. Previous methods (*30,42*) resulted in MA with unstable microtubules (*43,44*). In the studies of Rebhun *et al.* (*19*), Inoué *et al.* (*18*), and Cande *et al.* (*39*) exogenous tubulin was added to stabilize the MA microtubules. If one wishes to isolate a functional MA and subsequently evaluate the mechanism and biochemical parameters controlling mitosis, it would be desirable to isolate MA in quantities suitable for biochemical study in which exogenous tubulin would not complicate analysis. Filner and Behnke (*45*) reported a method for the isolation of microtubules from porcine brain using a DMSO–glycerol solution in which the microtubules were stable and did not lose their lability even after long periods of storage [see Behnke (*46*)]. The method has also been applied to other cellular systems for isolation of microtubules (*47–49*). This DMSO–glycerol method was adapted for the mass isolation of MA from sea urchin zygotes (*5,6,20,21,50*) and is described below.

A. Methodology for DMSO–Glycerol MA Isolation

The sea urchin eggs were collected using standard procedures (*51*), namely, 2-ml intracoelomic injections of 0.53 *M* KCl. The shed eggs were washed three times with fresh sea water. Sperm were obtained from excised testes and stored dry at 4 °C. Sperm suspensions were prepared immediately prior to use. Fertilization was achieved by mixing eggs and sperm in a test tube.

Fertilization membranes need to be removed in order to isolate MA. Several methods are available for removal of the fertilization membrane of sea urchin eggs, including chemical treatment of the newly formed fertilization membrane, enzymic digestion of the vitelline membrane, and mechanical shearing of the fertilization membrane. We used the method described by Kane (*30*) in which urea is used to weaken the newly forming fertilization membrane. Insemination is achieved by mixing the washed eggs with a large amount of freshly prepared sea urchin sperm suspension. An aliquot is removed and retained as a control of the fertilization procedure. Approximately 30 seconds later the inseminated eggs are collected with a hand-powered centrifuge. The supernatant is removed, and 1 M urea (pH 7.8) is added to the packed cells which are resuspended in the urea solution. (It is essential to add the urea within 60 seconds after insemination.) Kane (30) recommends the use of urea at acid pH and states that urea at basic pH causes the S. purpuratus zygotes to be very fragile. However, this fragility is necessary in order to isolate MA in DMSO–glycerol. If acid urea is used, the zygotes do not lyse. We have had successful isolations even when the zygotes fragment due to urea treatment.

After 1 minute of treatment with urea, the cells are collected by a hand-powered centrifuge (using gentle centrifugation because the eggs are very fragile at this time), and the supernatant is discarded. Fresh sea water is added, the cells are resuspended gently, and then the cells are collected by gentle centrifugation. Additional sea water (3–4 volumes) is added to the centrifuge tubes, and the cells are transferred to flat dishes and permitted to develop to metaphase of the first division at the species' optimum development time.

The time between insemination and metaphase varies for different species of sea urchin and is also dependent upon the temperature of the sea water. For example, fertilized eggs of Strongylocentrotus purpuratus will reach metaphase approximately 75–85 minutes after insemination, at their optimum temperature of 18 °C, whereas Lytechinus pictus zygotes require 110 minutes and develop best at 21 °C. The optimum temperature is maintained by placing the zygotes in flat dishes in a constant-temperature water bath. In addition, the sea water is precooled to the temperature of the water bath.

The developmental stage of the cells is established by observing an aliquot of cells using low-power bright-field microscopy or low-power polarization microscopy. In the former the MA can be distinguished against the darker background of the cell especially when the condensor is very much removed from Kohler illumination. In the latter, one sees a birefringent spindle. It is essential that the dishes in which the zygotes are placed after urea treatment have only a small amount of water and a maximal amount of surface area. To facilitate oxygen exchange the depth of the sea water should not exceed 1 cm.

When metaphase is reached, the zygotes are gently collected in 15-ml centrifuge tubes using a hand-powered centrifuge; the zygotes in the pellet are very gently resuspended in isolation medium containing 50% (v/v) glycerol–10% (v/v) DMSO–5 mM MgCl$_2$–5 mM Sorensen's phosphate buffer (final pH 6.8)–0.1 mM EGTA. The mixing should be performed carefully so that no cell lysis occurs. The resuspended cells are then centrifuged, and the supernatant is discarded. (This centrifugation requires much more force than the first because of the viscosity of the glycerol.) Fresh isolation medium is added to the pellet to about 10 times the volume of packed zygotes. Then the mixture is strongly agitated for about 2 minutes using a Vortex mixer. This agitation results in cell lysis and consequent liberation of the intact mitotic apparatus. A sample of the mixture is examined under a microscope to see if lysis is complete. If insufficient lysis occurs, the suspension is agitated again until more zygotes have lysed. The isolation procedures are conducted at room temperature. The suspension of lysed zygotes is then centrifuged at 225 g for 2 minutes. The resultant pellet contains unlysed eggs, large pieces of cytoplasm, and some MA. The supernatant contains MA and small pieces of cytoplasm. This supernatant is then centrifuged at 3000 g for 15 minutes. The resulting pellet contains mostly MA, with little contaminating material; this pellet is resuspended in a small volume of isolation medium and centrifuged at 3000 g to further clean up the MA. If the pellet from the first centrifugation at 3000 g is still contaminated with unbroken cells, a further centrifugation at 225 g for 2 minutes is required in order to separate MA from cells. Then subsequent centrifugation at 3000 g results in pellets of clean MA.

B. Special Methodology for Preparation of MA for Electron Microscopy (EM) Studies

It might be relevant to discuss some of the special techniques used in making preparations for electron microscopy.

MA are routinely fixed for several hours to days in glutaraldehyde (2% in whichever medium the MA are in) and postfixed for 1 hour in 1% osmium tetroxide. The MA are flat-embedded so that individual MA can be located, picked up, and mounted using light microscopy. Hence, the fixed MA are placed in molten 2% agar on a flat surface so that the agar forms a thin layer (with MA oriented parallel to the flat surface). Because the heat from the molten agar can alter glutaraldehyde-fixed microtubules (20), the MA are placed in agar after fixation with osmium tetroxide. After the agar solidifies small pieces of agar are transferred for dehydration and embedding in the same fashion as one ordinarily transfers tissue (52). This eliminates the need to continually centrifuge the MA to change solutions during dehydration and embedding.

Once the MA are embedded and located, the circled regions are cut out from the Epon using a tool resembling a cork borer with toothed rim; the tool is mounted on a drill press and the marked region of plastic (Epon, or Araldite) is drilled out. The disk-shaped region is then mounted on the end of an Epon block using epoxy cement.

The MA can be identified easily in the flat embedment using transmitted light and phase-contrast microscopy. One also can see the MA in the flat embedment using transmitted light and a stereo microscope (a standard dissection microscope). The MA cannot be seen when the drilled disk is mounted on the block because there is no transmitted light. In order to achieve the effect of transmitted light, a piece of filter paper is cemented directly to the bottom of the drilled-out disk; the disk with attached filter paper is mounted together on the end of an Epon block. The result is that light shining onto the surface of the block reflects up from the white filter paper giving a viewing quality comparable to that with transmitted light. This greatly facilitates trimming of the block.

One last point which may be relevant is the effect of some PCB-free immersion oils. These oils can be used to give improved optical properties when viewing cells either with phase-contrast microscopy or with a stereo microscope (they "smooth out" irregularities in the surfaces). It should be noted that Carl Zeiss PCB-free immersion oil dissolves or softens the Epon (or Araldite) so that it cannot be sectioned. Nikon PCB-free immersion oil is quite satisfactory in this regard.

C. Characteristics of the DMSO–Glycerol MA

The MA isolated by the DMSO–glycerol method have stable microtubules for a period of at least 2 weeks, and they also have stable birefringence and stable solubility properties in 0.5 M KCl over the same period (5). In addition these MA lose birefringence after treatment with increased hydrostatic pressure or low temperature (6,20), and these responses are stable for at least 1–2 weeks after isolation. Hence, in this regard the MA have *in vivo*-like properties, although the birefringence loss due to pressure or cold is not the same as *in vivo* (6,20).

D. Advantages and Disadvantages

One of the main advantages of this isolation procedure is its ability to produce large amounts of isolated MA which maintain their birefringence and their microtubules after periods of storage. In many of the other procedures the isolated MA lose birefringence and lose microtubules on storage. Another important feature is the fact that exogenous tubulin need not be added to obtain an isolated MA with stable birefringence and stable microtubules. In the current procedure all the tubulin present derives from the MA since no other tubulin is added.

The greatest disadvantage of the procedure is that the zygotes do not lyse as well as they do with the Kane (31) hexylene glycol method. There appear to be more unlysed zygotes in DMSO–glycerol than in parallel isolations with hexylene glycol. In addition, there is some variability in the procedure so that in some preparations it is difficult to separate the MA from contaminating cytoplasm even with differential centrifugation.

ACKNOWLEDGMENT

This work was supported by grants from the National Research Council of Canada.

REFERENCES

1. Mazia, D., *in* "The Cell" (J. Brachet and A. E. Mirsky, eds.), Vol. III, p. 77. Academic Press, New York, 1961.
2. Mazia, D., *in* "Cell Cycle Controls" (G. M. Padilla, I. L. Cameron and A. M. Zimmerman, eds.), p. 265. Academic Press, New York, 1974.
3. Rappaport, R., *Int. Rev. Cytol.* **31**, 169 (1971).
4. Mazia, D., and Dan, K., *Proc. Natl. Acad. Sci. U.S.A.* **38**, 826 (1952).
4a. Harris, P., *Exp. Cell Res.* **94**, 409 (1975).
5. Forer, A., and Zimmerman, A. M., *J. Cell Sci.* **16**, 481 (1974).
6. Forer, A., and Zimmerman, A. M., *J. Cell Sci.* **20**, 329 (1976).
7. Rebhun, L. I., and Sanders, G., *J. Cell Biol.* **34**, 859 (1967).
8. Kiefer, B., Saki, H., Solari, A. J., and Mazia, D., *J. Mol. Biol.* **20**, 75 (1966).
9. Tilney, L. G., Bryan, J., Bush, D., Fujiwara, K., Mooseker, M. S., Murphy, D. B., and Snyder, D. H., *J. Cell Biol.* **59**, 267 (1973).
10. Inoué, S., *in* "Primitive Motile Systems in Cell Biology" (R. D. Allen and N. Kamiya, ed.), p. 549. Academic Press, New York, 1964.
11. Inoué, S., and Sato, H., *J. Gen. Physiol.* **50**, 259 (1967).
12. Marsland, D., *in* "High Pressure Effects on Cellular Processes" (A. M. Zimmerman, ed.), p. 259. Academic Press, New York, 1970.
13. Salmon, E. D., *J. Cell Biol.* **65**, 603 (1975).
14. Salmon, E. D., *J. Cell Biol.* **66**, 114 (1975).
15. Zimmerman, A. M., *in* "High Pressure Effects on Cellular Processes" (A. M. Zimmerman, ed.), p. 235. Academic Press, New York, 1970.
16. Zimmerman, A. M., *Int. Rev. Cytol.* **30**, 1 (1971).
17. Zimmerman, A. M., and Marsland, D., *Exp. Cell Res.* **35**, 293 (1964).
18. Inoué, S., Borisy, G. G., and Kiehart, D. P., *J. Cell Biol.* **62**, 175 (1974).
19. Rebhun, L. I., Rosenbaum, J., Lefebvre, P., and Smith, G., *Nature (London)* **249**, 113 (1974).
20. Forer, A., and Zimmerman, A. M., *J. Cell Sci.* **20**, 309 (1976).
21. Forer, A., Kalnins, V. I., and Zimmerman, A. M., *J. Cell Sci.* **22**, 115 (1976).
22. Forer, A., *in* "Handbook of Molecular Cytology" (A. Lima-de-Faria, ed.), p. 553. North-Holland Publ. Amsterdam, 1969.
23. Hartmann, J. F., and Zimmerman, A. M., *in* "The Cell Nucleus" (H. Busch, ed.), Vol. 2, p. 450. Academic Press, New York, 1974.
24. Nicklas, R. B., *in* "Advances in Cell Biology II" (D. M. Prescott, L. Goldstein, and E. H. McConkey, eds.), p. 225. Appleton, New York, 1971.
25. Zimmerman, A. M., *in* "The Cell and Mitosis" (L. Levine, ed.), p. 159. Academic Press, New York, 1963.
26. Mazia, D., *Symp. Soc. Exp. Biol.* **9**, 335 (1955).
27. Mazia, D., Mitchison, J. M., Medina, H., and Harris, P., *J. Biophys. Biochem. Cytol.* **10**, 467 (1961).

28. Sakai, H., *Biochim. Biophys. Acta* **112**, 132 (1966).
29. Mazia, D., Petzelt, C., Williams, R. O., and Meza, I., *Exp. Cell Res.* **70**, 325 (1972).
30. Kane, R. E., *J. Cell Biol.* **12**, 47 (1962).
31. Kane, R. E., *J. Cell Biol.* **25** (No. 1, Pt. 2), 137 (1965).
32. Rebhun, L. I., and Sharpless, T. K., *J. Cell Biol.* **22**, 488 (1964).
33. Bryan, J., and Sato, H., *Exp. Cell Res.* **59**, 373 (1970).
34. Sato, H., Ellis, G. W., and Inoué, S., *J. Cell Biol.* **67**, 501 (1975).
35. Müller, W., *Chromosoma* **30**, 316 (1970).
36. Sisken, J. E., *Methods Cell Physiol.* **4**, 71 (1970).
37. Sisken, J. E., Wilkes, E., Donnelly, G. M., and Kakefuda, T., *J. Cell Biol.* **32**, 212 (1967).
38. Milsted, A., and Cohen, W. D., *Exp. Cell Res.* **78**, 243 (1973).
39. Cande, W. Z., Snyder, J., Smith, D., Summers, K., and McIntosh, J. R., *Proc. Natl. Acad. Sci. U.S.A.* **71**, 1559 (1974).
40. Sakai, H., and Kuriyama, R., *Dev. Growth Differ.* **16**, 123 (1974).
41. Sakai, H., Hiramoto, Y., and Kuriyama, R., *Dev. Growth Differ.* **17**, 265 (1975).
42. Kane, R. E., *J. Cell Biol.* **15**, 279 (1962).
43. Goldman, R. D., and Rebhun, L. I., *J. Cell Sci.* **4**, 179 (1969).
44. Kane, R. E., and Forer, A., *J. Cell Biol.* **25** (No. 3, Pt. 2), 31 (1965).
45. Filner, P., and Behnke, O., *J. Cell Biol.* **59**, 99a (1973).
46. Behnke, O., *Cytobiologie* **11**, 366 (1975).
47. Pipeleers, D. G., Pipeleers-Marichal, M. A., and Kipnis, D. M., *Science* **191**, 88 (1976).
48. Rubin, R. W., and Weiss, G. D., *J. Cell Biol.* **64**, 42 (1975).
49. Stearns, M. E., Connolly, J. A., and Brown, D. L., *Science* **191**, 188 (1976).
50. Forer, A., and Zimmerman, A. M., *Ann. N. Y. Acad. Sci.* **253**, 378 (1975).
51. Harvey, E. B., "The American *Arbacia* and Other Sea Urchins." Princeton Univ. Press, New Jersey, 1956.
52. Kellenberger, E., Ryter, A., and Sechaud, J., *J. Biophys. Chem Cytol.* **4**, 671 (1958).

Chapter 24

Isolation of the Mitotic Apparatus

JOSEPH L. TURNER AND J. RICHARD McINTOSH

Department of Molecular, Cellular, and Developmental Biology,
University of Colorado,
Boulder, Colorado

1. Introduction

The mitotic spindle is functionally defined as the cellular machinery that segregates the chromosomes at cell division. It forms for the occasion of division and disappears when its task is done. Morphologically the spindle shows considerable diversity from organism to organism, but it always has two poles with an array of fibers running from pole to pole and a fiber connecting each chromatid to each pole. With the light microscope, spindle fibers can be seen in living cells by using sensitive phase or polarization optics. The latter reveal that spindle fibers possess a weak, positive form birefringence (*1*). In the electron microscope a spindle fiber is identified as a bundle of microtubules. The spindle of fixed cells contains many ribosomelike particles, some vesicles, some microfilaments, and considerable ill-defined material of uncertain function, often called matrix.

The preparation of isolated mitotic spindles was first achieved by Mazia and Dan in 1952 (*2*). Since then several procedures for isolating spindles from different organisms have been described. The isolation of the spindle has received considerable attention because the potential rewards from a good isolate would be enormous. The process of chromosome motion is interesting and important in its own right, and a biochemical dissection of the spindle would obviously help to develop our understanding of the process. Further, many mechanisms for the change of cellular shape seem to be based on components similar to the spindle. It is therefore reasonable to hope that a characterization of the mitotic machinery and the control systems that regulate its assembly would help in our understanding of cellular morphogenesis in general.

Isolation procedures yield what is called a mitotic apparatus (MA), which is defined as a mitotic spindle, the chromosomes, the poles, astral fibers (if there are any in the cell type), and associated matrix components. No one has yet succeeded in producing a preparation of isolated MA which will move chromosomes in an apparently normal fashion. This has been a serious problem for students of mitosis, for isolated spindles are morphologically and biochemically complex, and without a functional assay, it is difficult to assess the importance of any particular component in the isolate. Nevertheless, isolated MA have been used to study microtubule structure (3,4), microtubule polymerization (5), microtubule birefringence (1,6), the thermodynamics of microtubule assembly (7), effects of D_2O on microtubules (8,9), synthesis of MA precursors (10,11), changes in mass distribution through mitosis (12), MA-associated ATPases (13–19), and MA-associated proteins (e.g., 20–26). Especially in these last two areas, the dangers from cytoplasmic contamination have already been mentioned in the literature (16,27).

II. Methods

Most of the published studies of isolated spindles have used marine eggs due to the ease with which large populations of highly synchronous dividing cells can be obtained. Use of eggs, however, makes removal of the tough fertilization membrane necessary; this may be accomplished by suspending the eggs in 1 M glycerol [for Spisula; (28)], or 0.7 mg/ml mercaptoethylglu-conamide (29), or 1 M urea (30), or 10 μg/ml pronase–10 mM ethylenediami-netertraacetic acid (EDTA)–1 mM dithiothreitol in Ca^{2+}-free sea water at pH 8.8 (17).

The initial problem in isolating an MA is stabilizing the spindle outside the cell. Spindles in situ are labile to cold (31), high pressure (32), certain drugs, such as colchicine (33), and to dilution of the cytoplasmic pool of spindle subunits (34,35). To maintain the MA outside the cell and purify it away from cytoplasmic contaminants, the medium used must support the structural stability of the MA in a cell-free environment. The following sections describe four ways in which this has been achieved.

A. Hexylene Glycol

Kane (36) reported the use of 1 M hexylene glycol (2-methyl-2,4-pentane-diol) (pH 6.4) for sea urchin MA isolations. This glycol was chosen for its ability to preserve the MA and for its apparent mildness (see urchin embryos can tolerate 0.1 M hexylene glycol in sea water and will develop to form normal plutei).

Sea urchins (*A. punctulata*) are grown in Ca^{2+}-free sea water, washed twice with a 19:1 mixture of 0.53 *M* NaCl and KCl, and then transferred to isolation medium [1 *M* hexylene glycol–10 m*M* KH_2PO_4 (pH 6.4 with KOH)]. Cells burst osmotically and the MA are then released by gentle agitation. The MA can be pelleted (500 *g*, two minutes) and then resuspended in fresh isolated medium. They should be handled at 0°C from lysis on to avoid loss of birefringence and irreversible stabilization of the MA (*37,38*).

Numerous variations on this method have been used to adapt it to clam eggs (*6,7*), star fish eggs (*39*), and mammalian cells (*40–42*). Problems with the method have been discussed by Forer and Goldman (*43,44*).

B. Cold Ethanol

A variation on Mazia and Dan's original isolation procedure has recently been developed (*17*). About 40 minutes after fertilization (i.e., 40 minutes before use), the eggs are washed several times with artificial sea water in which LiCl is substituted for NaCl. Li^+ does not prevent MA formation, but it does block disassembly. Isolation is accomplished by first resuspending the settled eggs in at least 10 volumes of 30% ethanol at –10°C and storing them for at least 24 hours. The medium is then changed to 30% ethanol–0.1% Triton X-100 (–10°C). Under these conditions the eggs will keep for several months. To isolate the MA, the eggs are allowed to settle and are resuspended in 2–3 volumes of cold water (10°C). They are then vortexed for a few seconds. Mazia *et al.* (*17*) report that this temperature is quite important. The MA suspension is now cooled to 0°C and the MA pelleted at 100 *g* for 10 minutes (0°C). They should now be washed in 30% ethanol (–10°C) to remove the Triton, and they may be washed in cold water. This method has been used chiefly to Ca^{2+}-activated ATPase in sea urchins (*17–19*).

C. MTM

The remaining methods discussed here represent applications of techniques developed for the handling of microtubules *in vitro* to the isolation of the MA. Forer and Zimmerman (*38*) have used Filner and Behnke's (*45*) microtubule medium [MTM: 50% (v/v) glycerol–10% (v/v) dimethyl sulfoxide (DMSO)–5 m*M* $MgCl_2$–5 m*M* phosphate buffer (pH 6.8)] with 0.1 m*M* EGTA to isolate sea urchin MA. Zygotes with developed MA were rinsed in the MTM and EGTA and gently pelleted. They were then resuspended in about 10 times their volume of MTM + EGTA and vortexed strongly. The MA are pelleted by spinning at 500 *g* for 5 minutes and rinsed in isolation medium. All steps may be at room temperature.

The birefringence of these MA is stable for at least 3 weeks (at room temperature or at 0 °C). The microtubules in the MA are stable for at least 2 weeks, but their morphology is notably altered from microtubules fixed *in vivo*. The MA (except for asters) is rapidly soluble in 0.5 M KCl, and it remains so for at least 10 days. The tubules are cold-labile if the MTM + EGTA is reduced to one-quarter strength (*38*; however, also see *46,47*). Sakai and Kuriyama (*48*) and Sakai et al. (*49*) have reported a similar method for isolating sea urchin MA in high concentrations of glycerol.

D.   Weisenberg Buffers

Recently, buffers that allow the cold-reversible polymerization of brain microtubules have been developed (*50,51*). Two reports have appeared describing MA isolations employing these buffers. Rebhun et al. (*28*) have isolated the MA of the surf clam, *Spisula*, in 0.1 M MES[1] or PIPES[2] (*52*) at (pH 6.85,–0.15–1.0 mM MgCl$_2$, 1–5 mM EGTA, 10 mM TAME[3] (an inhibitor of trypsin-type proteolysis), and the nonionic detergent Triton X-100 (1%). The eggs are suspended at room temperature in 2–3 volumes of isolation medium (room temperature). Vortexing or vigorous shaking are used to lyse the eggs, liberating the MA, which can then be centrifuged (1.5–2 minutes at 400–500 g) into a pellet.

The MA thus obtained are cold-labile initially, but are increasingly stable as they sit for 1 minute to several hours. In the presence of 1 mM guanosine 5′-triphosphate (GTP) they are capable of incorporating chick brain tubulin (at least at >9 mg/ml). They will grow in length and birefringence if resuspended in high concentrations of brain tubulin (15–20 mg/ml). Isolation in tubulin-containing buffers helps preserve cold lability.

Similar results have been reported by Inoué et al. (*53*) using the marine worm *Chaetopterus*. These authors have used 0.1 M PIPES (pH 6.94)–1 mM EGTA–2.5 mM GTP (PEG). Oocytes were perfused in PEG at 18–20 °C on a microscope slide. The birefringence was observed to fade until the oocyte lysed due to the hypotonicity of the PEG. The birefringence will fade away unless polymerization-competent tubulin is included in the lysis buffer. In 10 mg/ml tubulin, the birefringence will recover its original value (or more), and the spindle may grow both longer and thicker. At least for the first few minutes after lysis the birefringence is readily cold-labile. Interestingly, the birefringence is stable to even high concentrations (1 mM) of colchicine and colcemid and to dilution of the tubulin with PEG.

[1] MES = 2(*N*-morpholino)ethanesulfonic acid.
[2] PIPES = piperazine-*N*, *N*′-bis (2-ethanesulfonic acid).
[3] TAME = *p*-tosyl-L-arginine methyl ester HCl.

E. General Comments and Conclusions

In all of the above methods it is likely that small changes in pH can influence the amount and the properties of the protein in the isolate (e.g., *30, 36,41,43,44,54*). It has been suggested that better buffering than normally used is in order (*44*). The relative complexity of the egg cytoplasm can cause problems due to contamination of the isolate with relatively large amounts of cytoplasmic proteins (e.g., *16,27*). With the possible exception of the method described in Section II, D, the isolation solutions seem to function in part by being simply bad protein solvents (*36*), and this heightens potential problems with cytoplasmic contamination. With the exception of two reports (*55,49*), none of the spindles isolated thus far have been found to accomplish anaphaselike motion. Until such spindles are obtained, researchers will not be sure that the isolate contains all of the necessary components, and interpretation of data bearing on the mechanism of motion (e.g., enzymology) will be ambiguous.

III. Model Systems

A few studies have approached spindle chemistry from a different perspective. Although isolated MA have the virtue that their analysis is straightforward, as mentioned above, interpretation of the results is difficult due to the loss of spindle function. In two cases, a method has been developed which preserves chromosome motion through cell lysis, but there is no attempt made to remove the MA from the cell (*56,35*). These lysed "cell models" have proven difficult to work with, and little has yet been learned from them.

Hoffman-Berling (*56*) found that he could cause an increase in the anaphase separation of the chromosomes in glycerinated fibroblasts in the presence of ATP. Since other polyanions would work, the interpretation of this study is problematical.

The Weisenberg-type buffer system has been applied to gently lysed cells by Cande *et al.* (*35*). Mammalian tissue-culture cells (PtK$_1$) are lysed by flowing PEG (see above) containing 0.05% Triton X-100 and approximately 1 mg/ml brain tubulin across the growth surface. At lower tubulin concentrations, the spindles decrease in size and birefringence; at higher concentrations, no increase in length is seen, although the birefringence does increase. High-molecular-weight polyethylene glycol (Carbowax 20M) at concentrations of 2–5% can replace the polymerizable tubulin in maintaining the spindle after lysis. Cells in anaphase at the time of lysis continue chromosome separation at approximately the rate of untreated cells (57). These workers are trying to use the stopping and starting of chromosome motion after lysis to identify the enzymes important for chromosome movement (58).

We hope that studies of this kind will complement studies on completely isolated MA. Development of an understanding of the conditions necessary to support anaphse motion and eventually the metaphse-to-anaphase transition may well require the gentler model systems. This understanding should contribute to the development of functional criteria for evaluating and modifying existing isolation techniques and, ultimately, to the interpretation of the studies on the isolated MA.

REFERENCES

1. Sato, H., Ellis, G. W., and Inoué, S., J. Cell Biol. 67, 501 (1975).
2. Mazia, D., and Dan, K., Proc. Natl. Acad. Sci. U.S.A. 38, 826 (1952).
3. Cohen, W. D., and Gottlieb, T., J. Cell Sci. 9, 603 (1971).
4. Kiefer, B., Sakai, H., Solari, A. J., and Mazia, D., J. Mol. Biol. 20, 75 (1966).
5. Goode, D., J. Mol. Biol. 80, 531 (1973).
6. Rebhun, L. I., and Sander G., J. Cell Biol. 34, 859 (1967).
7. Sato, H., and Bryan, J., J. Cell Biol. 39, 118a (1968).
8. Marsland, D., and Zimmerman, A. M., Exp. Cell Res. 30, 23 (1963).
9. Marsland, D., and Zimmerman, A. M., Exp. Cell Res. 38, 306 (1965).
10. Bibring, T., and Cousineau, G. H., Nature (London) 204, 805 (1964).
11. Wilt, F. H., Sakai, H., and Mazia, D., J. Mol. Biol. 27, 1 (1967).
12. Rustad, R. C., Exp. Cell Res. 16, 575 (1959).
13. Mazia, D., Chaffee, R. R., and Iverson, R. M., Proc. Natl. Acad. Sci. U.S.A. 47, 788 (1961).
14. Miki, T., Exp. Cell Res. 29, 92 (1963).
15. Dirksen, E. R., Exp. Cell Res. 36, 256 (1964).
16. Weisenberg, R. C., and Taylor, E. W., Exp. Cell Res. 53, 372 (1968).
17. Mazia, D., Petzelt, C., Williams, R. O., and Meza, I., Exp. Cell Res. 70, 325 (1972).
18. Petzelt, C., Exp. Cell Res. 70, 333 (1972).
19. Petzelt, C., Exp. Cell Res. 74, 156 (1972).
20. Miki-Noumura, T., Embryologia 9, 98 (1965).
21. Miki-Noumura, T., Exp. Cell Res. 50, 54 (1968).
22. Sakai, H., Biochim. Biophys. Acta 112, 132 (1966).
23. Borisy, G. G., and Taylor, E. W., J. Cell Biol. 34, 535 (1967).
24. Bibring, T., and Baxandall, J., Science 161, 377 (1968).
25. Bibring, T., and Baxandall, J., J. Cell. Biol. 48, 324 (1971).
26. Cohen, W. D., and Rebhun, L. I., J. Cell Sci. 6, 159 (1970).
27. Bibring, T., and Baxandall, J., J. Cell Biol. 41, 577 (1969).
28. Rebhun, L. I., Rosenbaum, J., Lefebvre, P., and Smith, G., Nature (London) 249, 113 (1974).
29. Mazia, D., Mitchison, J. M., Medina, H., and Harris, P. J., Biophys. Biochem. Cytol. 10, 467 (1961).
30. Kane, R. E., J. Cell Biol. 12, 47 (1962).
31. Inoué, S., Rev. Mod. Phys. 31, 402 (1959).

32. Salmon, E. D., *J. Cell Biol.* **65**, 603 (1975).
33. Inoué, S., *Exp. Cell Res. Suppl.* **2**, 305 (1952).
34. Inoue, S., and Sato, H., *J. Gen. Physiol.* **50** Suppl., 259 (1967).
35. Cande, W. Z., Snyder, J., Smith, D., Summers, K., and McIntosh, J. R., *Proc. Natl. Acad. Sci. U.S.A.* **71**, 1559 (1974).
36. Kane, R. E., *J. Cell Biol.* **25**, 137 (1965).
37. Kane, R. E., and Forer, A., *J. Cell Biol.* **25**, 31 (1965).
38. Forer, A., and Zimmerman, A. M., *J. Cell Sci.* **16**, 481 (1974).
39. Bryan, J., and Sato, H., *Exp. Cell Res.* **59**, 371 (1970).
40. Sisken, J. E., Wilkes, E., Donnelly, G. M., and Kakefuda, T., *J. Cell Biol.* **32**, 212 (1967).
41. Sisken, J. E., *Methods Cell Physiol.* **4**, 71 (1970).
42. Wray, W., and Stubblefield, E., *Exp. Cell Res.* **59**, 469 (1970).
43. Forer, A., and Goldman, R. D., *Nature (London)* **222**, 689 (1969).
44. Forer, A., and Goldman, R. D., *J. Cell Sci.* **10**, 387 (1972).
45. Filner, P., and Behnke, O., *J. Cell Biol.* **59**, 99a (1973).
46. Forer, A., and Zimmerman, A. M., *J. Cell Sci.* **20**, 309 (1976).
47. Forer, A., and Zimmerman, A. M., *J. Cell Sci.* **20**, 329 (1976).
48. Sakai, H., and Kuriyama, R., *Dev., Growth Differ.* **16**, 123 (1974).
49. Sakai, H., Hiramoto, Y., and Kuriyama, R., *Dev. Growth Differ.* **17**, 265 (1975).
50. Weisenberg, R. C., *Science (N Y)* **177**, 1104 (1972).
51. Borisy, G. G., and Olmsted, J. B., *Science (N Y)* **177**, 1196 (1972).
52. Good, N. E., Winget, G. D., Winter, W., Connolly, T. N., Izawa, S., and Singh R. M. M., *Biochemistry* **5**, 467 (1966).
53. Inoué, S., Borisy, G. G., and Kiehart, D. P., *J. Cell Biol.* **62**, 175 (1974).
54. Goldman, R. D., and Rebhun, L. I., *J. Cell Sci.* **4**, 179 (1969).
55. Goode, D., and Roth, L. E., *Exp. Cell Res.* **58**, 343 (1969).
56. Hoffmann-Berling, H., *Biochim. Biophys. Acta* **15**, 226 (1954).
57. McIntosh, J. R., Cande, W. Z., Snyder, J. A., and Vanderslice, K., *Ann. N. Y. Acad. Sci.* **253**, 407 (1975).
58. McIntosh, J. R., Cande, W. Z., and Snyder, J. A., *In* "Molecules in Cell Movement" (S. Inoué, and R. E. Stephens, eds.), p. 31. Raven, New York, 1975.

Chapter 25

Fractionation of Nonhistone Chromosomal Proteins Utilizing Isoelectric Focusing Techniques

A. J. MacGILLIVRAY*

*Beatson Institute for Cancer Research,
Wolfson Laboratory for Molecular Pathology,
Bearsden, Glasgow, Scotland*

I. Introduction

An important finding of the recent studies of chromatin nonhistone proteins has been their apparent lack of tissue specificity, evidence which has lead to the conclusion that the major species observed are probably enzymes or structural components common to many chromatins (*1*). Hence in eukaryotes proteins that regulate gene expression may be present in the nonhistone protein fraction of chromatin in insufficient quantities for many current analytical procedures to detect them (*1*). There is, therefore, considerable interest in the fractionation of nonhistone proteins both to isolate minor components as well as perhaps to obtain the major polypeptides in purified form.

One of the approaches adopted in this laboratory has been to attempt to obtain a separation of nonhistone proteins similar to that observed by analytical two-dimensional gel electrophoresis, but on a preparative scale. An essential part of this protocol concerns the separation of polypeptides by isoelectric focusing, a technique that has demonstrated the wide range of isoelectric points of the nonhistone proteins (*2–5*). Due to the ability of these proteins to aggregate during focusing in vertical pH gradients stabilized by sucrose, we adapted the horizontal thin-layer isoelectric focusing technique of Radola (*6*) to suit our requirements (*7*). Our evidence is that the proteins which control the expression of the globin genes in chromatin *in vitro* are present in the hydroxyapatite HAP2 fraction of the nonhistone

Present address: Biochemistry Laboratory, School of Biological Sciences, University of Sussex, Falmer, Brighton, Sussex BN1 9QG, England.

proteins (8). Hence most of our interests have centered in the subfractiona-
tion of this group of proteins. Previous work (3) has shown that the HAP2
proteins are a complex mixture, and our initial studies (7) using thin-layer
isoelectric focusing showed the potential of this technique in their separation. Details of the basic technique of focusing in a thin layer are given in the
following sections using mouse liver HAP2 proteins as an example.

II. Preparation of Sample Proteins

Mouse liver nuclei are prepared by sucrose and Triton X-100 procedure as
previously described (9). Nuclear proteins are phosphorylated by incubation of the nuclei with γ-^{32}P-adenosine 5'-triphosphate (ATP) (10), after
which the nuclei are washed and chromatin is prepared by extraction with
0.14 M NaCl–0.05 M Tris-HCl (pH 7.5)–5 mM ethylenediaminetetraacetic
acid (EDTA) (11). Chromatin protein fractions are prepared by the hydroxyapatite procedure (11), and the HAP2 fraction is concentrated by dialysis
against Carbowax[1] to a concentration of 1–2 mg/ml. The concentrate is
then dialyzed against 8 M urea–0.3 M Tris-HCl(pH 8.3) and reduced by
the addition of dithioerythritol to a final concentration of 50 mM followed
by incubation at 37° for 3 hours. Finally the sample is dialyzed against
several changes of 8 M urea containing 2.5 mM dithioerythritol.

III. Thin-Layer Isoelectric Focusing

A stiff slurry is prepared by adding 3.5–4 gm of Sephadex G-50 Superfine to 40 ml of 8 M urea–1 mM dithioerythritol to which is added 1 ml of
pH 3–10 Ampholines (40% solution, LKB Instruments Ltd.). After mixing
and allowing the Sephadex to swell for a few minutes the slurry is evacuated
for 1 minute prior to being spread with a spatula on a clean 20 × 20 cm glass
plate. The layer is made even by repeatedly lifting one edge of the plate
5–10 cm and allowing the plate to fall back onto the bench (6). The layer is
then allowed to dry at room temperature for about 30 minutes until it loses ·
some 3% of its weight, after which it is transferred onto a cooled metal
block at 10°.

[1]Carbowax (MW 15,000–20,000) can be obtained from G. T. Gurr, High Wycombe,
Bucks, England.

The sample of proteins (5 mg in 3–5 ml) is made 1% (w/v) with respect to pH 3–10 Ampholines and sufficient Sephadex G-50 Superfine is added to make a workable slurry. A section (18 × 1 cm) is removed from the center of prepared layer and replaced with the slurry containing the sample proteins. The cathode electrode is prepared by soaking 20 × 1 cm strips of Whatman 3 MM chromatography paper in 0.4 M ethylenediamine in 8 M urea. After blotting to remove excess liquid two strips are placed on top of each other along one edge of the plate parallel to the sample application zone. A piece of platinum wire is then placed on top followed by a third strip of wetted chromatography paper. The anode is prepared in the same way at the opposite edge of the plate except that the electrode solution in 0.2 M H_2SO_4–8 M urea. Focusing is carried out on the cooled metal plate which is housed in a covered chamber containing reservoirs of water to prevent the layer drying out. The platinum electrodes are attached to a constant voltage supply and 7.5 V/cm is applied to the layer for 16 hours, followed by 15 V/cm for a further 2 hours.

The range of the pH gradient is determined by means of a spear-pH electrode. A print of the layer is then taken by placing a 20 × 3 cm strip of dry Whatman 3 MM paper in the middle of the layer along the line of the pH gradient. After the strip has become wetted it is dried and stained with Coomassie Blue as described by Radola (6). ^{32}P radioactivity is then located either by using a strip scanner[2] or by cutting the strip into 0.5-cm pieces followed by scintillation counting. Once the print has been taken the layer is then stored at $-20°$.

After the areas of interest have been located on the layer using the print or the scan of ^{32}P radioactivity as reference, they are scraped from the appropriate section of the frozen layer. Proteins are recovered by incubation of the Sephadex in several volumes of 6 M guanidine hydrochloride–0.3 M Tris-HCl (pH 8.3)–10 mM dithioerythritol at 37° for 3 hours, followed by removal of the Sephadex by filtration or centrifugation.

IV. Application to HAP2 Proteins

Figure 1 shows the results of focusing ^{32}P-labeled HAP2 proteins under the conditions described above. The experiment illustrated was part of a scheme designed to isolate the acidic and basic low-molecular-weight components designated "a" and "b" in Fig. 2A. Hence the thin layer was

[2]For example, Nuclear Chicago Actigraph.

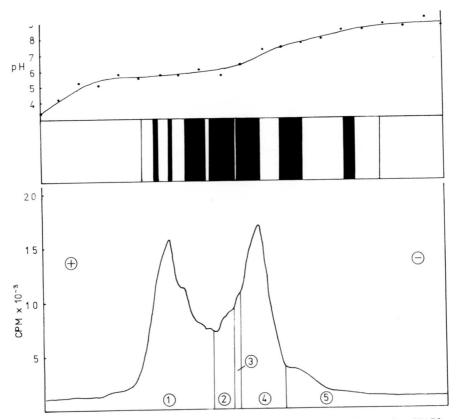

FIG. 1.   Isoelectric focusing in a thin layer of Sephadex G-50. ^{32}P-labeled mouse liver HAP2 proteins were prepared and applied to the layer as described in the text. The upper panel shows the pH gradient obtained after 18 hours. A print of the layer was prepared using a 20 × 3 cm strip of chromatography paper (see text). The middle panel is a representation of the protein bands detected by staining the print with Coomassie Blue. The lower panel shows the profile of ^{32}P radioactivity in the strip. The layer was divided into five sections as indicated, and the proteins were extracted from the Sephadex.

divided into five sections based on the distribution of ^{32}P radioactivity in the pH gradient. After recovery of the proteins from the Sephadex aliquots were analyzed by two-dimensional gel electrophoresis (3), the proteins being detected by staining with Coomassie blue and the phosphoproteins by autoradiography. Reproductions of these gels are given in Fig. 2B–F, where it can be seen that all of the components were recovered in sections 1–4. Note that section 1 contained only acidic proteins, while section 4 contained component "b" together with a range of other polypeptides (Fig. 2B and F).

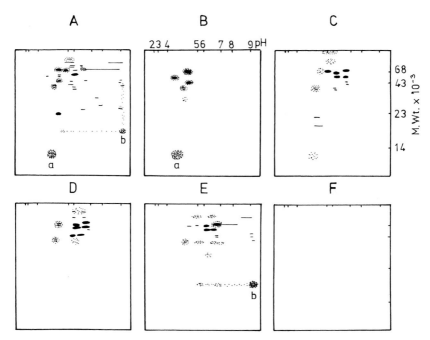

FIG. 2. Two-dimensional gel electrophoresis of proteins extracted from sections of the isoelectric focusing–thin-layer system described in Fig. 1. The horizontal axis represents isoelectric focusing in the first dimension, the pH gradient being indicated as shown in (B). The vertical axis shows separation in the second dimension using gel electrophoresis in sodium dodecylsulfate (SDS), the position of marker proteins of known molecular weight being given in (C). Proteins stained with Coomassie Blue are shown as lines or spots while the position of ^{32}P-labeled polypeptides is given by grains. (A). Unfractionated HAP2 proteins; (B–F) proteins extracted from sections 1–5 respectively of the layer. The two low-molecular-weight phosphoproteins "a" and "b" were found in extracts of sections 1 and 4 respectively and were subsequently isolated from these (see text).

Because of the size of the sections there was considerable overlap from one to another, a situation which can be averted, if necessary, by taking smaller sections of the layer (7).

The remainder of the proteins from section 1 were then applied to a gel filtration column [Sepharose 6B run in 6 M guanidine hydrochloride (3)] in order to isolate the low-molecular-weight phosphorylated protein "a" in virtually pure form. In the case of the extract of Section 4 this required further processing to remove nonbasic low-molecular-weight phosphoproteins. This was achieved by chromatography on QAE-Sephadex (3). In order to remove high-molecular-weight contaminants gel filtration was used as the final purification step.

V.   Conclusion

The efficiency of this procedure in separating proteins of differing iso-electric points obviously depends on the complexity of the sample. In the case of a heterogeneous group such as the HAP2 nonhistone proteins most sections of the gradient would require refocusing in order to obtain groups of polypeptides banding at distinct pH values. Alternatively, the pH gradient may be divided into a few sections each of which can be purified by other means, e.g., ion-exchange chromatography and gel filtration. We our-selves have preferred the latter approach, using the thin-layer system as part of a fractionation scheme to isolate individual nonhistone proteins such as the low-molecular-weight components described above (12).

REFERENCES

1. MacGillivray, A. J., In "The Organization and Expression of the Eukaryotic Genome" (E. M. Bradbury and K. Javaherian, eds). Academic Press, New York, 1977 (in press).
2. Barrett, T., and Gould, H. J., Biochim. Biophys. Acta 294, 165 (1973).
3. MacGillivray, A. J., and Rickwood, D., Eur. J. Biochem. 41, 181 (1974).
4. Suria, D., and Lieu, C. C., Can. J. Biochem. 52, 1143 (1974).
5. Peterson, J. C., and McConkey, E. H., J. Biol. Chem. 251, 555 (1976).
6. Radola, B. J., Biochim. Biophys. Acta 295, 412 (1973).
7. MacGillivray, A. J., and Rickwood, D., In "Isoelectric Focusing" (J. B. Arbuthnott and J. A. Beeley, eds.), p. 254. Butterworth, London, 1975.
8. Gilmour, R. S. and MacGillivray, A. J. in "The Molecular Biology of Hormone Action" (J. Papaconstantinou, ed.), p. 15. Academic Press, New York, 1976.
9. MacGillivray, A. J., Cameron, A., Krauze, R. J., Rickwood, D., and Paul, J., Biochim. Biophys. Acta 277, 384 (1972).
10. Rickwood, D., Riches, P. G., and MacGillivray, A. J., Biochim. Biophys. Acta 299, 162 (1973).
11. MacGillivray, A. J. This volume, page 329.
12. MacGillivray, A. J., Johnston, C., and Rickwood, D. (1977). In preparation.

Chapter 26

The Rapid Isolation, High-Resolution Electrophoretic Characterization, and Purification of Nuclear Proteins

WALLACE M. LeSTOURGEON AND ANN L. BEYER

Department of Molecular Biology,
Vanderbilt University,
Nashville, Tennessee

I. Introduction

Current attempts to understand the functions of the cell nucleus are varied and often highly complex. Frequently studies are focused on specific nuclear subfractions such as replicating versus nonreplicating sections of the genome, transcriptionally active versus quiescent chromatin fractions, ribosomal precursor particles, native versus reconstituted chromatin, the nuclear pore complex, metaphase chromosomes, RNP particles, etc. While each study requires unique methodological approaches, often it is very important to monitor the results of various manipulations by examining the protein components of subnuclear fractions. As a starting point a "fingerprint" of total chromatin or nuclear proteins is also very important. In many cases the overall amount or presence or absence of specific proteins can be diagnostic. Also the presence of degraded protein might reflect a procedural problem. In these respects attempts to characterize total nuclear protein or the protein components of subnuclear fractions have at least three procedural "problems" in common.

First, methods must be available for quantitatively solubilizing total protein in a nondegraded, nonaggregated form from various types of nuclear preparations. In comparative studies it is preferable that a single highly reproducible procedure should be used regardless of the starting material, and the extraction procedure should yield protein free of other chromatin components. Second, methods must be employed that allow the rapid examination of individual proteins present in submicrogram amounts in highly heterogeneous mixtures. Finally, for comparative purposes it is important to obtain certain forms of basic information concerning individual proteins. For example, it can be very useful to know the molecular weight, amino acid composition, isoelectric point, period of synthesis during developmental events, biological half-life, intranuclear concentration, and immunological specificity of specific proteins. In many cases this information requires the purification of individual polypeptides.

Described first in this article is a three-step 20-minute procedure for solubilizing "total" protein either from whole nuclei, from subnuclear preparations, or from various chromatographic or density-gradient fractions. The procedure yields nondegraded protein free of nucleic acid and polysaccharide contaminants, and following a single dialysis it is ready to load on high-resolution polyacrylamide gels. Described second are details of the electrophoretic procedures used routinely in the authors' laboratory. The procedures allow the direct visualization and quantitation of hundreds of polypeptides in submicrogram amounts, and less than 5 hours are required from loading samples to observing destained gels. Finally, selected procedures are described for obtaining purified proteins so that more complete characterizations of individual proteins can be obtained.

II. Extraction of "Total" Protein from Various Nuclear Preparations

The procedure described below for the complete extracting of protein from various preparations has evolved somewhat independently over the past few years in various laboratories (1–6). A detailed discussion of the evolution of the procedure has been presented elsewhere (6). Briefly, the the procedure capitalizes on the ability of hot sodium dodecyl sulfate (SDS) to quantitatively dissociate protein from nucleic acids, polysaccharide, and lipids. The author has previously reported that powerful protein solvents such as 9.0 M guanidine hydrochloride, 8.0 M urea, aqueous phenol, and deoxycholate will not solubilize all protein from chromatin (5–7). These findings have been observed by others (8). The procedure also takes advantage of the ability of aqueous phenol to specifically and rapidly concentrate protein in a phase separation from aqueous solutions containing protein, DNA, RNA, and polysaccharide. The absence of chemical or enzymic degradation of primary protein structure is particularly fortuitous. In order to achieve crisp banding in polyacrylamide gels with little or no background smear, protein degradation must be continuously kept at a minimum. Since most studies are conducted on relatively small amounts of starting material the procedure outlined below will concentrate on milligram or less amounts of starting material.

A. Step 1

To a small amount of starting material [10^8 nuclei or nucleoli, 0.5–2 mg wet weight chromatin, a trichloroacetic acid (TCA) or alcohol precipitate of protein from chromatographic fractions, etc.] pelleted by centrifugation in the bottom of a 2.0-ml Potter–Elvehjem tissue homogenizer (Tri-R, S-30, 8.0 mm inside diameter (ID) × 14.8 cm) add 1–1.5 ml of protein extraction solution (solution 1, see below). Suspend the starting material gently with the matching Teflon pestle and place in boiling water for 3 minutes with occasional gentle agitation. Remove and cool to room temperature under tap water for 1 minute.

Important Notes Concerning Step 1

a. The small Potter–Elvehjem is very convenient for working with small samples. It is a straight tube and can be fitted with a cork spacer so that it can be centrifuged at 1000 g in an International Model PRJ or similar centrifuge. This is particularly useful since at no time is it necessary to transfer solution from tube to tube and risk loss of material.

b. If the starting material contains high amounts of DNA it is frequently helpful to reduce the viscosity by shearing. This is effectively achieved by attaching a motor drive to the Teflon pestle and homogenizing briefly.

c. If the starting material is an acid precipitate [TCA, perchloric acid (PCA)] the acid should be removed by washing the precipitate twice with absolute methanol or ethanol. These washings can be performed directly in the small Potter–Elvehjem. Slight wetness from residual alcohol does not interfere with protein extraction and need not be removed before adding solution 1.

d. If the starting material is a salt precipitate or contains high salts incompatible with SDS (i.e., potassium salts, guanadine–HCl, ammonium sulfate), the salts should be removed by washing 2 or 3 times with 70% cold methanol or ethanol before adding solution 1.

e. If a considerable amount of nonprotein material remains visible after heating in solution 1 it should be removed by centrifugation. Protein present in the starting material will be soluble in the supernatant.

f. The high 2-mercaptoethanol in solution 1 has been observed to yield the same effects (unpublished) as reducing and alkylating the protein following standard procedures (9). With time 2-mercaptoethanol in solution will lose some of its potency. In practice solution 1 is good for 2–3 weeks at room temperature.

g. If the starting material is impacted (i.e., TCA precipitate, solid tissue) it may be necessary to use a motor drive on the Teflon pestle. Any sudsing induced by such a procedure can be removed by centrifugation at 1000 g for 1–2 minutes.

h. If the starting material is pure protein (alcohol or TCA precipitate or lyophilized sample) skip step 2 below.

B. Step 2

To the Potter–Elvehjem containing the solubilized and reduced starting material add 1.0 ml aqueous phenol (solution 2)/ml of aqueous phase (solution 1). Make an emulsion by homogenizing with the Teflon pestle either manually or with the motor drive. After 5 minutes centrifuge the emulsion (1000 g) to separate the phases; draw out the lower phenol phase with a long 23-gauge needle attached to a 1.0-ml tuberculin syringe. Add fresh phenol and repeat the extraction procedure. Combine the phenol phases in a second small Potter–Elvehjem.

Important Notes Concerning Step 2

a. Aqueous phenol is a powerful protein solvent which excludes DNA, RNA, and polysaccharide. Dilute protein in large volumes of aqueous solutions can be concentrated many-fold by their selective partitioning into a much smaller phenol phase. Phenol will pull histone and nonhistone out of aqueous solution (6) without discrimination, and it does not selectively solubilize only phosphorylated protein as once thought.

b. Solution 1 for solubilizing protein contains ethylenediaminetetraacetic acid (EDTA) and high levels of 2-mercaptoethanol and SDS. These agents inhibit many enzymes; however, some proteases (Proteinase-K, Merck) are not denatured in SDS, and it is advisable to add the phenol solution without delay. However, in the authors' experience proteolytic activity has not been detected in preparations where nuclear extracts have been allowed to stand at room temperature several minutes before adding the phenol solutions.

c. Usually a small white interface will form between the lower phenol phase and upper aqueous phase. This interface contains some DNA, RNA, polysaccharide, and SDS micelles of the above. If a small amount of this interface material contaminates the withdrawn phenol phase no deleterious effects result. Considerable amounts of nucleic acid or polysaccharide will cause crescent bands, band smearing, and high background in gels.

d. If a very large amount of starting material is being extracted (2–5 mg protein) it is advisable to extract at least twice with phenol. If a very small amount of starting material is extracted (500 μg or less) usually only one phenol extraction is required to effectively concentrate the protein in the phenol phase. Additional extractions will dilute the sample.

e. If phase separation is achieved in a refrigerated centrifuge (which is recommended) the lower phenol phase and upper aqueous phase may appear cloudy. This cloudiness disappears in a few minutes on warming to room temperature.

C. Step 3: Preparing Extracted Protein for Electrophoresis

Following step 2, the protein originally present in whole nuclei, chromatin, subnuclear fractions, etc. will be completely solubilized in the phenol solution. The only further requirement is to reconstitute the protein in a suitable "sample" buffer for electrophoresis. The easiest way to reconstitute protein for electrophoresis is to dialyze the phenol phase against "sample buffer" (solution 3) for a minimum of 4 hours or overnight. The phenol phase should be placed in previously prepared dialysis tubing (A. H. Thomas Co., 3787-D12, $\frac{1}{4}$-in. tubing or similar) and dialyzed against two changes of 200 volumes of solution 3. During the first 3 hours of dialysis it is recommended that the contents of the dialysis tubing be occasionally mixed by passing through pinched fingers. Gels can be run the next morning, or the samples can be stored frozen for later analysis. Ideal protein concentration for both slab and disc gels is 1 mg/ml sample buffer.

IMPORTANT NOTES CONCERNING STEP 3

a. Protein reconstituted in "sample buffer" is stable for years when stored frozen. However, frequent thawing and refreezing will lead to fuzzy bands and high background.

b. Protein solubilized in the phenol phase is stable for several days if stored in the refrigerator.

c. If it is expected that considerably less than 1 mg/ml protein is reconstituted in sample buffer the preparations can be rapidly concentrated before electrophoresis by embedding the dialysis tube containing the sample in Sephadex G-100. Phenol and 2-mercaptoethanol will interfere in the Lowry et al. (10) assay for protein. For quantitation an aliquot of the phenol phase or reconstituted protein should be dialyzed extensively against 0.1% SDS.

d. If one has in hand a preparation of pure protein (alcohol or TCA precipitate, lyophilized sample, etc.) it is not necessary to use the phenol purification step. Simply solubilize directly in sample buffer (solution 3) and load on gels.

e. Through the procedures outlined above, the extraction of protein, free of nucleic acid, is achieved without the use of nucleases and extended incubation periods at elevated temperatures which could activate endogeneous proteases.

D.   Required Solutions

Solution 1 (Protein Extraction Solution):
2% sodium dodecyl sulfate from BDH Biochemicals distributed by Gallard-
   Schlesinger Chemical Mfg. Corp., Carle Place, Long Island, New York.
20 mM EDTA
20 mM Tris-HCl (pH 8.2)
0.1% 2-mercaptoethanol
Note: The EDTA aids in dissociating nucleic acid–bound proteins. The high mercaptoethanol assures complete reduction of disulfide bonds.

Solution 2 (Buffer-saturated Phenol for Protein Concentration and Purification): Using a separatory funnel of suitable size saturate redistilled phenol three times with equal volumes of solution 1 not containing SDS. The pH of the phenol solution should stabilize near 7.6 after buffer saturation. Appropriate caution should be exercised in phenol distillation and buffer saturation. Saturation with buffer is facilitated by melting the phenol in a hot water bath before adding to buffer previously placed in the separatory funnel. The saturated phenol is collected as the bottom phase after separation on standing at room temperature and should be stored in the refrigerator.

Solution 3 (Sample Buffer):
500 ml 1.0 M sucrose
20 ml 2-mercaptoethanol
20 ml 10% SDS solution
19 ml 0.2 M NaH_2PO_4
81 ml 0.2 M Na_2HPO_4
Bring to 2000 ml with distilled water. The pH should be consistently near 7.4. The high mercaptoethanol maintains all protein in a completely reduced state. Potassium buffers cannot be used due to their incompatibility with SDS.

III.   Electrophoretic Separation of Protein for Analytical or Comparative Purposes

The procedure described above for rapidly solubilizing total protein from various sources is intended for use primarily as a "screening" procedure for low levels of nuclear protein. It is important therefore that an efficient and highly sensitive system be used for rapidly resolving the mixtures of protein into as many individual components as possible. The discontinuous polyacrylamide gel system discussed here is basically that described by Ornstein (11) and Davis (12) as modified by Laemmli (13) with the addition of SDS. The system has been used successfully in thin-slab, tubular, and preparative gel electrophoresis (largest gel so far cast is 2.5 cm ID) and as the second dimension of a two-dimensional gel electrophoresis system. The gel system has previously been used successfully by O'Farrell (14) as the second dimension of a two-dimensional electrophoresis system. In the author's experience single bands in urea-containing gels or Weber–Osborn gels (15) often are resolved into several components when the gel system described below is used. The stock solutions and gel recipes are listed below with relevant comments concerning each solution. Presented first are several suggestions which may lead to consistently better gels. Some knowledge of SDS–polyacrylamide gel electrophoresis is assumed.

A. Analytical Thin-Slab Electrophoresis

For routine analysis of nuclear, subnuclear, chromatographic, or density-gradient fractions the technique of thin-slab SDS–polyacrylamide gel electrophoresis has several advantages over cylindrical or "disc" gels as follows. (Comments pertain to studies using the Hoeffer model SE 520–0.75-mm slab gel unit, Hoeffer Scientific Instruments, 2609 California Street, San Francisco; also sold by Bio-Rad Laboratories.)

1. From 1 to 40 samples can be analyzed at one time.

2. The average time required for electrophoresis is 3 hours.

3. Gel slabs are easily removed and can be stained in 30 minutes (0.05% Coomassie blue–10% MeOH–10% acetic acid) and diffusion destained for visual inspection in 30 minutes (complete destaining in 4–6 hours). No special equipment is required for staining or destaining.

4. Individual polypeptide bands containing as little as 0.2 μg protein can be detected visually and recorded photographically and spectrophotometrically. Linear dye binding occurs between 0.2–2.5 μg per polypeptide band.

5. A heterogeneous sample (load of 30 μg in 30 μl) of total nuclear protein will allow direct visualization of individual proteins present at an intranuclear concentration as low as 40,000 molecules per nucleus (faintest bands).

6. Thin-slab gels consistently yield increased resolution with crisp boundaries of bands.

7. Increased reproducibility can be expected even when variation occurs in sample load volume.

8. Identical polypeptides in separate fractions migrate "in line," and if the sample application solutions contain very different concentrations of electrolytes or SDS such that the "ion front' is not straight, identical polypeptides still "point" to each other in the gel.

9. The thin-slab gels can be dried *in situ* on hard fine filter paper supports following the general procedure of Fairbanks *et al.* (*16*) for autoradiography or for storage in notebooks. No glycerol or other softening agents are required to prevent gel cracking.

When assembling the slab gel equipment all components should be clean and dry. Assembly should follow manufacturer's recommendations although it is sometimes necessary to use small amounts of silicone grease to prevent leaks where glass plates and polyvinylchloride (PVC) spacers mesh against the rubber seals. It is not necessary to siliconize the glass plates. The lower-resolving gel should be added from the top by capillary wetting of the outer plate using a narrow-tip 20-ml transfer pipette instead of through the bottom as suggested by the manufacturer. In this way adding the resolving gel solution can be achieved in about 20 seconds and no needle and syringe are required. When casting 0.7-mm slabs, straight interfaces are best achieved with the plastic inserts designed for this purpose. It is necessary to overfill the unit around the inserts to prevent polymerization below the desired level. Small bubbles or dust particles under the insert will create large depressions in the resolving gel interface. When casting thick slabs (2–3 mm) better interfaces result when an overlay of 0.1% SDS or anhydrous isobutanol is used. Stacking gels and loading wells should be formed with the various-width combs supplied by the manufacturer. The more narrow the sample application well, the better the resolution and sensitivity. Band broadening does not occur unless samples are drastically overloaded (>4 μg/band) or unless DNA or RNA is present in the sample.

The lower electrode buffer should not cover more than 1.0 cm of the bottom gel plates, and as much upper electrode buffer as the upper chamber will hold should be used. The lower electrode buffer solution can be reused at least once without adverse effects. All electrophoresis should be performed at room temperature and cooling water need not be used. When two 0.75-mm slabs are used a constant current of 20 mA is applied until the marker dye is just into the resolving gel, and the current is then adjusted to 40 mA. Higher currents must be used for thicker slabs.

Sample buffer (solution 3 above) contains 0.25 M sucrose for increased density. The samples are loaded by underlay into the wells after the upper electrode buffer is added. Before loading samples but after adding upper electrode buffer each well should be flushed out with electrode buffer using a tuberculin syringe and 2 in. $\times$ 24 gauge needle. Any unpolymerized stacking gel solution in the wells will prevent clean sample application. A sufficient amount of bromphenol blue solution should be added to each sample to facilitate visual observation during loading and to monitor the ion front during electrophoresis. Turn off the current when the ion front is 1.0 mm from the bottom of the gel. Figures 1 and 2 illustrate typical slab-gel separations of total nuclear protein from mammalian cells in tissue culture and from avian erythrocytes. Various known proteins are labeled in the figures.

B. Analytical Cylindrical-Gel or "Disc"-Gel Electrophoresis

Laboratories with equipment for cylindrical-gel electrophoresis and spectrophotometric equipment for scanning cylindrical gels can expect opti-

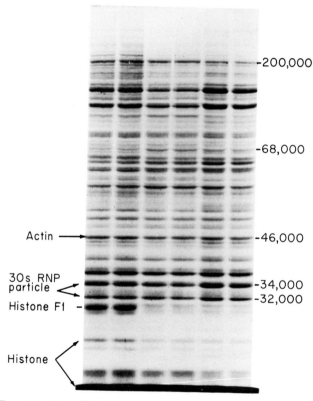

FIG. 1. Example of "total" nuclear protein extracted, purified, and resolved in thin-slab SDS-containing 8.75% polyacrylamide gels as described in text. Nuclei were isolated from three strains of Chinese hamster ovary (CHO) cells in tissue culture. Samples were run in duplicate to show banding fidelity. Gels containing 8.75% acrylamide are suitable for resolving the nonhistones. See Fig. 2 for histone resolution in thin slabs.

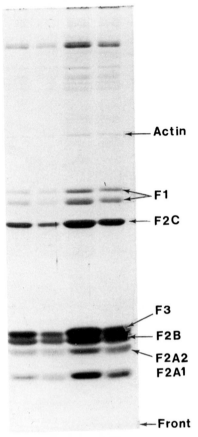

FIG. 2. Example of "total" chicken reticulocyte nuclear proteins extracted, purified, and resolved in thin-slab SDS-containing 12.5% polyacrylamide gels as described in text. Endogenous actin (MW 46,000 in this system) was labeled to provide a reference for comparing to the separation in Fig. 1. The resolved histones were labeled following the pattern of resolution achieved by Panyim and Chalkley (19) and through comparative studies in this laboratory.

mum results using the same gel solutions and recipes described above for slab-gel electrophoresis. In general casting 12 or 24 perfect 5- or 6-mm diameter gels is a bit more tricky and much more time-consuming than casting two 40-place slabs. Since in tubular gels each sample runs independently, more attention must be paid to constant gel length, diameter, loading volume, electrolyte concentration in sample load solution, and protein load. In the authors' laboratory a Hoeffer model DE 101 12-place electrophoresis unit is used. However, if one has access to the slab-gel equipment described previously it can be converted for tube-gel work by simply drilling 12 or more 13-mm holes in the upper reservoir chamber and in the plastic plate which serves as the bottom plate to the unit core. In slab-gel work these holes can be plugged with small rubber stoppers so that the upper chamber will again hold electrode buffer. Usually 5-mm ID by 15-cm glass tubes are used for casting the discontinuous cylindrical gels. As in slab-gel work, the smaller the gel diameter the better the resolution. The same equipment and glass tubes are used when resolving RNA species in 3% gels and for analytical isoelectric focusing of protein in 4% polyacrylamide gels containing 2% ampholyte and 8.0 M urea.

The most frequent problem encountered in cylindrical-gel work is band dishing (crescent-shaped bands convex toward the anode, see gel 1, Fig. 3). The following suggestions should help minimize this problem.

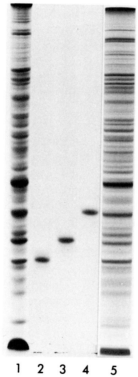

1 2 3 4 5

FIG. 3. Examples of various proteins purified by preparative gel electrophoresis and band "dishing" in tubular gels. Gels 1 and 5 show a fraction of residual nuclear proteins from the lower eukaryote *Physarum polycephalum* resolved in tubular 8.75% acrylamide gels. In gels 1–4 protein was resolved in 4.5 mm × 10 cm gels and photographed lying in a grooved Plexiglas plate. In gel 5 protein was resolved in a 6 mm × 10 cm gel and photographed while supported in a glass tube filled with gel destaining solution. Gels 2–4 show the excellent purification achieved with the preparative gel equipment described in Fig. 5. These purified bands yield single spots when subjected to isoelectric focusing in 4% polyacrylamide gels as described by O'Farrell (*14*).

In casting cylindrical gels some apparatus must be used to support all the glass tubes parallel to each other and vertical. A rectangle of Plexiglas 1.5 in. thick, 2 in. wide, and 6 in. long with 12 holes precision-drilled perpendicular to the long axis serves this purpose well. For more accuracy the Plexiglas block can be fitted with adjustable screw-type legs for leveling on the lab bench. The diameter of the holes should be very near to the outside diameter of the gel tubes to provide proper parallel and vertical support. Flat gel bottoms after polymerization are important in obtaining straight bands, and Parafilm serves better than rubber cups for sealing the bottom of the gel tubes.

In comparative studies or for more consistent reproducibility it is important that the internal diameter of the glass tubes be as similar as possible. Precision-wall glass tubes are recommended. In the same context it is important that resolving and stacking gel length be identical among all gels. This can be facilitated by using identical-length glass tubes premarked in the glass shop at an appropriate distance (10.5 cm) from the bottom. Filling resolving gel solution to the mark can be easier and faster than measuring each aliquot. If gel tubes have slightly different internal diameters measured volumes of gel solution will always yield gels of different length.

The glass tubes should be silanized to help prevent wall effects. For silanization, the tubes should be washed in chromic acid or strong hot Alconox, rinsed well in distilled and deionized water, dried at 160°C overnight, and while still warm dipped in a solution of reagent-grade carbon tetrachloride containing hexamethyl disilazane [add 3.0 ml of HMDS (Pierce, 84770, Box 117, Rockford, Illinois) to 50 ml of carbon tetrachloride in a 50-ml glass graduated cylinder]. The tubes can be dipped directly into the 50-ml graduated cylinder two or three times, transferred upright into a glass beaker, and heated for 30–60 minutes in a dry-heat sterilizer. Silanization should be performed in a fume hood.

After filling all tubes with lower resolving gel solution to the mark, an overlay of 0.1% SDS or anhydrous isobutanol should be applied. Better interfaces can be obtained if a small-lumen (30 gauge) 3–4 in. needle is used attached to a 1.0-ml tuberculin syringe with smooth plunger action. Sometimes, for unknown reasons, straighter bands can be obtained if after filling the tubes with resolving gel solution they are capped with small stoppers and all the tubes at once are tipped over 2 or 3 times. After removing the caps, let stand for 2 minutes to allow for wall drainage, then add a small amount (0.1 ml) of isobutanol to yield a straight interface following polymerization. During polymerization a small cardboard box covered with aluminum foil should be placed over the gel tubes to prevent cool or warm drafts or radiant heat from nearby equipment from causing nonuniform polymerization. Before adding a constant volume of stacking gel solution (0.2 ml for 5-mm ID tubes) the isobutanol or 0.1% SDS should be aspirated off and the lower gel surface and tube rinsed with a small volume of stacking gel solution not containing TEMED (formula for the latter is given in Section III,C). After adding complete stacking gel solution a straight interface should be made again either with 0.1% SDS or isobutanol. Stacking gels are not required for good separation in either cylindrical gels or slab gels; however, on an empirical basis gels with stacking gels seem to yield slightly sharper bands after staining. Figures 3 and 4 illustrate the results of protein electrophoresis in tubular gels.

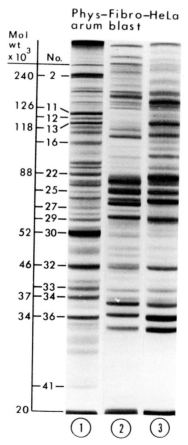

FIG. 4. Example of protein resolution in 6 mm × 12.5 cm tubular gels. All gels were subjected to electrophoresis simultaneously in a single 12-place unit. While some protein homologies seem to exist among the mammalian cell types better comparisons would be possible if the samples had been run parallel in thin-slabs.

C.　Required Stock Solutions, Recipes, and Important Notes in Making Polyacrylamide Gels

STOCK SOLUTIONS

1. 10% SDS solution (w/v) (500 ml) (BDH Chemicals Ltd., 30176).
2. 0.1% SDS solution (w/v) (100 ml).
3. Acrylamide solution (store in refrigerator): acrylamide, 30.0 gm (put into solution first); bisacrylamide, 0.8 gm; distilled water qs (total volume) 100.0 ml, then filter or centrifuge to remove lint.
4. Resolving gel buffer (pH 8.8): Tris base, 18.0 mg (Eastman, 4833); 10% SDS solution, 4.0 ml; 6 N hydrochloric acid, about 5 ml to pH 8.8; distilled water qs 100.0 ml.
5. Stacking gel buffer (pH 6.8): Tris base, 6.0 gm; 10% SDS solution, 4.0 ml; 6 N hydrochloric acid, about 7 ml to pH 6.8; distilled water qs 100.0 ml.
6. Electrophoresis buffer (electrode buffer, reservoir buffer). Can be made as concentrate and stored frozen. If so, leave out SDS in concentrate and add after thawing. Tris base, 6.0 gm; glycine, 28.7 gm; 10% SDS solution, 20.0 ml; distilled water qs 2000.0 ml.
7. Ammonium persulfate solution (make fresh): ammonium persulfate, 1.0 gm (Matheson Coleman and Bell, AX1340-CB151); distilled water, add 10.0 ml.
8. N,N,N',N'-tetramethyl-ethylenediamine (TEMED) (Eastman, 8178).
9. Isobutanol (2-methyl-1-propanol).

RECIPES FOR SDS–POLYACRYLAMIDE GELS

The following recipe will provide sufficient resolving gel solution for making 12 cylindrical gels 10.5 cm by 6 mm; one 2-cm ID by 8-cm prep gel; or two 0.7-mm thin-slab gels. The recipe must be doubled to make one 3-mm thick-slab gel.

For 8.75% acrylamide resolving gels:

a. In a 1000-ml round-bottom flask or other device suitable for evacuation combine the following: distilled water, 18.0 ml; resolving gel buffer, 10.0 ml; acrylamide solution, 11.7 ml.

b. Deaerate by evacuating the flask with a standard water aspirator, and heat gently over a Bunsen burner until the solution just begins to boil. With a good vacuum boiling should begin at about 30 to 35 °C.

c. Cool to room temperature under reduced vacuum; then add: ammonium persulfate solution, 150 mm³; TEMED, 10 mm³.

d. Mix; then follow suggestions in text for casting gels.

Note: in 8.75% acrylamide gels proteins with molecular weights less than 16,000 will usually migrate with the ion front and not be resolved. For resolving proteins with lower molecular weights the acrylamide concentration can be increased by simply altering the ratio of water and acrylamide solution used. See Fig. 2 for resolution of histones in 12.5% acrylamide gels.

For making 3% stacking gels:

a. In a 50-ml Erlenmeyer flask combine: distilled water, 6.5 ml; stacking gel buffer, 2.5 ml; acrylamide solution, 1.0 ml; ammonium persulfate solution, 30.0 mm³.

b. Mix; then use a few drops of this solution to rinse off the upper surface of the polymerized resolving gels.

c. Add TEMED, 10.0 mm³.

d. Mix; then follow suggestions in text for casting stacking gels.

Note: Stacking gel solution does not need to be deaerated. Stacking gels are not required for good protein separation; however, slightly sharper bands are more consistently obtained when stacking gels are used. For cylindrical gels apply a constant current of 1 mA/tube until the marker dye is 1 mm into the resolving gel; then adjust the current to 2 mA/tube. If the sample is contaminated with DNA, RNA, or polysaccharide the tops of the stacking gel will invaginate and cause band distortion. Distortion of the stacking gel interface will also result if too high a current is applied during electrophoresis.

IV. Protein Purification through Preparative SDS–Gel Electrophoresis and Isoelectric Focusing

As stated previously it is often important in many studies to precisely identify protein components of various subnuclear fractions. Much useful information can be obtained on purified proteins even though they may have been exposed to SDS during purification. Described below are two procedures where SDS–gel electrophoresis is used to purify individual polypeptides from highly heterogeneous preparations. As in analytical work, proteins are separated as a function of their differing molecular weights. Also described is a procedure for purifying individual proteins in the absence of SDS through preparative isoelectric focusing in urea and ampholite containing sucrose gradients. In this procedure proteins are separated as a function of differing isoelectric points. The various procedures can yield up to milligram quantities of protein which is sufficient to obtain several amino acid analyses, more accurate molecular weights, isoelectric points, and antisera to specific polypeptides. In many cases classical procedures for purifying specific proteins are less than satisfactory when applied to nuclear

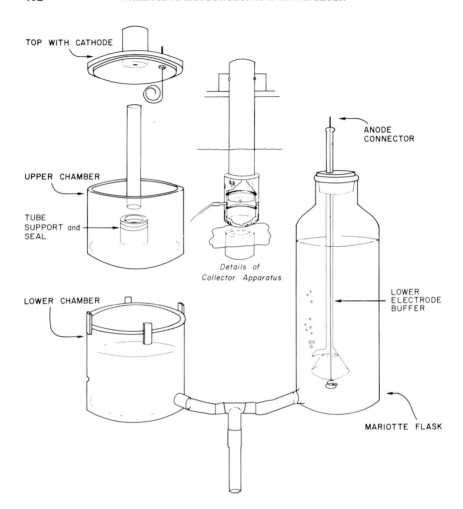

FIG. 5. Equipment for preparative gel electrophoresis and preparative isoelectric focusing. On assembly the lower chamber is hermetically sealed against the upper chamber with silicone grease. In this way a constant hydrostatic pressure can be maintained through use of the Mariotte flask. Gas generated by the lower electrode in the Mariotte flask does not change the hydrostatic pressure because it is vented up and out the bubble tube. By positioning the Mariotte flask the flow rate of lower electrode buffer over the bottom of the gel can be carefully controlled and maintained. As protein molecules elute off the bottom of the gel, lower reservoir buffer (flowing in through three small holes in the Plexiglas collector apparatus) rinses the protein into a 1-mm ID tubing attached to a fraction collector. The dead volume below the gel bottom is kept at about 1.5 mm by inserting into the collector apparatus a machined Plexiglas cylinder covered with dialysis tubing. The collector apparatus, dialysis membrane, and the insert are held firmly in place by "O" rings. An "O" ring inside the tube support collar in the upper reservoir chamber holds the tube in place and prevents upper reservoir buffer from

proteins. The techniques discussed here have been proved successful in purification of specific nonhistone proteins.

A. Large-Diameter Cylindrical-Gel Electrophoresis and Equipment

There is no fundamental difference between preparative gel electrophoresis and analytical gel electrophoresis. The basic objective is to collect various populations of molecules as they migrate out the bottom of large-diameter cylindrical gels. The equipment diagrammed in Fig. 5 for preparative gel electrophoresis using large gels was designed by the author and can be custom-built for about $100. Details of its operation appear in the figure legend. An example of proteins purified through this procedure is presented in Fig. 3.

In preparative electrophoresis the same gel solutions and recipes are used as previously described for analytical work. However, when casting large-diameter gels (2 cm) the volume of ammonium persulfate solution and TEMED is doubled to induce more rapid polymerization. Large gels which polymerize slowly will frequently show uneven polymerization which can be observed as swirling refractive lines running through the gel matrix. As much as 20 mg of a heterogeneous mixture of protein can be loaded on a 2-cm diameter gel. For best resolution more than 2 mg protein per band should not be loaded for each separation. Larger-diameter gels can be used for higher concentrations of protein per band. For 2-cm diameter gels a constant current of 20 mA should be applied until the marker dye is just into the resolving gel; then the current should be adjusted to 40 mA.

B. Thick-Slab Preparative Gel Electrophoresis

Other than differences in equipment design there is no fundamental difference between protein purification in large-diameter cylindrical gels or thick-slab gels. Thick-slab preparative gel electrophoresis can be performed using the same equipment described for analytical slab-gel work. In preparative work the 3-mm PVC spacers have inlet and outlet ports for pumping a suitable solution (electrode buffer) through a narrow gap between the bottom of the resolving gel and a lower polyacrylamide plug. This is more

---

leaking into the lower chamber. The flow rate should be adjusted so that no marker dye escapes into the lower chamber. See additional comments in text for use of this equipment and also for converting the equipment for preparative isoelectric focusing in sucrose gradients containing ampholytes and 8.0 M urea. Figure 3 shows various proteins purified using this equipment.

clearly described in Fig. 6. The procedural changes described above for forming large-diameter gels should be followed in forming 3-mm thick-slabs. In both procedures protein purification must be monitored by performing analytical gel electrophoresis on various fractions. The ability to examine 40 samples at one time greatly simplifies the problem of locating fractions containing purified proteins of interest. Using 3-mm thick-slabs, proteins with molecular weights differing as little as 300 can be clearly separated and purified. Occasionally it is necessary to first use the procedure of preparative isoelectric focusing described below to reduce the number of polypeptides which have similar molecular weights and which migrate closely in SDS-containing gels.

C. Preparative Isoelectric Focusing of Proteins in Sucrose Gradients

In cases where proteins of interest are not well separated through molecular-weight sieving (presence of other proteins with similar molecular weight), or in cases where it is desirable to purify proteins in the absence of SDS, the procedure of preparative isoelectric focusing can be very helpful.

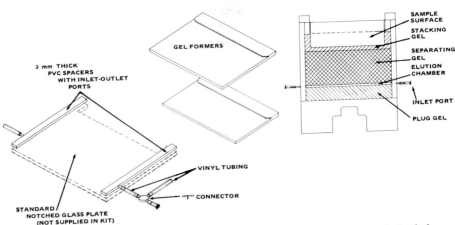

FIG. 6.   Equipment for slab-gel preparative electrophoresis. The essential step in the design is to form a channel for elution buffer to carry off samples in a continuous flow. This is achieved by introducing a narrow sucrose band in the preparative-slab gel. The operator prepares the slab by layering in, first, a dense plug gel at the base of the glass plates; then, a narrow band of sucrose; and finally, the resolving gel. After the gels polymerize and electrophoresis begins, the sucrose in the narrow channel is forced out one elution port as elution buffer is pumped in at the other end. The attractive feature of this design is that the same equipment for analytical thin-slab electrophoresis can be used for purifying individual polypeptides from heterogeneous mixtures. Courtesy of Hoefer Scientific Instruments, 650 Fifth St., San Francisco, California.

The equipment diagrammed in Fig. 5 for preparative SDS–gel electrophoresis can also be used for isoelectric focusing of proteins in solution. The only difference is that the gel tube bottom is covered with tightly stretched dialysis tubing and a 1-cm plug of acrylamide gel is polymerized in the bottom of the tube. The solution inside the tube should contain 8.0 M urea–2% ampholites–0.01% mercaptoethanol and a linear sucrose gradient from 30% on the bottom (resting on the gel plug) to 10% on the top. The 10% acrylamide plug polymerized over the dialysis tubing should contain 30% sucrose and all other components of the column solution except ampholites and 2-mercaptoethanol. In a 2-cm glass tube the volume of ampholite–urea solution should be 22 ml if the tube length is 5 in. This will allow sufficient space on top to overlay up to 5 ml of sample. Protein should be solubilized in 8.0 M urea adjusted to pH 9 with KOH. The lower electrolyte solution should contain 8.0 M urea–30% sucrose–0.01% 2-mercaptoethanol, and the pH should be adjusted to 2.5 with H_3PO_4. The upper electrolyte solution is 0.02 N NaOH.

Up to 30 mg of a heterogeneous protein preparation can be focused per run. In the beginning a current of 15 mA should be applied using a constant-voltage power supply; after 6 hours the current should drop to about 1 mA. The current should be readjusted to 6 mA, and after 10–14 hours the run can be terminated. Fractions are collected by puncturing the bottom gel plug with a suitable syringe needle and dripping. The contents of the column containing the separated protein can also be pumped through a fraction collector, but resolution efficiency is sacrificed due to mixing in the connecting tubing.

V.   Overview

The procedures outlined in this chapter are intended for use in general survey studies or in detailed comparative studies concerning specific proteins where sensitivity, reproducibility, simplicity, and speed are required. The methods described for purifying specific polypeptides are reliable but generally are considered to yield denatured protein. However, in many cases much useful information can be obtained on such proteins. Evidence also exists that SDS-treated proteins can be used in immunological studies (*17, 18*). Methods for purifying specific nuclear proteins through more classical procedures are described in a second article in this book by the senior author. It should be emphasized that the procedures outlined above can be instrumental in monitoring the progress of attempts to characterize nuclear proteins through more classical biochemical techniques. As the field of nuclear protein research becomes more sophisticated so must the procedures become more sophisticated and especially more reliable. Considerable enthusiasm has recently been generated by the potentials for two-parameter polyacrylamide gel electrophoresis in nuclear protein research (isoelectric focusing and molecular-weight seiving). In many initial studies however, the two-parameter system may not be the most useful procedure to start with. In comparative studies the most essential features of an electrophoretic system are reproducibility, lack of artifact, and potential for valid quantitation. When these problems are solved the two-parameter system should be of significant value in increasing the investigator's ability to monitor and characterize more minor nuclear constituents.

ACKNOWLEDGMENTS

The authors thank Kenneth J. Hardy for providing isolated chicken reticulocyte nuclei for studies. Supported by NSF grant BMS 75-03105.

REFERENCES

1. Steele, W. J., and Busch, H., *Cancer Res.* **23**, 1153 (1963).
2. Shelton, K. R., and Allfrey, V. G., *Nature (London)* **228**, 132 (1970).
3. Teng, C. S., Teng, C. T., and Allfrey, V. G., *J. Biol. Chem.* **246**, 2597 (1971).
4. Helmsing, P. J., *Cell Differ* **1**, 19 (1972).
5. LeStourgeon, W. M., and Rusch, H. P., *Arch. Biochem. Biophys.* **155**, 144 (1973).
6. LeStourgeon, W. M., and Wray, W., *In* "Acidic Proteins of the Nucleus" (I. L. Cameron and J. R. Jeter, eds.), p. 59. Academic Press, New York, 1974.
7. LeStourgeon, W. M., Forer, A., Yang, Yeu-Zu, Bertram, J. S., and Rusch, H. P., *Biochim. Biophys. Acta* **379**, 529 (1975).
8. Pederson, T., and Bhorjee, J. S., *Biochemistry* **14**, 3238 (1975).
9. Crestfield, A. M., Moore, S., and Stein, W. H., *J. Biol. Chem.* **238**, 622 (1963).
10. Lowry, D. H., Rosebrough, N. J., Farr, A. L., and Randall, R. J., *J. Biol. Chem.* **193**, 265 (1951).
11. Ornstein, L., *Ann. N. Y. Acad. Sci.* **121**, 321 (1964).
12. Davis, B. J., *Ann. N. Y. Acad. Sci.* **121**, 404 (1964).
13. Laemmli, U. K., *Nature (London)* **227**, 680 (1970).
14. O'Farrell, P. H., *J. Biol. Chem.* **250**, 4007 (1975).
15. Weber, K., and Osborn, M., *J. Biol. Chem.* **244**, 4406 (1969).
16. Fairbanks, G., Levinthal, C., and Reeder, R. H., *Biochem. Biophys. Res. Commun.* **20**, 393 (1965).
17. Silver, L. M., and Elgin, S. C. R., *Proc. Natl. Acad. Sci. U.S.A.* **73**, 423 (1976).
18. Tjian, R., Stinchcomb, D., and Losick, R., *J. Biol. Chem.* **250**, 8824 (1974).
19. Panyim, S., and Chalkley, R., *J. Biol. Chem.* **246**, 7557 (1971).

Chapter 27

Two-Dimensional Polyacrylamide Gel Electrophoretic Fractionation

PATRICK H. O'FARRELL AND PATRICIA Z. O'FARRELL

Department of Biochemistry and Biophysics,
University of California, San Francisco,
San Francisco, California

I. Introduction

Electrophoresis has been used extensively in the analysis of chromosomal proteins. Although a number of procedures successfully resolve the various species of histones, the NHC (nonhistone chromosomal) proteins have proved to be so heterogeneous that most electrophoretic techniques can only resolve and detect the major components. The recent development of a high-resolution, two-dimensional electrophoresis system (1) provides a technique suitable for the analysis of this complex group of proteins. The technique combines isoelectric focusing in the first dimension and sodium dodecyl sulfate (SDS) slab-gel electrophoresis in the second dimension. As an example of the resolution afforded by this technique, Fig. 1 displays the electrophoretic separation of total proteins derived from a rat liver hepatoma cell line (HTC). Up to 1600 components have been resolved and proteins at levels as little as one part in ten million have been detected. Denaturing agents are present in both dimensions so that the tendency of NHC proteins to aggregate is diminished. To date the procedure has been used by Peterson and McConkey (2) to analyze the protein composition of chromatin in HeLa Cells. These studies revealed that there are at least 450 distinct NHC protein species.

This chapter describes the two-dimensional electrophoresis procedure and some problems that might be encountered with samples prepared from chromatin. For a complete discussion of all the electrophoretic parameters and for some variations in the technique not described here see O'Farrell (1).

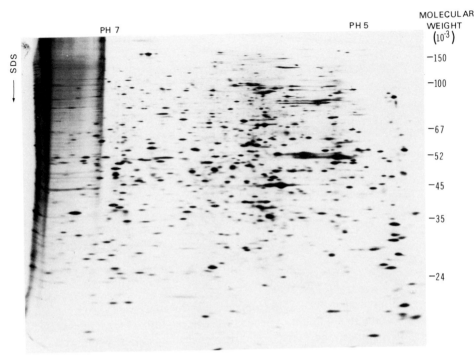

FIG. 1. The separation of total cell protein from a rat hepatoma cell line (HTC). The cells were labeled in microwells as described in the first labeling protocol of Section II,A. 5×10^5 cpm of trichloroacetic acid–precipitable radioactivity were loaded on the first-dimension gel and electrophoresed through both dimensions as described in Sections III and IV. The autoradiogram was obtained from a 10–day exposure. The upper-left corner is the top (basic end) of the first-dimension gel, and isoelectric focusing is from left to right. The SDS dimension is from top to bottom. The SDS gel is an exponential gradient of 10 to 16% acrylamide which yields the indicated molecular-weight distribution of proteins. The high background of unresolved material along the basic (left) side of the figure is typical of the separation obtained with crude extracts from mammalian cells.

II.   Preparation of Samples for Electrophoresis

The best results are obtained when the proteins to be separated are radio-actively labeled and detected by autoradiography. Below, we describe the conditions for labeling cell protein to a high specific activity with radio-active methionine. These are general conditions meant to serve as guidelines for application to particular systems. Radioactive precursors other than [^{35}S]methionine can be useful for some experiments; e.g., $^{32}PO_4$ can be used to detect phosphorylation of specific proteins (3). We first describe the conditions for preparation of radioactive cell extracts, and then we provide some recommendations for the purification of chromosomal proteins so that they are suitable for detection by two-dimensional electrophoresis.

A. Radioactive Labeling of Cells Growing in Tissue Culture

The following protocol is used to label cells growing on surfaces for the preparation of total cellular extracts. Cells are grown at 37°C in multiwell dishes (containing microwells with an area of 2 cm²; obtained from Falcon Plastics) to a density of about 5×10^4 cells/cm². The medium is removed, and the cells are washed with 2 ml of methionine-free medium containing 10% whole serum. This medium is removed and replaced with 200 μl of fresh methionine-free medium supplemented with serum and containing 80 μCi of [³⁵S]methionine (specific activity between 100 and 400 Ci/mM; obtained from New England Nuclear or Amersham-Searle). Labeling times ranging in length from 15 minutes to 4 hours at 37°C have been used. Following the labeling period, the radioactive medium is removed. To remove serum proteins, cells are washed with 2 ml of phosphate-buffered saline at room temperature. To prepare a total cell extract, the cells are lysed by flooding the surface of the microwell with 100 μl of lysis buffer [9.5 M urea–2% Nonidet P-40–5% β-mercaptoethanol–1.6% Ampholines (pH 5–7)–0.4% Ampholines (pH 3.5–10)]. The lysed cells in 100 μl of lysis buffer are transferred to small vials and stored at -70°C. To avoid possible artifactual modifications of proteins (1), the time that the samples are at room temperature is kept to a minimum and the samples are never heated.

Cells growing in suspension culture can be labeled using a similar procedure. The cells are grown to a density of about 25% of their saturation density and collected by centrifugation. The cells are washed by resuspension in methionine-free medium and centrifuged. The cells are resuspended in methionine-free medium containing 400 μCi of [³⁵S]methionine per milliliter, at a final cell density about 4 times the initial cell density. After the labeling, the cells are centrifuged and washed with phosphate-buffered saline as above. The cell pellet is lysed by the addition of lysis buffer (approximately 100 μl/10⁵ cells).

For the preparation of subcellular fractions, a larger number of cells are labeled with radioactive methionine. For example, cells in 100-mm diameter petri dishes are labeled in 1–2 ml of methionine-free medium containing 200–400 μCi/ml [³⁵S]methionine. The plates should be agitated occasionally to prevent desiccation of the cells. After labeling and washing the cells with phosphate-buffered saline, the cells are removed with a rubber policeman, lysed, and subcellular fractions are prepared.

The optimal labeling conditions will obviously depend on the cell type, growth conditions, and experimental requirements. The described procedures have been used to label a wide range of cell types in a number of tissue-culture media. In all cases, methionine-free medium refers to a methionine-free medium supplemented with 10% whole serum, and due to the methionine content of the serum, the medium actually contains approximately 3 μM methionine (4). Mammalian tissue-culture cells show large changes in the synthesis of different proteins when the growth conditions are altered (5); thus it is essential that the growth and labeling conditions be carefully standardized for any set of experiments involving a comparison of different samples.

The number of proteins detected depends on the specific activity of the protein sample to be electrophoresced; thus, in order to compare accurately two or more samples, the same number of acid-precipitable cpms should be subjected to electrophoresis. Nonionic detergent (Nonidet P-40) in the lysis buffer can cause artifacts in the determination of the amount of trichloroacetic acid (TCA)–insoluble radioactivity. The addition of TCA to a solution of NP-40 causes the detergent to separate into a different phase which can pass through a filter and can carry protein with it. To avoid this problem, a small aliquot of a sample in lysis buffer is used to determine the acid-precipitable radioactivity. Added to 0.5 ml of water are 1 drop of 0.2% bovine serum albumin, 1 μl of sample, and 0.5 ml of 10% TCA. The precipitate is allowed to form for 20 minutes at room temperature and is collected on Millipore filters washed with 5% TCA. With a 3-hour labeling period we generally obtain $1-5 \times 10^6$ TCA-precipitable cpm/10^5 cells corresponding roughly to 100,000 cpm/μg of protein.

B. Preparation of Subcellular Fractions

A large number of methods for preparation of chromatin and extraction of chromosomal proteins have been published. Although it is not the purpose of this chapter to review these methods, this electrophoretic technique has some specific requirements and a number of precautions must be taken. Some suggestions for chromosomal protein isolation are given below.

The electrophoretic technique has sufficient resolution that a single charge change in a protein will result in an easily detectable change in electrophoretic position. It is possible to generate artifactual charge heterogeneity of proteins by a number of different modification reactions. Since heterogeneity would destroy the usefulness of the separation, it is essential that it be avoided. Thus, during sample preparation the proteins should not be subjected to extremes of pH, stored as a lyophylized powder exposed to air, or treated with any reagent capable of generating charge heterogeneity.

These restrictions do not represent an all-inclusive list. Researchers should learn the chemical reactivity of the various groups in the protein(s) of interest and be aware that even extremely low levels of modification are a severe detriment.

Nucleic acids are precipitated by basic Ampholines and will form a smear at the top (basic end) of the isoelectric focusing gel. This highly ionic precipitate can bind many proteins and produce artifacts in their separation. The severity of the effect is proportional to the amount of nucleic acid present in the sample, and small amounts of whole cell extract can be loaded on a gel without regard for its nucleic acid content. However, in the analysis of chromatin protein, it is essential that a large proportion of the nucleic acid be removed.

For the preparation of samples for two-dimensional electrophoresis there is no advantage gained by the separation of chromosomal proteins into histone and nonhistone fractions (see Section V, A). Suitable fractionation procedures are ones in which total chromsomal proteins are solubilized using conditions which will not modify the proteins, followed by the removal of nucleic acid. These methods include: the solubilization of proteins in 7 M urea–3 M NaCl and the removal of DNA by centrifugation as described by Shaw and Huang 6); the solubilization of proteins by 5 M urea–2 M NaCl and removal of DNA by hydroxyapatite, as described by MacGillivray and Rickwood (7); the solubilization of proteins with urea and digestion of DNA with S1 nuclease as outlined by Peterson and McConkey (2). In these cases samples for electrophoresis can be prepared directly (from the supernatant after centrifugation, from the effluent of the hydroxyapatite column, or from the enzyme digestion mixture) by the addition of solid urea to give a final concentration of about 9.5 M and the addition of 1 volume of lysis buffer. The high salt present in these samples will not be detrimental to the isoelectric focusing gels; however, to ensure reproducible results the amount of salt in all samples should be kept constant.

Two other precautions should be mentioned regarding the preparation of chromosomal proteins. First, protease inhibitors such as DIFP (diisopropylfluorophosphate) and sodium bisulfite are frequently used during the isolation of nuclei and chromatin. DIFP is a reagent that inactivates serine proteases at a high rate, but it also will react nonspecifically with proteins to produce artifactual charge heterogeneity. Sodium bisulfite is a weak oxidizing agent, and oxidizing conditions produce heterogenity. For these reasons, it is suggested that these reagents be avoided and further that a reducing agent be present in all buffers used for the preparation of chromatin. Second, in the procedures suggested above urea is used to solubilize the protein. Urea spontaneously hydrolyzes to produce isocyanate which can react with proteins to produce charge heterogeneity (8, 9). Urea solutions should be deionized immediately before use or freshly prepared from high-quality crystalline urea. Furthermore, to compete for the reaction of isocyanate with protein, lysine should be included in all the urea buffers. Finally, the isolation should be accomplished as rapidly as possible and elevated temperatures avoided.

III. Isoelectric Focusing Dimension

A. Preparation of the Isoelectric Focusing Gel

Materials and solutions:

1. Pyrex tubing with internal diameter of 2.5 mm and accurately cut into 13-cm sections. (It is important that the isoelectric focusing gel is firmly attached to the walls of the gel tubes during electrophoresis. Thus these tubes are cleaned well by soaking in chromic acid and rinsed in water. The tubes are then soaked in a saturated solution of KOH in ethanol, rinsed in water, and air-dried.)

2. Tube stand (can be made of corrugated cardboard wrapped around a bottle so that tubes will stand vertically in the corrugations when held in place with a rubber band).

3. Parafilm.

4. Vacuum flask.

5. Loading syringe, 2.5–10 ml syringe with 6-in. blunt end 20-gauge hypodermic.

6. Urea (Schwarz Mann "ultrapure").

7. Acrylamide stock for isoelectric focusing, 28.38% acrylamide and 1.62% bis-acrylamide (stored at 4°C).

8. 10% (w/v) Nonidet P-40 (Particle Data Laboratories Ltd., 115 Hahn St., Elmhurst, Illinois) (stored at 4°C).

9. 40% Ampholines (LKB)[1] pH 5–7 and pH 3.5–10 (stored at 4°C).

10. 10% ammonium persulfate (stored at 4°C).

11. TEMED (N,N,N',N'-tetramethylethylenediamine) (stored at 4°C).

12. Lysis buffer [9.5 M urea–5% β-mercaptoethanol–2% NP-40–1.6% Ampholines (pH 5–7)–0.4% Ampholines (pH 3.5–10)] (stored in aliquots at minus 70°C).

One end of the gel tubes is sealed with Parafilm, and the tubes are placed in the stand. To make 10 ml of gel mixture (sufficient to pour 18 gels), weigh out 5.5 gm of urea into a 125-ml vacuum flask, add 1.33 ml of acrylamide stock, 2 ml of 10% NP-40, 1.95 ml of H_2O, 0.4 ml of Ampholines (pH 5–7), and 0.1 ml of Ampholines (pH 3.5–10). The urea is dissolved by

[1] Ampholines from other sources give different results.

swirling gently in a 37 °C water bath. Briefly vacuate this solution to remove dissolved gasses. Add 5 μl of TEMED and 10 μl of 10% ammonium persulfate, and immediately fill the loading syringe. To fill the gel tubes without trapping bubbles, the tip of the long hypodermic is inserted to the bottom of the tubes and withdrawn slowly as the acrylamide mix runs into the tubes. The tubes are filled to a mark made 5 mm from the top so that all gels are the same height. The gel mixture is overlayered with water and the gels allowed to sit for 1 hour. The overlay and unpolymerized gel mixture is aspirated from the top of the gels and replaced with 25 μl of lysis buffer. A small amount of water is layered over the lysis buffer to prevent urea from crystallizing at the top. The upper surface of the gel is allowed to equilibrate with lysis buffer for 1 hour.

B. Isoelectric Focusing

Materials and solutions:
1. 10 mM H$_3$PO$_4$.
2. 20 mM NaOH degassed.
3. Overlay buffer [9 M urea–0.8% Ampholines (pH 5–7)–0.2% Ampholines] (pH 3.5–10) (stored as aliquots at −70 °C).
4. Cylindrical gel tank.
5. Rubber stoppers with holes to adapt the tank to the narrow tubes.
6. Power supply with adjustable constant-voltage output capable of giving 800 V output.

The Parafilm is removed from the gel tubes, and the gels are loaded in the electrophoresis tank [the dialysis membrane used previously (1) is unnecessary]. The lower chamber is completely filled with 10 mM H$_3$PO$_4$. The solution above the gels is removed by aspiration and is replaced by 25 μl of lysis buffer which is overlayered with 20 mM NaOH. The upper reservoir is filled with 20 mM NaOH and connected to the negative terminal of the power supply, while the lower reservoir is connected to the positive terminal. The gels are prerun for 15 minutes at 200 V (constant voltage), 300 V for 30 minutes, and 400 V for 30 minutes. Do not attempt to cool the gels because the urea will come out of solution. After the prerun, the NaOH is removed from the upper chamber and discarded. The liquid above the gels is removed by aspiration, and the samples are loaded.

For total cell extracts, the optimal sample volume loaded is less than 25 μl containing 5 × 10^5 cpm and less than 20 μg of protein. There is, however, a great deal of flexibility: if the gel is made 1 cm shorter, 100 μl can be loaded; although the resolution will be lower, 100 μg or more of protein can be applied to the gel; and even if as few as 500 cpm are applied to the gel, some proteins will be detected but the sensitivity will be 1000-fold less than under optimal conditions. Of course, samples of subcellular fractions containing a limited number of protein species will require fewer cpm for detection.

The sample is overlayered with 10 μl of overlay buffer which in turn is overlayered with 20 mM NaOH. The upper reservoir is refilled with 20 mM NaOH and reconnected to the power supply. The focusing is continued overnight either for 13.5 hours with the power supply at 400 V or for 18 hours with the power supply at 300 V. The next day the voltage is increased to 800 V and focusing is continued for 1 additional hour.

C.   Equilibration and Treatment of First-Dimension Gel

Materials and solutions:
1. Syringe with tubing connected to outlet.
2. SDS sample buffer [2.3% SDS–5% β-mercaptoethanol–10% glycerol–62.5 mM Tris-HCl (pH 6.8)]. (Use only high-quality SDS, such as that manufactured by British Drug House (distributed by Gallard–Schlesinger Chemical Mfg. Corp., Carl Place, L. I., New York), for all solutions containing SDS.)
3. Screw-capped 15-ml tubes.
4. Dry Ice–ethanol bath.

A syringe is connected to the gel tube by means of a short piece of tubing, and the gel is slowly extruded by air pressure from the syringe into a tube containing 5 ml of SDS sample buffer. The gels are equilibrated with the SDS sample buffer by shaking at room temperature for 1 hour. At this time the gels can be frozen by laying the tightly stoppered tube on its side in a Dry Ice–ethanol bath. The frozen gels can be stored indefinitely at −70°C. Frozen gels are thawed quickly by running warm tap water over the tubes. The 5-ml SDS sample buffer is replaced with 5 ml of fresh buffer either after the first hour of equilibration or, if the gels were frozen, immediately after thawing. Equilibration is continued for 1 additional hour during which time the preparation of the second-dimension slab gels should be completed (see Section IV).

IV.   SDS Electrophoresis Dimension

A.   Apparatus for the Second-Dimension Slab Gel

The electrophoresis tank is essentially that described by Studier (*10*) with dimensions appropriate for the size of the gel slab used.

Two different types of glass plates have been used for casting the gels.

The first, referred to as beveled glass plates, is more convenient to use but more difficult to construct [see Fig. 2A and B as well as (1)]. The second type of plate used is the same as that normally used with a Studier apparatus for one-dimensional electrophoresis (see Fig. 2C). The glass plates are assembled with spacers along the bottom and side edges. These spacers must make a watertight seal, which is usually accomplished by three latex rubber spacers (16 × 6 × 0.8 mm) lightly greased with silicone grease. The assembled plates are held together with binder clamps. In the following description the term "base of the notch" refers to the positions marked in Fig. 2B and C, and the term "groove" refers to the groove in which the first-dimension gel is placed. For the beveled plates the groove is formed

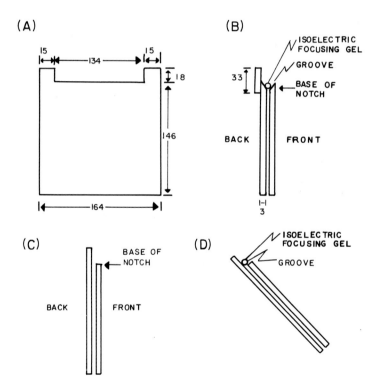

Fig. 2. Illustration of the glass plates used for the second SDS dimension. (A) The dimensions of a notched plate. (B) and (C) two different ways of assembling plates so that they can subsequently be used to load the first-dimension isoelectric focusing gel. In (B), two notched plates are beveled at a 45 degree angle at their notched edges. The back plate is formed by gluing a rectangle of glass (164 × 33 mm) to the edge of the notch. The groove for the isoelectric focusing gel is formed by the two beveled edges. In (C), one notched (unbeveled) plate is combined with one square plate (164 × 164 mm). In (D), the plates in (C) have been tilted to a 45 degree angle to form the groove for the isoelectric focusing gel.

by the two beveled edges (Fig. 2B). As shown in Fig. 2D a groove can be formed with unbeveled plates by leaning the assembly at a 45° angle.

B.   Preparation of the SDS Separating Gel

Materials and solutions:
1. Assembled glass plates.
2. Gradient maker.
3. Magnetic stirrer.
4. Vacuum flasks.
5. Lower gel buffer: 0.4% SDS–1.5 M Tris-HCl (pH 8.8) at 23 °C (stored at 4 °C).
6. SDS acrylamide stock: 29.2% (w/v) acrylamide–0.8% (w/v) bis acrylamide (stored at 4 °C).
7. 10% ammonium persulfate (stored at 4 °C).
8. 75% glycerol (w/v).
9. TEMED.

The lower separating gel is poured the day prior to running the second dimension. Gels of different porosities can be made to optimize the separation of proteins in different molecular-weight ranges. The best resolution of complex mixtures of proteins has been obtained with a concave exponential gradient (10 to 16%) acrylamide gel made as follows. To make the light solution for one gel add to a vacuum flask 4 ml of lower gel buffer, 5.3 ml of acrylamide stock, and 6.7 ml of water. To make the dense solution add to a second vacuum flask 2 ml of lower gel buffer, 4.3 ml of acrylamide, and 1.7 ml of 75% glycerol. Degas both solutions, add 25 μl of ammonium persulfate to the light solution and 10 μl to the dense solution, and add 8 μl of TEMED to the light solution and 4 μl to the dense solution. Pipette 5 ml of the dense solution into the mixing chamber of a gradient maker containing a magnetic stirring bar. Convert the mixing chamber to a constant-volume chamber by stoppering it. Load all 16 ml of the light solution into the reservoir chamber, start the magnetic stirrer, open the channel connecting the chambers, and start pouring the gradient (either by gravity or with a peristaltic pump) at a rate of 3 ml/minute. The gradient is poured from the top of the glass plates, and the flow is stopped when the level reaches a mark 1 in. below the base of the notch. The mixing chamber of the gradient maker should still be full, and this remaining acrylamide solution is discarded. The gel solution is overlayered by a spray of atomized water and allowed to set until the gel interface can be seen as a sharp straight boundary (about 1 hour). The overlay and unpolymerized acrylamide is removed by aspiration and replaced with a 4-fold dilution of the lower gel buffer. The gel is allowed to stand overnight.

C. Preparation of the Stacking Gel

Materials and solutions:
1. Teflon strip (0.8 × 130 × 20 mm).
2. Stacking gel buffer: 0.4% SDS–0.5 M Tris-HCl (pH 6.8) at 23°C (stored at 4°C).
3. SDS acrylamide stock (as above, Section IV, B).
4. 10% ammonium persulfate.
5. TEMED.
The stacking gel of the second-dimension slab must be prepared during the last hour of equilibration of the first-dimension gel (see Section III, C). At the start of the hour add to a vacuum flask 1.25 ml of stacking gel buffer, 0.75 ml of acrylamide stock, and 3 ml of H_2O. While this solution is degassing under vacuum, remove by aspiration the solution covering the interface of the separating gel. Remove the stacking gel mixture from the vacuum, add 15 μl ammonium persulfate and 5 μl TEMED, and pipette the mixture on top of the separating gel, filling to the base of the notch. Taking care that no bubbles are trapped, insert the edge of the Teflon strip 2 mm below the base of the notch. Allow the stacking gel to set for about 30 minutes, remove the Teflon strip, and rinse the surface of the gel with water, removing all excess liquid by aspiration. Load the first-dimension gel on top of the second as described below (Section IV, D).

D. Loading and Running of the Second-Dimension Gels

Materials and solutions:
1. 1% agarose (Seakem obtained from Marine Colloids Inc.) in SDS sample buffer (stored as aliquots at 4°C).
2. Boiling H_2O bath.
3. Parafilm.
4. Gel tank for slab gels.
5. 0.1% Bromphenol blue.
6. SDS running buffer: 0.192 M glycine–25 mM Tris–0.1% SDS. *Caution: Do not titrate this buffer, and do not add any salt.*
7. Syringe with a bent cannula.
8. Power supply capable of delivering 20 mA constant current to each slab gel.
The agarose is melted in the boiling water bath 5 minutes before the equilibration of the first-dimension gel is complete. The first-dimension gel is removed from the equilibration tube, placed on a sheet of Parafilm, and aligned parallel to one edge of the parafilm. All excess liquid is removed

from around the gel. If unbeveled glass plates are used, lean the plates against a test-tube rack to form a 45 degree angle to horizontal. The angle between the back plate and the base of the notch forms the "groove" in which the agarose and the first-dimension gel will be placed (see Fig. 2D). For the beveled plates described in Section III,A, the groove is formed by the two beveled plates standing upright. One milliliter of the melted agarose is pipetted into the groove and, using the parafilm as a spatula, the first-dimension gel is quickly transferred into the groove. The agarose is allowed to set for 5 minutes. The gasket sealing the bottom of the glass plates is removed, and the gel is clamped to the electrophoresis tank.

Add 2 drops of 0.1% Bromphenol blue and sufficient running gel buffer to the upper reservoir so that the gel is covered. Fill the lower buffer reservoir with running buffer and remove air trapped under the gel with a stream of running buffer ejected from a syringe connected to a cannula which is bent upward at the tip. Start electrophoresis at 20 mA constant current per gel and continue until the Bromphenol blue front reaches the bottom of the gel (approximately 4 hours).

E. Staining, Drying, and Autoradiography of the Slab Gels

Materials and solutions:
1. Pyrex dishes.
2. Fixing and staining solution: 50% TCA–0.1% Coomassie Brilliant blue.
3. Destaining solution: 7% acetic acid.
4. Gel dryer.
5. Vacuum pump or very good aspirator.
6. Whatman 3 MM filter paper.
7. Kodak no-screen x-ray film.

After completion of electrophoresis, the gel is removed from the plates and placed in a Pyrex dish containing the fixing and staining solution. After 30 minutes, the gel is removed into a Pyrex dish containing 7% acetic acid. The gel is destained overnight by several changes of the acetic acid. The gel is destained overnight by several changes of the acetic acid. On a fully destained gel it should be possible to detect all compact protein spots containing more than a few hundreths of a μg of protein. The destained gel is transferred to a piece of wet 3 MM paper and dried on a drying apparatus (such as that manufactured by Hoeffer Scientific). The dried gel is exposed to 5 × 7 in. x-ray film. Normally a gel containing 5×10^5 cpm of total cell protein is exposed for 1 day, and to obtain a high-sensitivity autoradiogram a second exposure of 10 days is made.

V. Limitation of the Technique and Difficulties in Interpretation

A. Loss of Basic Proteins

In the first-dimension isoelectric focusing gel high concentrations of denaturants are used, and prolonged running times are required to reach equilibrium. Under these conditions we have been unable to produce stable pH gradients which extend significantly above pH 7. Thus, all proteins with basic isoelectric points will be excluded from the first-dimension gel. Although a large fraction of the cellular proteins have isoelectric points within the range encompassed by this procedure, it is clear that a number of chromosomal proteins will not enter the described gels: histones certainly will not be fractionated and some of the nonhistone chromosomal proteins will also be too basic for this system (7). The inability to detect basic proteins may in some cases be a severe drawback. Current efforts are directed toward the design of a complementary gel system for the analysis of basic proteins.

Because the histones fail to enter this gel system there is no necessity to fractionate chromosomal proteins into histone and nonhistone preparations.

B. Requirement for Isotopically Labeled Proteins for High-Sensitivity Analysis

Although unlabeled proteins can be fractionated and detected in this system, the results obtained are far inferior to those obtained with radiolabeled proteins. A large number of proteins are present in such minute quantities that they can only be detected by systems capable of very high sensitivity. The sensitivity of autoradiography of protein labeled to high specific activity is about four orders of magnitude higher than the sensitivity of staining of unlabeled protein.

Since labeling of whole tissues to a high specific activity is extremely difficult, it is not possible to obtain optimum sensitivity and resolution in all cases, and the approach is best suited for tissue-culture cells. Furthermore, to prepare samples of high specific activity, small numbers of cells must be used; thus chromatin preparation procedures must be scaled down.

C. Definition of a Chromosomal Protein

A major problem in the analysis of chromosomal proteins is one of definition. A chromosomal protein is defined as a protein found in a subcellular fraction called chromatin (11). However, the composition of chromatin is dependent on the method of isolation (12). Since the fraction "chromatin" cannot be precisely defined, neither can a chromosomal protein. An additional problem is encountered if an attempt is made to detect the minor components of chromatin or to determine the total number of chromosomal proteins. In a preparation of chromatin which is 95% pure, the residual 5% of the protein of nonchromosomal origin produce a severe complication. Since the gel technique is capable of detecting proteins that comprise as little as 1 part in ten million (10^{-5}%), and since a large number of the proteins present are in a low abundance (2), an additional criterion is needed to determine whether low-abundance proteins are due to contamination or are really low-abundance chromosomal proteins. This might be established by comparing gel patterns from (1) purified chromosomal proteins, (2) total cellular proteins, and (3) a mixture of total cellular proteins and purified chromosomal proteins. These gels are compared to identify the proteins corresponding to "chromosomal proteins" in the total cell extract. If these proteins are highly enriched in the chromatin fraction then they are (by definition) chromosomal proteins. It is hoped that the sensitivity of detection of the described technique will be helpful in analysis of the proteins involved in the structure of chromatin and the control of its expression.

ACKNOWLEDGMENTS

Patrick O'Farrell is supported by a fellowship from the Jane Coffin Childs Memorial Fund for Cancer Research. Patricia O'Farrell is supported by funds from the USPHS.

REFERENCES

1. O'Farrell, P. H., *J. Biol. Chem.* **250**, 4007 (1975).
2. Peterson, J. L., and McConkey, E., *J. Biol. Chem.* **251**, 548 (1976).
3. O'Farrell, P. Z., and Goodman, H. M., *Cell* **9**, 289 (1976).
4. Eagle, H., *Science* **122**, 43 (1955).
5. Ivarie, R. D., and O'Farrell, P. H., Submitted for publication.
6. Shaw, L. M. J., and Huang, R. C. C., *Biochemistry* **9**, 4530 (1976).
7. MacGillivray, A. J., and Rickwood, D., *Eur. J. Biochem.* **41**, 181 (1974).
8. Stark, G. R., *in* "Methods in Enzymology," Vol. 11: Enzyme Structure (C. H. Werner Hirs, ed.), p. 590. Academic Press, New York, 1967.
9. Bobb, D., and Hofstee, B. H. J., *Anal. Biochem.* **40** 209 (1971).
10. Studier, F. W., *J. Mol. Biol.* **79**, 237 (1973).
11. Elgin, S. C. R., and Weintraub, H., *Ann. Rev. Biochem.* **44**, 725 (1975).
12. MacGillivray, A. J., Cameron, A., Krauze, R. J., Rickwood, D., and Paul, J. *Biochim. Biophys. Acta* **277**, 384 (1972).

Part E.   Chromatin Fractionation. I

Chapter 2 8

Methods for Fractionation of Chromatin into Transcriptionally Active and Inactive Segments

JOEL M. GOTTESFELD

MRC Laboratory of Molecular Biology,
Cambridge, England

I.   General Introduction

Cellular differentiation in eukaryotes can be traced, in part, to differential gene activity. This statement rests on three important observations: first, the various cell types of an organism posses the same genetic complement (*1*); second, differentiated cells of metazoan creatures transcribe a limited fraction of their full genetic potential (*2,3*); and, third, the RNA populations of cells are, to a certain degree, tissue-specific (*3–6*). When compared to the transcription of protein-free DNA, the *in vitro* transcription of isolated chromatin is highly restricted (*7*). Even though this observation was made well over a decade ago, the specific chromosomal components responsible for the control of transcription have yet to be identified. Both RNA (*8, 9*) and nonhistone proteins (*10–13*) have been suggested as potential modulators of gene activity. It appears that the regulatory components, whatever they might be, remain intact in isolated chromatin; *in vitro* transcription of the globin gene has been found to be highly tissue-specific (*14,15*).

There is a vast genetic and cytological literature which demonstrates that the structure of transcriptionally active chromatin is markedly different from that of transcriptionally inactive regions of chromatin. First of all, RNA synthesis appears to be restricted to the extended euchromatic regions of thymocyte nuclei; no transcription can be detected by electron microscope autoradiography in the condensed heterochromatic regions of the nucleus (*16*). Second, during the pachytene stage of amphibian oogenesis, transcription takes place in the extended loops of the lampbrush chromosomes (17). Similarly, prior to transcription in polytene cells of Drosophila, the specific chromomeres to be transcribed extend and become puffs (18). On the other hand, chromatin condensation is associated with a loss of genetic activity. The morphological differences between transcriptionally active and inactive regions of chromatin must have a biochemical basis, and hence, separation of chromatin into active and inactive components should be possible. Successful methods for chromatin fractionation will allow a direct comparison of the structure and composition of these chromatin species. This information will undoubtedly increase our understanding of the mechanisms of eukaryotic gene control. Furthermore, techniques for fractionation will provide a means for the purification of the gene regulatory elements themselves.

In developing methods for chromatin fractionation, two main technical difficulties are encountered. First, chromosomal DNA must be reduced in size to lengths small enough to allow the subsequent separation of active and inactive regions. Second, a physical means of fractionation is required. In the following pages many of the methods currently in use for chromatin fractionation will be outlined, and a brief account of the properties of the isolated fractions will be given. In addition, methods for the fragmentation of chromosomal DNA will be discussed.

II.   Criteria for Fractionation

Before discussing the actual methods for chromatin fractionation, it is essential to set down reasonable criteria to test the validity of any fractionation method. First of all, since the fraction of the genome which is transcribed in a given cell type is limited, the amount of DNA recovered in the active fraction should be small and should correspond to the template activity of the particular cell type under study. Second, the DNA sequence complexity of the active fraction should be small compared to the whole genome and should correspond to the fraction of the genome which is active in the particular cell type under investigation. In other words, the active fraction must be a specific subset of the genome rather than a random sample of DNA sequences. Third, the active fraction must be enriched in DNA sequences complimentary to cellular RNA. Fourth, specific transcribed DNA sequences should be found in the active fraction of transcribing cells but not in the active fraction of nontranscribing cells. For example, globin sequences should be found in high concentration in the active fraction of reticulocyte chromatin, but not in the active fraction of, let's say, brain chromatin. Fifth, and finally, nascent RNA chains should be isolated with active chromatin. Many of these criteria have been applied to chromatin fractions produced by a variety of methods. To date, however, no investigator has reported the isolation of an active fraction that satisfies all of the criteria listed above.

III. Methods for the Fragmentation of Chromosomal DNA

A. Introduction

To minimize the cross-contamination of active and inactive chromatin species, chromosomal DNA must be fragmented to a length less than that of the average unit of transcription. In higher eukaryotes, the major product of nuclear transcription is hnRNA (for review, see *11, 19*). The size of these molecules ranges from a few thousand nucleotides to giant transcripts up to 30,000–50,000 nucleotides in length. This size heterogeneity introduces problems in terms of yields of pure active-fraction chromatin. It is obvious that the amount of active-fraction chromatin which can be separated from inactive regions of chromatin will increase as the chromatin DNA size is reduced. This amount should first increase and then level off at an amount corresponding to the fraction of DNA which is transcriptionally active in the particular cell type or tissue under investigation.

B. Sonication

The earliest attempts at chromatin fractionation used sonication as a means for the disruption of nuclei and the fragmentation of chromatin DNA (*20*). Chromatin was sonicated for 5 seconds at 20,000 Hz and 7 A in the Branson S–75 Sonifier. More recently, Reeck *et al.* (*21*) have used the Branson Sonifier–Cell Disruptor (Model W185). Chromatin in 1 mM Tris-Cl (pH 8) is sonicated at a concentration of 0.4–0.6 mg DNA/ml. Eighteen ml of the chromatin suspension is placed in the aluminum fluted cell and the microtip of the sonifier horn is extended 4.5 cm into the sample. The sonication cell is suspended in a vigorously stirred ice-water bath. Temperature is maintained at less than 2 °C during the 120 seconds of sonication at 70 W. This procedure results in the generation of chromatin fragments with a DNA molecular weight of 7×10^5 (500 base pairs). Reeck *et al.* report that the melting temperature (T_m) of sonicated chromatin is lowered by 1.5 °C relative to that of unsonicated chromatin. Sonication of chromatin does not result in the release of detectable amounts of free protein in solution.

Arnold and Yound (22) have utilized the Branson S125 Sonifier. Chromatin in 10 mM Tris-Cl (pH 8)–2 mM β-mercaptoethanol was sonicated at a concentration of 0.5 mg DNA/ml. Ten-milliliter aliquots were sonicated at 20,000 Hz at 96 W for a total of 8 seconds (two bursts of 4 seconds with 15-second cooling between). Temperature was maintained at 0 °C during sonication. After sonication the chromatin was centrifuged at 5000 g for 10 minutes. Ninety-five percent of the chromatin DNA was found in the supernatant.

Rickwood *et al.* (23) have used the Dawe Soniprobe with a $\frac{1}{4}$-in. tip. Samples were sonicated at 3.5 A at a power setting of 4. Fifteen-second bursts were followed by 15-second periods of cooling for a total sonication time of 2 minutes.

C.   French Pressure Cell

The French pressure cell has been utilized by many workers. Chromatin is generally dissolved in 10 mM Tris-Cl (pH 8). Pressures of 3000–20,000 psi have been used. After shearing, the chromatin suspension is usually clarified by centrifugation (16,000 g for 20 minutes). Janowski *et al.* (24) report that when chromatin at 0.7–1 mg DNA/ml is sheared at 4000 psi the resulting fragments contain DNA of 2×10^6 molecular weight (3000 base pairs).

D.   Homogenization

Soluble chromatin as prepared in many laboratories is sheared by high-speed blending or homogenization. Again chromatin is sheared in 10 mM Tris-Cl (pH 8), and the resulting suspension is clarified by centrifugation at 16,000 g for 10–20 minutes. After shearing at 40 V for 90–120 seconds in the VirTis "45" homogenizer (at 0–2 °C in an ice-water bath), the resulting soluble chromatin contains DNA $6–12 \times 10^6$ molecular weight (10,000–20,000 base pairs). McCarthy *et al.* (25) report that 5 minutes of homogenization at 40 V results in chromatin DNA fragments of 3×10^6 molecular weight (5000 base pairs).

E.   Nuclease Digestion

1.   Deoxyribonuclease II

Spleen acid deoxyribonuclease (DNase) (Worthington, HDAC) has been used for cleavage of chromosomal DNA in purified chromatin (26–28) and in crude nuclear lysates (J. M. Gottesfeld, unpublished). The first method given is used with purified chromatin (29). After sedimentation through 1.7 M sucrose, the chromatin pellet is washed once with 10 mM Tris-Cl (pH 8) and dialyzed overnight at 4 °C against 200 volumes of 25 mM sodium acetate (pH 6.6). Alternatively, the chromatin is washed once with sodium acetate and resuspended in this buffer. The chromatin concentration is adjusted to about 0.4 mg DNA/ml and DNase II is added to 100 units/ml. Incubations are carried out at 24 °C, and the reaction is terminated by the addition of 0.1 M Tris-Cl (pH 11) to give a final pH of 7.5. With rat liver chromatin, a 5-minute incubation has been found to be optimal for subsequent fractionation.

DNase II digestions have been carried out on nuclear lysates in the following manner: nuclei are prepared by standard techniques and lysed in 0.2 mM ethylenediaminetetraacetic acid (EDTA) (pH 8) (see next section). Sodium acetate (1 M, pH 6.6) is added to the suspension with rapid stirring to give a final concentration of 25 mM. If necessary, the pH is carefully adjusted to 6.6. Incubations with DNase II are carried out as described above.

Note that DNase II as supplied (Worthington, HDAC) contains both RNase and protease activities. Incubations should be as brief as possible to avoid difficulties.

2. MICROCOCCAL NUCLEASE

The method of Noll, Thomas, and Kornberg (30) is given. Nuclei, prepared by standard methods, are suspended in 0.34 M sucrose–60 mM KCl–15 mM NaCl–15 mM Tris-Cl (pH 7.4)–15 mM β-mercaptoethanol–0.15 mM spermine–0.5 mM spermidine–1 mM CaCl$_2$. The concentration of nuclei is about 1.5×10^8/ml (0.44 ml of buffer per gm of starting tissue). Micrococcal nuclease (Worthington) is added to 15 units/ml, and digestions are carried out at 37 °C for 30 seconds. The reaction is terminated by the addition of 0.02 ml of 0.1 M EDTA (pH 7)/ml of nuclei and chilling on ice. The nuclei are centrifuged at 4000 g for 5 minutes and lysed by suspension in 0.2 mM EDTA (pH 7) with the use of a Pasteur pipette. After centrifugation for 2 minutes at 4000 g, the opalescent supernatant contains 75% of the input nuclear DNA. Ninety-five percent of the chromatin DNA molecules are 3000 to 30,000 nucleotide pairs in length. The weight average length was reported to be 9000 base pairs. Smaller fragments (weight average length of DNA = 1600 base pairs) are obtained by incubation at 150 units of nuclease per milliliter for 30 seconds.

3. RESTRICTION ENDONUCLEASES

Bacterial restriction endonucleases have been used for the cleavage of chromosomal DNA in intact nuclei (31) and in isolated chromatin (J. M. Gottesfeld, unpublished; R. S. Gilmour, personal communication). The enzymes Eco RI, Eco RII, Bsu, and Hind III have been used successfully. The method of Pfeiffer et al. for digestion in nuclei is as follows: rat liver nuclei prepared by the method of Hewish and Burgoyne (32) were treated with the enzymes *Bsu* and *Eco* RII in aliquots containing 0.5 A_{260} units of DNA in 50 μl. Incubations were carried out at 37 °C for 3 hours in a buffer containing 60 mM KCl–15 mM NaCl–0.5 mM spermine–0.5 mM spermidine–0.2 mM EDTA–0.2 mM EGTA–15 mM β-mercaptoethanol–15 mM Tris-Cl (pH 7.4)–10 mM MgCl$_2$. Pfeiffer *et al.* report that large amounts of enzyme are required for *in situ* digestions. Incubation with the *Bsu* enzyme, at a concentration 80-fold higher than that required for the complete digestion of protein-free DNA, results in the production of chromatin fragments containing DNA molecules ranging in size from about 400 nucleotide pairs to about 12,000 nucleotide pairs. The average length was about 2400 base pairs. Note that EGTA must be used to inhibit endogenous nuclease activity during the incubations.

A method for digestion of isolated chromatin with restriction nucleases is now given (J. M. Gottesfeld, unpublished). Nuclei are prepared from rat liver by standard methods (32) and lysed in the following manner: the final nuclear pellet is taken up in 10 mM Tris-Cl (pH 8) and diisopropylphosphofluoridate (DFP) is added to 8 mM. The solution is stirred at room temperature for 20 minutes. EGTA (pH 8) is added to 0.1 mM at the end of this time. DFP treatment is used here to prevent proteolysis during the long incubations with restriction nucleases. NOTE: Extreme caution should be used in handling solutions containing DFP as this compound is a lethal nerve poison. The solution is centrifuged at 16,000 g for 10 minutes, and the resulting chromatin pellet is taken up in digestion buffer. Two buffer systems have been used. In the first [6 mM Tris-Cl (pH 7.5)–50 mM NaCl–6 mM MgCl$_2$–0.1 mM EGTA], the chromatin is highly condensed and large amounts of enzyme are required for digestion. The enzymes *Eco* RI and *Hind* III have been used. Incubations are carried out at 37° for 2–3 hours. The second buffer system is that of Polisky *et al.* (33). The chromatin pellet is taken up in 25 mM Tris-Cl (pH 8.5)–150 μM MnCl$_2$. The enzyme *Eco* RI is added (10 units/500 μl of chromatin at 2–5 A_{260} units/ml), and incubations are carried out for 2-3 hours at 37°. The latter method has the advantage that the chromatin is soluble under the digestion conditions and much less enzyme is required for cleavage. Note, however, that base sequence specificity is altered under these latter conditions (33).

F.   Evaluation of Methodology — Comments and Prospects

Care should be taken that the method chosen for fragmentation of chromosomal DNA does not result in changes in chromatin structure and/or changes in specific DNA–protein interactions. Control experiments should be performed to test whether protein migration or dissociation occurs during shearing. Noll *et al.* (*30*) report that native chromatin structure is lost upon shearing in the Sorvall Omni-Mixer. Chromatin dissolved in buffers of low ionic strength is highly extended, and, under these conditions, it is most susceptible to shear damage. On the other hand, Woodhead and Johns (*34*) have reported that condensed chromatin (in 0.14 *M* NaCl) still retains a regular DNA repeat pattern after shearing in the M.S.E. blender. It is advisable to perform the "micrococcal test" of Noll *et al.* (*30*) on all preparations of sheared chromatin to insure that nucleosome structures are preserved.

Methods of DNA cleavage in intact nuclei (micrococcal nuclease and restriction enzymes) are currently being applied to the problem of chromatin fractionation. These methods have the advantage that mechanical shear damage is avoided. The nuclease methods have further advantages: the procedures are gentle, simple, and quite rapid; no expensive equipment is required; and the enzymes are now readily available.

IV. Methods of Chromatin Fractionation

A. Differential Centrifugation

1. Sucrose Gradient Sedimentation

The first reported attempt at a biochemical fractionation of active and inactive regions of chromatin was that of Frenster *et al.* (*20*). Thymocyte chromatin, prepared from nuclei disrupted by sonication, was separated into fractions corresponding by ultrastructural criteria to eu- and heterochromatin. Fractionation was achieved by simple differential centrifugation. The euchromatin fraction contained only about 20% of the nuclear DNA but most of the rapidly labeled RNA (*20*). The same approach has been taken by many workers—with greater or lesser degrees of success and evidence for fractionation. Chalkley and Jensen (*35*) have used sucrose gradients to fractionate chromatin. Again, the basis for fractionation is the suspected hydrodynamic difference between repressed and active chromatin. Extended active chromatin should have a lower sedimentation coefficient than condensed, repressed regions of chromatin. Indeed, many workers have found that the slowly sedimenting chromatin species have a higher template activity than the more rapidly sedimenting chromatin molecules (*35–38*). Doenecke and McCarthy (*39*) have concluded that the basis for fractionation by the sucrose gradient method is the differential aggregation of heterochromatin fibers.

The methods given here are those of McCarthy *et al.* (*25*) and Berkowitz and Doty (*38*). In both instances linear 6–60% (w/v) sucrose gradients containing 10 mM Tris-Cl (pH 8) are used. Chromatin fragmented by sonication is centrifuged in the Beckman SW40 rotor at 25,000 rpm (78,000 g) for 17 hours at 4 °C (*38*). If chromatin is sheared by VirTis homogenization (40 V for 5 minutes), optimal separation is obtained after centrifugation at 30,000 rpm (113,000 g) for 6 hours in the SW40 rotor. Alternatively, chromatin sheared in the French pressure cell (3000 psi) is centrifuged for 14 hours at 30,000 rpm (113,000 g).

The slowly sedimenting fractions from sucrose gradients have many of the properties expected of transcriptionally active chromatin. Nascent RNA cosediments with this fraction; furthermore, Berkowitz and Doty (*38*) have found that this RNA is enriched in globin nucleotide sequences in a hemoglobin-producing cell population (reticulocytes). Further hybridization experiments are needed to demonstrate that the DNA of the slowly sedimenting fraction actually codes for the RNA sequences of the cell type under investigation. Nonetheless, the slowly sedimenting fraction isolated by Berkowitz and Doty has several interesting characteristics: this fraction displays a 6° lower melting profile and is highly susceptible to DNase I relative to the fast sedimenting template-inactive fraction. The chemical compositions of the fractions were also found to be quite different; the active fraction is enriched in nonhistone protein and lacks the very lysine-rich histones and some of the arginine-rich histones.

2.   GLYCEROL GRADIENT CENTRIFUGATION

Sheared or sonicated chromatin can be resolved into rapidly and slowly sedimenting species by centrifugation in linear glycerol gradients. Chromatin is prepared by standard methods and fragmented by sonication, VirTis shearing, or passage through the French pressure cell. After centrifugation at 10,000–12,000 g for 15–30 minutes, the chromatin supernatant (3 ml at 0.5–1 mg DNA/ml) is layered on 7.6–76% (v/v) linear glycerol gradients containing 10 mM Tris-Cl (pH 8). Magee *et al.* (*40*) have used the Beckman SW27 rotor and centrifuged for 15 hours at 22,500 rpm. The same conditions have been used by Murphy *et al.* (*36*) except the SW25.1 rotor was employed. The slowly sedimenting fractions from these gradients were found to exhibit very high template activities when assayed with both prokaryotic (*E. coli*) and eukaryotic RNA polymerases (yeast and homologous polymerases I and II) (*36,40*).

B.   Gel Filtration Chromatography

In addition to density gradient centrifugation, the chromatographic method of Janowski *et al.* (*24*) relies on the hydrodynamic properties of repressed and active chromatin (see above). The chromatographic resin is agarose (Bio-Rad A-50 or A-150), and the column is prepared for use in the following manner: the agarose beads are packed under low pressure (pressure head <50 cm) in a column (2.6 $\times$ 40 cm) fitted with upward-flow plungers. The column is equilibrated with either 10 mM Tris-Cl (pH 8)–1 mM β-mercaptoethanol or with the same buffer containing 0.15 M KCl–0.1 M MgCl$_2$. Chromatin is soluble in either of these solvents, and fractionation may be achieved in either buffer system. Chromatin is prepared by standard procedures and fragmented by VirTis homogenization (5 minutes at 40 V) or by shearing in the French pressure cell (at 3000 psi). The chromatin sample (1–3 ml at 0.7–1 mg DNA/ml) is applied to the resin in the chosen buffer, and the chromatogram is developed by elution at 15 ml/hour. Under the low-salt conditions, the active chromatin fractions emerge at the end of the elution profile while in high-salt active chromatin appears at the front. It shoult be noted that histone H1 is dissociated under the high-salt conditions. Nonetheless, Janowski et al. recommend the high-salt buffer as superior resolution is obtained.

Janowski et al. have identified the active fractions by association of nascent RNA chains, in vitro template activity, RNA polymerase binding, and association of hormone–receptor complexes in chromatin from hormone target tissues. Appropriate DNA reassociation and DNA–RNA hybridization experiments are needed to verify this method.

C. Thermal Chromatography on Hydroxyapatite

McConaughy and McCarthy (41) have described a method for the fractionation of chromosomal DNA by thermal elution from hydroxyapatite. The method is based upon the observation that the thermal denaturation profile of chromatin extends over a very much broader range of temperatures than that for protein-free DNA. This is presumably due to the differential stabilization of particular sections of the DNA against thermal denaturation by associated chromosomal proteins.

Hydroxyapatite is prepared for use by boiling for 5 minutes in 10 mM sodium or potassium phosphate (pH 6.8). The resin (1 gm) is packed into waterjacketed columns and equilibrated with 0.12 M phosphate buffer (pH 6.8) at 60°C. Potassium phosphate is recommended over sodium since elution of DNA is shifted to lower temperatures. Further temperature reduction may be achieved by substituting 10 mM cesium phosphate as the elution buffer (42). Higher elution temperatures were found with Clarkson hydroxyapatite than with the Bio-Rad product. As capacities of hydroxyapatite vary according to lot and manufacturer, it is wise to test each batch before use. Generally, 100–200 μg of double-stranded nucleic acid (DNA or chromatin) is retained by 1 ml of resin. In the study of McConaughy and McCarthy (41), chromatin was sheared in the French pressure cell at 12,000 psi. The sheared chromatin is absorbed to the resin, and the column is heated in 5° increments to 100 °C by means of a circulating water bath. At each temperature, 5 ml of buffer (in 2 aliquots) is forced through the column under air pressure. Alternatively, a peristaltic pump may be used to elute single-stranded DNA. Elution of DNA remaining on the column at 100 °C is accomplished by washing with 8 M urea–0.24 M sodium phosphate (pH 6.8)–10 mM EDTA.

DNA extracted from the chromatin fractions eluting from hydroxyapatite at the lowest temperatures is enriched several-fold in transcribed sequences. Furthermore, the fractions eluting at low temperature are not greatly enriched in sequences transcribed in nonhomologous tissues. Thermal chromatography of chromatin has the disadvantage that native DNA–protein interactions are destroyed; however, actively transcribed DNA may be purified by this method.

D. ECTHAM–Cellulose Chromatography

Simpson and his colleagues have reported a method for chromatin fractionation based on ion-exchange chromatography (21,43,44). The chromatographic resin is ECTHAM–cellulose, a weak cation-exchanger. The low content of ionizing groups of relatively low pK makes this absorbant well suited for use with chromatin since mild conditions are required for elution. ECTHAM–cellulose is prepared by coupling tris(hydroxymethyl) aminomethane to cellulose with epichlorohydrin (45). The resulting resin has a capacity of 0.11 meq/gm. ECTHAM–cellulose is prepared for use by extensive washing 0.5 N HCl, 0.5 N NaOH, water, and 0.01 M Tris-Cl (pH 7.2). The resin (2 gm) is packed into a column measuring 1.2 × 15 cm under air pressure (3–4 psi). The column is equilibrated to 0.01 M Tris-Cl (pH 7.2) at 4 °C. Ten milliliters of sonicated chromatin (A_{260} = 7.4) are applied to the column at 3 ml/hour. Elution is achieved by washing the column with 0.01 M Tris base–0.01 M NaCl at a flow rate of 2 ml/hour, and 1-ml fractions are collected. Titration of the ionizing groups of the resin results in the elution of the chromatin as a broad peak. The weakest-bound molecules are eluted first, followed by a gradual transition to the more strongly bound species. At about fraction 48 (48 ml of eluant) the ionizing groups are fully titrated and the pH of the eluant rises to about 9.2. This results in the elution of the most tightly bound chromatin molecules. Reeck et al. (21) report that only 80% of the applied chromatin is recovered by elution with 0.01 M Tris base. The remaining material can be recovered only under conditions that destroy chromatin structure. Alternatively, Reeck et al. (21) describe a method of displacement chromatography for the recovery of noneluted material. By exceeding the capacity of the resin 10-fold, only the tightest-binding 10% of the sample is retained on the column. Titration elution then results in 80% recovery of the bound material for an overall recovery of 98%.

The late-eluting fractions have many of the properties expected for transcriptionally active chromatin. Late-eluting fractions exhibit a markedly lower T_m than bulk chromatin or the early eluting fractions (21). This finding is consistent with the results of McConaughy and McCarthy (41). The late-eluting fractions are greatly enriched in RNA polymerase initiation sites (44). Moreover, the late-eluting chromatin molecules are more extended in conformation than the early eluting material—the late-eluting chromatin species exhibit lower sedimentation values and a circular dichroism spectrum more like that of protein-free DNA than the early eluting molecules. The DNA of the late fractions is more susceptible to nuclease attack than the DNA of the early fractions (46). The fractions exhibit differences in chemical composition: the late-eluting chromatin has a greatly increased ratio of nonhistone protein to DNA (up to 2.5:1) and a diminished content of histone HI. Recently, however, Howk *et al.* (47) failed to detect differences in the content of specific transcribed DNA sequences in the chromatin fractions produced by ECTHAM–cellulose chromatography. This finding casts serious doubt as to whether the late-eluting fractions correspond to transcriptionally active chromatin *in vivo*.

E.   Partition Chromatography

Turner and Hancock (48) have described a two-phase aqueous polymer system for partition chromatography of sheared chromatin. The aqueous polymer solutions are prepared using Dextran T500 (lot 17, Pharmacia, Sweden) and polyethyleneglycol 6000 (Carbowax 6000, Fluka, Switzerland). The volume ratio of the two phases is 1.0. Single-step partitions are carried out in 10-ml nitrocellulose tubes which are inverted mechanically at about 30 cycles/minute for 5 minutes; the tubes are then centrifuged at 1000 g for 20 minutes. Separations are also obtained with a countercurrent apparatus (49). Chromatin is first sheared in a French pressure cell to a DNA molecular weight of 2×10^6. Optimal separation is obtained in 0.1 mM Tris-Cl (pH 7.4)–0.2 mM EDTA–0.25 mM KCl, and the two phases contain 5% Dextran and 4% polyethyleneglycol, respectively. Chromatin partition is extremely sensitive to the ionic composition of the medium; increasing concentrations of salt (KCl) shift the chromatin progressively from the polyethyleneglycol phase (upper phase) into the Dextran phase (lower phase). The countercurrent method used with the solvent system described above results in the isolation of a population of chromatin molecules enriched in nonhistone protein and nascent RNA. This fraction represents about 30% of the input DNA.

F.   Buoyant Density Centrifugation in Metrizamide

Until recently isopycnic centrifugation of chromatin as a method for fractionation was complicated by the fact that protein–DNA complexes are dissociated by the high ionic stength of the centrifugation media (cesium salts, for example). This problem may be overcome, to a certain degree, by crosslinking of protein and DNA with formaldehyde (50). Cross-linking, however, leads to irreversible changes in chromatin; therefore, nonionic or weakly ionic media would be desirable for isopycnic banding of chromatin. Hossainy et al. (51) have used chloral hydrate for this purpose. Unfortunately, this compound has several undesirable properties; for example, purified DNA is denatured by chloral hydrate. The compound metrizamide has been used by Monahan and Hall (52) and by Rickwood et al. (23) for fractionation of sheared chromatin. Metrizamide [2-(3-acetamide-5-N-methyl acetamide-2,4,6-triiodobenzamide)-2-deoxy-D-glucose] dissolves in water to give solutions of high density, low ionic strength, and low viscosity. These solutions form density gradients when centrifuged in a fixed-angle rotor, and sheared chromatin appears to separate into two fractions of different buoyant density upon centrifugation (23,52).

The procedure given is that of Rickwood et al. (23). Chromatin is solubilized in 1 mM EDTA–1 mM N-2-hydroxyethylproperazine-N'-2-ethanesulfonic acid (HEPES) (pH 7.0) and sheared by sonication. The sheared chromatin is then mixed with a solution of metrizamide (Nyegaard and Co. A/S, Oslo, Norway) in the same buffer, and the metrizamide concentration is adjusted to 41.5% (w/v). Metrizamide concentrations are determined from the refractive index of the solution using the following relationship:

$$\rho = 3.350c - 3.462$$

The mixture is overlaid with paraffin oil and centrifuged in 5-ml portions in the M.S.E. 10 × 10 aluminum fixed-angle rotor at 30,000 rpm (72,000 g) or in 15-ml portions in the M.S.E. 8 × 40 titanium fixed-angle rotor at 27,000 rpm (76,800 g) for 44 hours at 2°C. The metrizamide–chromatin mixture contained 0.1–0.2 mg of chromatin DNA/per milliliter. The gradients are fractionated and densitites are calculated from the refractive index.

Extensively sheared chromatin (with a DNA molecular weight of 2.4×10^5) exhibits a clear bimodal distribution after centrifugation in metrizamide (23). The major difference between the two fractions is the proportion of protein complexed with DNA. RNA polymerase binds to both fractions, and no differences have been found in the concentration of globin genes in the density fractions from cells synthesizing hemoglobin and in the fractions from a tissue which does not synthesize hemoglobin. These studies suggest that the metrizamide fractions do not correspond to transcriptionally active and inactive regions of chromatin. It is likely that the density fractions arise from differences in the nonhistone protein composition of nucleosomes. The functional significance of this remains unclear.

G.   Differential Solubility

Chromatin may be fractionated on the basis of differential solubility in solutions containing salt or divalent cations. This approach is based on the observation that chromatin is highly aggregated at physiological ionic strength (0.14–0.2 M Na$^+$). Under these conditions, the small fraction of chromatin which is available to RNA polymerase *in vivo* would be predicted to be less aggregated and perhaps soluble in sheared chromatin preparations (*26*). Arnold and Yound (*22*) and Bonner and co-workers (*27,28*) have used Mg^{2+} ions to effect a fractionation. This method relies on the same expected solubility properties of repressed and active chromatin as does precipitation with salt. Precipitation with Mg^{2+} avoids salt conditions where chromosomal protein rearrangement might take place. In the procedure of Arnold and Young (*22*) chromatin DNA is fragmented by sonication. The DNase II method of Marushige and Bonner (*26*) is recommended over sonication as this latter method does not lead to detectable levels of protein rearrangement during enzyme treatment (see Section III,E,1). After incubation with nuclease, large chromatin aggregates are removed from solution by centrifugation at 27,000 g for 10 minutes at 4°. To the supernatant, 1/100th volume of 0.2 M MgCl$_2$ is added drop by drop with rapid stirring. The solution is maintained on ice with constant stirring for 20 minutes, and then it is centrifuged for 10 minutes at 27,000 g. The second supernatant is the active chromatin fraction.

Several criteria have been applied to test the validity of this fractionation method: (1) the amount of DNA recovered in the Mg^{2+}-soluble fraction varies according to the cell type or tissue under investigation but corresponds to the template activity of the particular chromatin studied (*27*); (2) the Mg^{2+}-soluble fraction contains a subset of DNA sequences rather than a random sample of the genome; (3) these DNA sequences are enriched about 5-fold in sequences transcribed into cellular RNA (*28*); and (4) nascent RNA is copurified with the Mg^{2+}-soluble fraction (*22,27,53,54,55*). The association of nascent RNA with the chromatin DNA of the Mg^{2+}-soluble fraction has been confirmed by gel filtration (*22*) and by sucrose gradient centrifugation (Gottesfeld, unpublished).

The Mg^{2+}-soluble fraction shares many of the properties of active fractions isolated by other procedures: it is enriched in nonhistone protein; it is most susceptible to DNase; and the DNA of this chromatin fraction melts at significantly lower temperatures than unfractionated chromatin or the inactive fractions.

V. Conclusions

Many of the methods currently in use for the fragmentation of chromosomal DNA and the subsequent fractionation of chromatin into active and inactive regions have been described in the previous pages. Most of the methods for chromatin shearing (sonication, pressure, homogenization) are harsh and undoubtedly cause irreversible changes in chromatin structure and specific DNA–protein interactions. Consequently, fractionation results with chromatin sheared in these ways should be viewed with caution. The more-gentle nuclease methods of chromatin fragmentation offer promise for the future.

All of the methods for chromatin fractionation listed above result in the isolation of a portion of the input chromatin with at least some of the properties expected of transcriptionally active chromatin. In many instances, however, the basis for fractionation is obscure. At the outset I listed five criteria to test the validity of the fractionation methods. No technique reported to date has satisfied all of these criteria. In most cases the appropriate DNA reassociation and DNA–RNA hybridization experiments have yet to be performed. In some cases the results of such experiments have been negative (*23,47,*). Clearly, more work is needed to determine whether the existing methods of chromatin fractionation are indeed satisfactory; new and gentle techniques for fractionation are also needed.

Acknowledgment

I wish to thank the Helen Hay Whitney Foundation for financial support.

Note Added in Proof

J. R. Tata and B. Baker (*Cell*, in press) have devised a new method for chromatin fractionation based on simple differential centrifugation following mild nuclease treatment. Nuclei are washed with 0.25 M sucrose–0.5 mM MgCl$_2$–0.25% Triton X-100 and then suspended in 100 mM Tris-Cl (pH 8.0)–100 mM NaCl–1.0 mM dithiothreitol–8% glycerol–0.25 mM CaCl$_2$. Micrococcal nuclease is added to 0.2 units per milliliter and digestion is carried out at 20°–29°C for 60 seconds. The reaction is stopped by the addition of EGTA to 20 mM, and the sample is centrifuged at 1200 g for 10 minutes. The low-speed pellet contains 87% of the input DNA in the form of heterochromatin and nucleoli. The supernatant is centrifuged at 150,000 g for 60 minutes. The high-speed pellet contains 9% of the input DNA in the form of polynucleosome (6–150 nucleosomes). Over 85% of endogenous template-engaged RNA polymerase II (B) is recovered in this fraction, as well as the bulk of *in vivo* labeled heterodisperse RNA. Appropriate hybridization experiments are currently under way to establish whether the high-speed pellet fraction contains DNA sequences which are transcribed *in vivo*.

REFERENCES

1. Gurdon, J. B., *Dev. Biol.* **4**, 256 (1962).
2. Brown, I. R., and Church, R. B., *Dev. Biol.* **29**, 73 (1972).
3. Grouse, L., Chilton, M. D., and McCarthy, B. J., *Biochemistry* **11**, 798 (1972).
4. Ryffle, G. U., and McCarthy, B. J., *Biochemistry* **14**, 1379 (1975).
5. Axel, R., Feigelson, P., and Schltz, G., *Cell* **7**, 247 (1976).
6. Galau, G. A., Klein, W. H., Davis, M., Wold, B., Britten, R. J., and Davidson, E. H., *Cell* **7**, 487 (1976).
7. Huang, R. C., and Bonner, J., *Proc. Natl. Acad. Sci. U.S.A.* **48**, 1216 (1962).
8. Huang, R. C., and Bonner, J., *Proc. Natl. Acad. Sci. U.S.A.* **54**, 960 (1965).
9. Britten, R. J., and Davidson, E. H., *Science* **165**, 349 (1969).
10. Paul, J., and Gilmour, R. S., *J. Mol. Biol.* **34**, 305 (1968).
11. Davidson, E. H., and Britten, R. J., *Quart. Rev. Biol.* **48**, 565 (1973).
12. Barrett, T., Maryanka, D., Hamlyn, P. H., and Gould H. J., *Proc. Natl. Acad. Sci. U.S.A.* **71**, 5057 (1974).
13. Stein, G. Park, W., Thrall, C., Mans, R., and Stein, J., *Nature (London)* **257**, 764 (1975).
14. Gilmour, R. S., and Paul, J., *Proc. Natl. Acad. Sci. U.S.A.* **70**, 3440 (1973).
15. Axel, R., Cedar, H., and Felsenfeld, G., *Proc. Natl. Acad. Sci. U.S.A.* **70**, 2029 (1973).
16. Littau, V. C., Allfrey, V. G., Frenster, J. H., and Mirsky, A. E., *Proc. Natl. Acad. Sci. U.S.A.* **52**, 93 (1964).
17. Miller, O. L. *Nat. Cancer Inst. Monogr.* **18**, 79 (1965).
18. Berendes, H., *Int. Rev. Cytol.* **35**, 61 (1973).
19. Lewin, B., *Cell* **4**, 11 (1975).
20. Frenster, J. H., Allfrey, V. G., and Mirsky, A. E., *Proc. Natl. Acad. Sci. U.S.A.* **50**, 1026 (1963).
21. Reeck, G. R., Simpson, R. T., and Sober, H. A., *Proc. Natl. Acad. Sci. U.S.A.* **69**, 2317 (1972).
22. Arnold, E. A., and Young, K. E., *Arch. Biochem. Biophys.* **164**, 73 (1974).
23. Rickwood, D., Hell, A., Malcolm, S., Birnie, G. D., MacGillivray, A. J., and Paul, J., *Biochim. Biophys. Acta* **353**, 353 (1974).
24. Janowski, M., Nasser, D. S., and McCarthy, B. J. *Gene Transcription Reprod. Tissue, Trans. Karolinska Symp. Res. Methods Reprod. Endocrinol. 5th 1972*, p. 112.
25. McCarthy, B. J., Nishiura, J. T., Doenecke, D., Nasser, D. S., and Johnson, C. B., *Cold Spring Harbor Symp. Quant. Biol.* **38**, 763 (1973).
26. Marushige, K., and Bonner, J., *Proc. Natl. Acad. Sci. U.S.A.* **68**, 2941 (1971).
27. Billing, R. J., and Bonner, J., *Biochim. Biophys. Acta* **281**, 453 (1972).
28. Gottesfeld, J. M., Garrard, W. T., Bagi, G., Wilson, R. F., and Bonner, J., *Proc. Natl. Acad. Sci. U.S.A.* **71**, 2193 (1974).
29. Marushige, K., and Bonner, J., *J. Mol. Biol.* **14**, 160 (1966).
30. Noll, M., Thomas, J., and Kornberg, R., *Science* **187**, 1203 (1975).
31. Pfeiffer, W., Horz, W., Igo-Kemenes, T., and Zachau, H. G., *Nature (London)* **258**, 450 (1975).
32. Hewish, D. R., and Burgoyne, L. A., *Biochem. Biophys. Res. Commun.* **52**, 504 (1973).
33. Polisky, B., Greene, P., Garfin, D. E., McCarthy, B. J., Goodman, H. M., and Boyer, H. W., *Proc. Natl. Acad. Sci. U.S.A.* **72**, 3310 (1975).
34. Woodhead, L. and Johns, E. *FEBS Lett.* **62**, 115 (1976).
35. Chalkley, R. G., and Jensen, R. H., *Biochemistry* **7**, 4380 (1968).
36. Murphy, E. C., Jr., Hall, S. H., Shepherd, J. H., and Weisen, R. S., *Biochemistry* **12**, 3843 (1973).
37. Chesterton, C. J., Coupar, B. E. H., and Butterworth, P. H. W., *Biochem. J.* **143**, 73 (1974).

38. Berkowitz, E. M., and Doty, P., *Proc. Natl. Acad. Sci. U.S.A.* **72**, 3328 (1975).

39. Doenecke, D., and McCarthy, B. J., *Biochemistry* **14**, 1366 (1975).

40. Magee, B. B., Paoletti, J., and Magee, P. T., *Proc. Natl. Acad. Sci. U.S.A.* **72**, 4830 (1975).

41. McConaughy, B., and McCarthy, B. J., *Biochemistry* **11**, 998 (1972).

42. Nasser, D. S., and McCarthy, B. J. in "Methods in Enzymology," Vol. 40: Hormone Action, Part E, Nuclear Structure and Function (B. W. O'Malley and J. G. Hardman, eds.), p. 93. Academic Press, New York, 1975.

43. Simpson, R. T., and Reeck, G. R., *Biochemistry* **12**, 3853 (1973).

44. Simpson, R. T., *Proc. Natl. Acad. Sci. U.S.A.* **71**, 2740 (1974).

45. Peterson, E. A., and Kuff, E. L., *Biochemistry* **7**, 2916 (1969).

46. Simpson, R. T., and Polacow, I., *Biochem. Biophys. Res. Commun.* **55**, 1078 (1973).

47. Howk, R. S., Anisowicz, A. Silverman, Y., Parks, W. P., and Scolnick, E. M., *Cell* **4**, 321 (1975).

48. Turner, G., and Hancock, R., *Biochem. Biophys. Res. Commun.* **58**, 437 (1974).

49. Albertsson, P. A., "Partition of Cell Particles and Macromolecules," 2nd ed. Wiley (Interscience), New York, 1971.

50. Brutlag, D., Schlehuber, C., and Bonner, J., *Biochemistry* **8**, 3214 (1969).

51. Hossainy, E., Zweidler, A., and Bloch, D. P., *J. Mol. Biol.* **74**, 283 (1973).

52. Monahan, J. J., and Hall, R. H., *Nucleic Acid Res.* **1**, 1359 (1974).

53. Bonner, J., Gerrard, W. T., Gottesfeld, J. M., Holmes, D. S., Sevall, J. S. and Wilkes, M., *Cold Spring Harbor Symp. Quant. Biol.* **38**, 303 (1973).

54. Pederson, T., and Bhorjee, J. S., *Biochemistry* **14**, 3238 (1975).

55. Kimmel, C. B., Sessions, S. K., and MacLeod, M. C., *J. Mol. Biol.* **102**, 177 (1976).

Chapter 29

Chromatin Fractionation by Chromatography on ECTHAM–Cellulose

ROBERT T. SIMPSON

*Developmental Biochemistry Section, Laboratory of Nutrition and Endocrinology,
National Institute of Arthritis, Metabolism, and Digestive Diseases,
National Institutes of Health, Bethesda, Maryland*

I. Introduction

Chromatin fractionation using chromatography on ECTHAM–cellulose was initiated in an attempt to reduce the physicochemical heterogeneity of chromatin and isolate and characterize the composition and structure of chromatin fractions which differed in functional properties from the bulk of the material, for example, constitutive heterochromatin and chromatin active as a template for DNA-dependent RNA polymerase. When these studies were begun, most available chromatin fractionation methods relied on the size and/or shape of the fragmented chromatin as a basis for separation (1–4). We hoped to devise a method that would be less sensitive to the size of the nucleoprotein moiety generated during the shearing which is a mandatory prerequisite to fractionation methods. While the elution of large polyanionic molecules from ion-exchange adsorbents is a function of their size, for the case of chromatin fractionation on ECTHAM–cellulose the separation appears to be based not on size but rather on the histone content of the various nucleoprotein species.

The adsorbent utilized, ECTHAM–cellulose (5), has tris(hydroxymethyl)aminomethane as an ionizing group, coupled to cellulose with epichlorhydrin. The adsorbent is prepared at a relatively low capacity, about 0.1–0.15 meq/gm cellulose. The pK of the ionizing group is lower than that of the more common anion-exchange celluloses. Together these two characteristics allow the chromatography of nucleoproteins under sufficiently mild conditions so that dissociation of proteins from RNA or DNA is not likely to be a problem. For example, the fractionation of chromatin is carried out in a pH range from 6.5 to 9.1 at ionic strength of about 0.01. The fractionation method is relatively rapid: the time from animal sacrifice to completion of separation is about 24 hours. Further, the fractions are obtained at reasonable concentrations (50–500 μg DNA/ml) in low ionic strength Tris buffers near neutral pH, enabling them to be directly used for a variety of further manipulations. Once chromatin has been prepared, the methodology is completely standard and has been used for a variety of tissues from several animal species and tissue-culture cell lines. The scale of the fractionation has been varied from input amounts of chromatin DNA of 200 μg to over 100 mg without effect on the ensuing resolution.

In consideration of potential chromatin fractionation schemes, it is important to distinguish methods that display a spectrum of types of nucleoproteins from those which divide the chromatin fragments into two populations. Since different structural and functional segments of the genome are almost certainly arrayed contiguously along a single polynucleotide chain, shearing of the chromatin is a prerequisite to fractionation. If shearing can be carried out specifically, as may be the case for the DNAse II digestion and fractionation described by Gottesfeld *et al.* (6), then separation of the sheared material into two populations by simple procedures such as differential sedimentation or solvent precipitation is a reasonable approach. On the other hand, most commonly employed shearing methods are likely to induce random chain breakage and thereby create a population of chromatin fragments that contains varying proportions of the different types of nucleoprotein segments. For this type of population a fractionation method which displays this *spectrum* of species in a population of sheared chromatin, such as the chromatographic method described here or sucrose gradient sedimentation (7), is mandatory.

II.   Preparation and Shearing of Chromatin

Nuclei are prepared by a method originally devised by Hymer and Knuff (8). Animals are exsanguinated and organs are removed and trimmed of connective tissue and blood vessels. The chosen organ is minced in an ice-cold beaker with scissors. Buffer A [0.25 M sucrose–3 mM CaCl$_2$–0.01 M Tris (pH 8.0)] is added, using 4–5 ml/gm tissue, and the mince is blended in the 1-quart glass jar of a Waring blender attached to a powerstat. Blending is carried out at high voltage (85–90 V) long enough to complete the mincing, i.e., until nearly no macroscopic tissue fragments are visible. This usually requires less than 30 seconds. The voltage is lowered to 45–50 V, and blending is carried out for an additional 2 minutes. The sample is filtered through three layers of cheesecloth, and a crude nuclear pellet is collected by centrifugation for 10 minutes at 2000 g. The pellet is resuspended in half the volume of buffer A initially employed and mixed gently in the blender jar, usually at a voltage of 15–20 V. Triton X-100 is added to a final concentration of 1%, and stirring is continued for 2 minutes. The centrifugation is repeated exactly as above. This step of Triton washing and sedimentation is repeated until the nuclear pellet is nearly white in color. For liver from older animals three or four repetitions may be required. In contrast, nuclei from young animal tissues and cultured cells generally are sufficiently pure after a single cycle of Triton washing. At this stage the nuclei, examined by phase-contrast microscopy, should have a well-preserved morphology and be nearly free of debris and cytoplasmic tags. The nuclei are washed twice by sedimentation from buffer A lacking detergent.

The nuclear pellet is suspended in 0.05 M Tris-Cl(pH 8.0), using 2 ml/gm wet weight of tissue. Homogenization is carried out with a motor-driven Teflon and glass Potter–Elvehjem homogenizer for three to five strokes. Chromatin is then sedimented by centrifugation at 5000 g for 15 minutes. Washing in similar fashion is then carried out twice more with 0.05 M Tris-Cl and once each with 0.01 M, 0.005 M, and 0.001 M Tris-Cl, each time centrifuging as above. During the lowering of the ionic strength the pellet should swell to become as expanded gel. When the pellet is suspended in mM Tris-Cl, an aliquot should be diluted in 1% sodium dodecyl sulfate (SDS) and the absorbance at 260 nm determined to estimate the amount of DNA in the preparation. The final pellet is suspended in 1 mM Tris-Cl (pH 8.0), at a concentration of about 10 A_{260}/ml, and sheared in a blender for 90 seconds at 60 V. Unsheared material is removed by centrifugation for 20 minutes at 7500 g, and the chromatin in the supernatant is utilized for fractionation.

Twenty-milliliter aliquots of chromatin are sonicated using a Branson Sonifier, Model W185, equipped with a microtip (Heat Systems–Ultrasonics, Inc., Plainview, New York). Samples are held in the aluminum "cold shoulders" cell (from the same manufacturer) suspended in a vigorously stirred ice-water bath. Sonication with this instrument proceeds for 2 minutes at a power output of 60–65 W. This time was selected since it represented the beginning of a plateau when chromatin size was measured during sonication by viscometric techniques (Fig. 1). When establishing conditions for sonication with other instruments, it would be wise to similarly follow

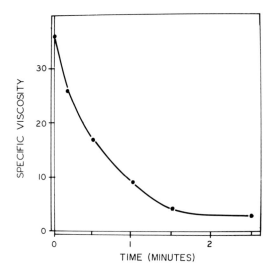

FIG. 1. Sonication of chromatin. The specific viscosity of chromatin samples was determined in a three-bulb Ostwald-type low-shear viscometer at 20° as a function of the time of sonication under conditions described in text.

the viscosity of the preparation and select sonication conditions at the onset of the plateau. Chromatin sonicated in this fashion has a weight-average molecular weight corresponding to about 900 base pairs of DNA. It melts less than 1° below unsonicated chromatin. No proteins are dissociated by the sonication procedure, and the template properties of the nucleoprotein are not altered by sonication. The chromatin is adjusted to contain 0.01 M Tris-Cl(pH 6.5) by slow addition with vigorous stirring of a 100-fold concentrated stock solution.

All pH values detailed are the values of stock solutions, 1 M for all Tris buffers and 0.25 M for ethylenediaminetetraacetic acid (EDTA) buffers, measured at 20°. Except where otherwise noted all operations involved in chromatin preparation and chromatography are carried out at 0–4°.

III.   Preparation of ECTHAM–Cellulose

The methodology used is exactly that of Peterson and Kuff (5) although we have scaled up the preparation to 10 times that described without any difficulties other than mechanical problems of adequately mixing the components. The need for obtaining homogeneous dispersion of the Tris and cellulose initially, and of the complete mixture later, cannot be over-emphasized. Any high or low local concentrations of reactants will produce correspondingly high or low local concentrations of ionizing groups on the product and lead to tight-binding sites and lowered recoveries from the columns.

Dissolve 60 gm of NaOH in 175 ml of water. In this solution dissolve 20 gm of Tris and cool the mixture in an ice bath. The solution is then added to 60 gm of cellulose (the original reference specifies 100–230 mesh cellulose sieved from Solka-Floc SW-40-B, Brown Co., Berlin, New Hampshire, but we have used several sources—Whatman CF11, and Bio-Rad Cellex N1, for example—without major differences in product), and the two components are mixed thoroughly with a strong rod or manually, wearing gloves. When as evenly dispersed as possible, the mixture is cooled in an icebath for 30–60 minutes. Thirty milliliters of epichlorhydrin (Eastman) are added, and the mixture is again mixed thoroughly. The reaction is allowed to proceed overnight at room temperature in a covered vessel in a hood. The adsorbent is then suspended in 500 ml of 2 M NaCl and transferred to a coarse-sintered glass filter attached to a vacuum line. The salt is removed by suction, and then the adsorbent is washed on the funnel successively with 1 N NaOH, water to neutrality, 1 N HCl, water to neutrality, 1 N NaOH, and water to neutrality. Fines are removed from the adsorbent by dilution to a large volume in water and washing by decantation. The final step is to wash the cellulose again on the glass filter and then, without disturbing the cake, wash thoroughly with ethanol. Traces of ethanol are removed in a rotatory evaporator or vacuum oven at 40°.

The degree of modification of the cellulose can be measured by nitrogen determination or estimated by titration. ECTHAM–cellulose prepared as described above has a content of ionizable groups of 0.11–0.14 meq/gm. The pK_a of the bound Tris group is 7.1 in water. Barring contamination by isotopes or microorganisms, the adsorbent can be reused extensively, for at least 100 chromatographic runs.

IV. Column Operating Parameters

Prior to use or after completion of a run, the adsorbent is allowed to soak for 30 minutes in 0.5 N NaOH using 10 ml/gm cellulose. The alkali is removed by vacuum filtration and the cellulose washed with water until the washings are neutral. The adsorbent is then suspended in water (10 ml/gm), and 1 M Tris-Cl (pH 6.5) is added to achieve a final buffer concentration of 0.1 M. After 10 minutes the cellulose is washed with the starting buffer [0.01 M Tris-Cl (pH 6.5)] several times and left as a slurry of about equal volumes of cellulose and buffer.

Any standard chromatography column with low dead space and a top fitting capable of being pumped, such as those made by Pharmacia Corp., Bio-Rad, or Chromaflex, may be used. The column is filled with the slurry with the outlet closed. When 1–2 cm of adsorbent have settled to the bottom, the outlet is opened wide and the column is allowed to pack by gravity or with low pressures of air (up to 4 lb/in.² g.). More slurry is added before the top of the cellulose has dried, and the operation is continued until the appropriate size has been attained. The column is then washed with two column volumes of starting buffer at a flow rate 1.5 times the eluting flow rate.

Generally the dimensions of columns employed lead to a ratio of height to diameter of 15–25. The relationships between column adsorbent volume, sample load, and elution flow rate are:

Adsorbent volume (ml) = 5 × sample load (mg DNA)

Flow rate (ml/hr) = 0.14 × adsorbent volume (ml)

The sample is loaded at a flow rate 1.5 times the elution flow rate, and the column is then washed with starting buffer at the elution flow rate until one column volume has been passed through. Elution is carried out by switching to a buffer of 0.01 M Tris (base)–0.01 M NaCl. This switch can be conveniently accomplished by use of a Gralab timer and two Flow Stopper solenoid valves (LKB Instruments). Fractions are collected on a time basis, usually one every 15 minutes.

The absorbance at 260 nm and pH is determined for each fraction. Particularly with small fraction volumes the use of spectrophotometer cells with a path length of 1 mm is expeditious avoiding the need for dilution and requiring only 0.25 ml to fill. Appropriate fractions are then pooled, dialyzed or otherwise manipulated prior to further analysis.

The elution profile is characteristically a broad featureless peak, having a sharp trailing edge except for some tailing of material with low optical density (Fig. 2). When comparing the extremes of the eluted material, as opposed to examining a particular property all across the elution profile, we routinely examine the characteristics of the first 10% of the eluted material and the last 10% of the eluate.

Recovery of applied material is generally 85–90%. Lower recoveries are obtained if the column is not loaded to 80–90% of capacity. Material remaining on the column may be eluted with 0.5 N NaOH or solutions of guanidine hydrochloride. It is likely that the material not eluted with Tris represents a random selection of the types of nucleoprotein in the input sample, as has been observed for other cellulosic adsorbents. Evidence

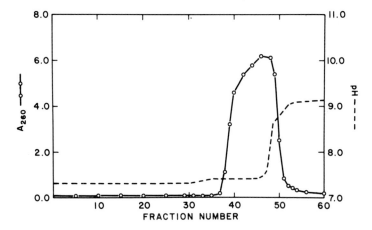

FIG. 2. Elution of sonicated chromatin from an ECTHAM–cellulose column. The 1.2 × 15 cm column contained 2 gm of adsorbent. Ten milliliters of sonicated chromatin (A$_{260}$ = 7.4) were applied at 3 ml/hour. The titrant, 0.01 M Tris–0.01 M NaCl, was applied at 2 ml/hour. Fraction size = 1 ml. The pH of the buffers here was measured at 4°.

for this derives from displacement chromatography of chromatin on ECTHAM-cellulose. A column was loaded to 5 times its capacity with chromatin and then eluted as normal. In this case, material with the properties characteristic of the last eluted segments in a normal column run began to elute midway through the elution profile, well before the pH change. The properties of the last eluted material from such a displacement column are virtually identical to those last eluted from a conventional chromatography experiment, suggesting that no particular species has been bound especially firmly to the adsorbent.

V. Properties of Chromatin Fractions

The thermal denaturation profiles of early and late-eluted chromatin samples are shown in Fig. 3 (9). This is the simplest and most readily available method for monitoring the efficacy of the fractionation, and so it will be described here. Appropriate samples are dialyzed for 24 hours versus two changes of 0.25 mM EDTA (pH 7.0). The samples are removed from dialysis bags and diluted to A_{260} = 0.25–0.30 with dialyzate. The solutions degassed briefly using a water aspirator and placed in cuvettes with Teflon

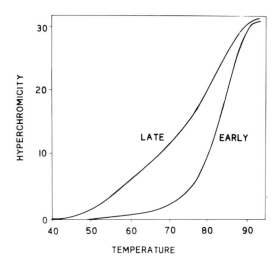

Fig. 3. Thermal denaturation of early and late-eluted HeLa cell chromatin fractions obtained by ECTHAM–cellulose chromatography. Samples in 0.25 mM EDTA, pH 7.0, were denatured using an ACTA III spectrophotometer equipped with multiple sample changer and thermostatted cell block. Temperature increases were obtained with a Neslab thermal programmer.

stoppers. Thermal denaturation profiles are recorded using a spectrophotometer with a 0.1 absorbance scale, a multiple sample changer, and a thermostatted cell block. Temperature increases are obtained with a programmable heating unit and usually are at about 0.3–0.5° minute. Several instruments which have these capabilities are available. The early eluted chromatin should have less than 10% of its total hyperchromicity developed at 70°, while the late-eluted chromatin segments usually have 35–40% of their total hyperchromicity evident by that temperature.

Some of the compositional, functional, and structural properties of chromatin which elutes early and late in the elution profile are compared in Table I. It is clear that the material on the tailing edge of the elution profile comprises what has been called the "template-active" portion of chromatin, is depleted in histone content, particularly H1, has more non-histone proteins than the remainder of the chromatin segments, and exists in an extended, more DNA-like structure. Certain of these features are those expected if these chromatin segments has been partially depleted of H1 by the action of the chromatin-associated protease. Evidence that this is not the case is as follows. Storage of chromatin for 24 hours under the conditions existing for the late-eluted material does not alter the circular dichroism spectrum, histone content, or thermal denaturation profile. Addition of 1 mM diisopropylphosphofluoridate (DFP) to the chromatin does not alter the results of chromatography. Rechromatography of a

TABLE I

COMPARISON OF SOME PROPERTIES OF EARLY AND LATE-ELUTED
ECTHAM–CELLULOSE CHROMATIN

Property	Early	Late	Reference
Histone, gm/gm DNA	1.2	0.75	*10*
Nonhistone protein, gm/gm DNA	1.0	2.8	*10,11*
Satellite DNA, mouse liver, % total	19	2	*12*
RNA polymerase binding sites, % protein-free	1	40–50	*13*
Sedimentation coefficient, Svedberg	21	9	*12*
Ellipticity at 280 nm, degrees/cm² dmol	2800	6000	*15*
Melting, % total hyperchromicity at 70°	<10	~40	*9*
Molecular weight, × 10^{-6}	2.6	2.0	*14*
Phosphate available to polylysine, %	25	70	*16*
Nuclease-susceptible DNA, %	40	65–70	*16*

fraction from the middle of the column eluate does not create low-melting, H1-depleted fractions.

Chromatin fractionation on ECTHAM–cellulose is not applicable to certain types of investigation. First, samples differing greatly in size will not resolve in the same fashion as those of a tighter size distribution—the overall charge on the polyanionic nucleoprotein determines elution position rather than the ratio of histone to DNA. Second, numerous small molecules chromatograph on ECTHAM–cellulose, either by ionic or nonspecific interactions. Among these are estradiol and triiodothyronine, thus precluding the application of this method of chromatin fractionation for experiments on the binding of these hormones to different types of nucleoproteins present in chromatin.

In summary, the method of ion-exchange chromatography on ECTHAM–cellulose appears to resolve a series of components which are present in sheared chromatin. There is no evidence to suggest that the fractionation procedure itself creates the disparate types of chromatin segments found in the elution profile. Indeed, all properties examined balance across the profile, with the early and late fractions differing in opposite direction from the property of the input material. The question has been raised whether any chromatin prepared by methods which involve shearing accurately reflects the distribution of components present in the cell *in vivo* (*17*). While this must currently remain moot, it must be noted that no functional properties have been reported yet for isolated nucleosomes or for chromatin prepared by nuclease digestion methods. In contrast, a large body of evidence suggests that the chromatin isolated by conventional means retains many of the structural features and the specificity of transcription characteristic of the chromatin *in vivo* (*18–23*).

REFERENCES

1. Frenster, J. H., Allfrey, V. G., and Mirsky, A. E., *Proc. Natl. Acad. Sci. U.S.A.* **50**, 1026 (1963).
2. Duerksen, J. D., and McCarthy, B. J., *Biochemistry* **10**, 1471 (1971).
3. Chalkey, R., and Jensen, R. H., *Biochemistry* **7**, 4380 (1968).
4. Marushige, K., and Bonner, J., *Proc. Natl. Acad. Sci. U.S.A.* **68**, 2941 (1971).
5. Peterson, E. A., and Kuff, E. L., *Biochemistry* **8**, 2916 (1969).
6. Gottesfeld, J. M., Garrard, W. T., Bagi, G., Wilson, R. F., and Bonner, J, *Proc. Natl. Acad. Sci. U.S.A.* **71**, 2193 (1974).
7. Charles, M. A., Ryffel, G. U., Obinata, M., McCarthy, B. J., and Baxter, J. D., *Proc. Natl. Acad. Sci. U.S.A.* **72**, 1787 (1975).
8. Hymer, W. C., and Kuff, E. L., *J. Histochem. Cytochem.* **12**, 359 (1964).
9. Reeck, G. R., Simpson, R. T., and Sober, H. A., *Proc. Natl. Acad. Sci. U.S.A.* **69**, 2317 (1972).
10. Simpson, R. T., and Reeck, G. R., *Biochemistry* **12**, 3853 (1973).
11. Reeck, G. R., Simpson, R. T., and Sober, H. A., *Eur. J. Biochem.* **49**, 407 (1974).
12. Simpson, R. T., *Biochem. Biophys. Res. Commun.* **65**, 552 (1975).
13. Simpson, R. T., *Proc. Natl. Acad. Sci. U.S.A.* **71**, 2740. (1974).
14. Simpson, R. T., *in* "Current Topics in Biochemistry 1973" (C. B. Anfinsen and A. M. Schechter, eds.) p. 135. Academic Press, New York, 1974.
15. Polacow, I., and Simpson, R. T., *Biochem. Biophys. Res. Commun.* **52**, 202. (1973).
16. Simpson, R. T., and Polacow, I., *Biochem. Biophys. Res. Commun.* **55**, 1078 (1973).
17. Noll, M., Thomas, J. O., and Kornberg, R. D., *Science* **187**, 1203 (1975).
18. Wilson, G. V., Steggles, A. W., and Niehuis, A. W., *Proc. Natl. Acad. Sci. U.S.A.* **72**, 4835 (1975).
19. Axel, R., Cedar, H., and Felsenfeld, G., *Proc. Natl. Acad. Sci. U.S.A.* **70**, 2029 (1973).
20. Gilmour, R. S., and Paul, J., *Proc. Natl. Acad. Sci. U.S.A.* **70**, 3440 (1973).
21. Astrin, S., *Biochemistry* **14**, 2700 (1975).
22. Swetly, P., and Watanabe, Y., *Biochemistry* **13**, 4122 (1974).
23. Stein, G. S., Park, W. D., Thrall, C. L., Mans, R. J., and Stein, J. L., *Biochem. Biophys. Res. Commun.* **63**, 945 (1975).

Chapter 30

Fractionation of Chromatin in a Two-Phase Aqueous Polymer System

A. J. FABER

Swiss Institute for Experimental Cancer Research,
Lausanne, Switzerland

1. Introduction

In the past few years several methods have been developed for the fractionation of chromatin. The primary objective of such experiments is to separate, from the bulk of the chromatin, those regions actively undergoing transcription. Chromatin has been shown by hybridization studies to direct the *in vitro* synthesis of RNA similar to that produced in the intact cell (*1–4*). This suggests that, within isolated chromatin, the necessary control elements for transcription are present. Therefore, a detailed study of the ultrastructure, protein, RNA and DNA components, and other properties of the transcribed and nontranscribed regions should contribute to a better understanding of the mechanism of gene activation and the regulation of gene activity that occurs sequentially during cellular differentiation.

Chromatin fractionation methods can be divided, in general, into two groups. The first relies on differences in the degree of condensation of regions within the chromatin structure. Thus chromatin, which has been sheared appropriately, can be separated into fractions by differential centrifugation (*5–10*) or column chromatography (*11–14*). Autoradiographic studies have established that the extended chromatin fibers ("euchromatin") represent the template-active regions whereas the condensed chromatin ("heterochromatin") represent inactive regions (*15–18*). The second group of methods for fractionation is based on the differential precipitation of sheared chromatin by monovalent or divalent cations (*19–21*).

In an earlier paper a method was described for the fractionation of sheared chromatin in a two-phase aqueous polymer system (*22*). This procedure allows one to isolate milligram quantities of the chromatin fraction which bears rapidly labeled RNA, from soluble chromatin prepared at low ionic strength (23). Evidence is presented here that this template-active fraction represents chromatin bearing nascent RNA attached to the template by RNA polymerase.

II.   Methods

A.   Chromatin Preparation

Chromatin is prepared from mouse P815 cells (24) (grown in suspension culture) by the low ionic strength method described by Hancock et al. (25). In brief, the method involves washing the cells in the presence of 0.1 M sucrose to remove serum proteins and growth medium salts, lysing the cells at low ionic strength [0.2 mM phosphate (pH 7.4)] by the nonionic detergent Nonidet P-40, followed by several centrifugations through sucrose. Dounce homogenization of the chromatin gives fragments with an average single-stranded DNA molecular weight of 9×10^5 [measured in alkaline isokinetic sucrose gradients by utilizing a marker of polyoma form II DNA in combination with Studier's (26) equation]. The sheared chromatin is then centrifuged (Spinco rotor SW65, 15,000 rpm, for 30 minutes at 4°) to remove any remaining nucleoli or nuclear envelope fragments, and the supernatant is dialyzed overnight against 0.2 mM ethylenediaminetetraacetic acid (EDTA) (pH 7.4).

In order to monitor the distribution of chromatin during fractionation in the two-phase system, cells are grown in the presence of radioactive precursors. DNA is labeled with [^{14}C]thymidine (6 μCi/liter with unlabeled thymidine 0.2 μg/ml) during two generations of cell growth (approximately 24 hours). Nascent RNA is labeled by a 3-minute pulse of [^{3}H]uridine (3 mCi/liter) followed by rapidly pouring the cells into crushed frozen growth medium.

B.   Single-Step Partition

Dextran T500 (Pharmacia, Uppsala, Sweden) and polyethylene glycol 6000 (Carbowax, Fluka, Switzerland) are prepared as 20% (w/w) and 40% (w/w) aqueous solutions, respectively. The dextran is dissolved by gentle stirring at 90°–95° and then sterilized by autoclaving. Its precise concentration can be determined by polarimetry (specific optical rotation $[\alpha]_D^{25} = + 199°$). The polyethylene glycol (PEG) should not be heated, and for this reason it is prepared using sterile technique. Both solutions can be stored for several weeks at 4°.

The theory and application of two-phase aqueous polymer systems for the partition of biological material has been studied in detail (27,28). When aqueous solutions of two different polymers (e.g., Dextran, PEG) are mixed above a critical concentration, an immiscible liquid two-phase system is formed. Partition of a soluble substance in this system is characterized by a partition coefficient K (ratio of the concentration of the partitioned substance in the upper phase relative to the lower phase). In an earlier paper (22) chromatin was shown to transfer progressively from the upper phase (PEG-rich) to the lower phase (Dextran-rich) as the concentration of KCl was increased in the phase mixture. In the experiments described here, the KCl concentration at which K equals 1.0 is found to vary from 0.2 mM to 0.5 mM KCl for different preparations of chromatin. This variation is probably due to differences in their endogenous ion content. Before each multistep fractionation, the KCl concentration to give a K equal to 1.0 is determined by single-step partition as shown in Fig. 1A. An average partition coefficient of 1.0 for total chromatin should give the best condition for a multistep fractionation of components with different affinities for the two phases (27).

Single- and multistep partitions are carried out in 13.5 ml cellulose nitrate

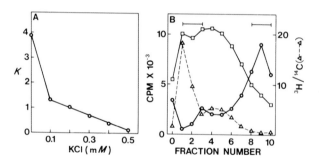

FIG.1.   (A) The partition coefficient (K) of chromatin in a single-step partition in the two-phase system at various concentrations of KCl. Chromatin (20 μg DNA) prepared from cells labeled with [¹⁴C]thymidine and pulse-labeled for 3 minutes with [³H]uridine was added to a 2-gm phase mixture containing 5% Dextran–4% PEG–0.1 mM Tris-HCl (pH 7.4)–0.2 mM EDTA and the indicated KCl concentration. After partition, the trichloroacetic acid (TCA)-precipitable radioactivity was determined for both the upper and lower phases of each sample, and the K value for [¹⁴C]DNA was determined. The K value for [³H]RNA (not shown) followed a parallel curve. (B) Countercurrent distribution of chromatin containing pulse-labeled RNA. The chromatin preparation used in Fig. 1A was fractionated by countercurrent distribution in 10 tubes each containing 10 gm of phase mixture. The composition of the phase system was as described for Fig. 1A, with 0.2 mM KCl. The chromatin sample (containing 1.8 mg DNA) was loaded in tube O. The bars indicate fractions pooled for removal of polymers (Fig. 2). (O)¹⁴C-labeled DNA; (□)³H-labeled RNA.

tubes (Beckman 50 Ti rotor tubes) sealed with silicone stoppers. The phase system found most suitable for chromatin partition contains 5% Dextran–4% PEG–0.1 mM Tris-HCl (pH 7.4)–0.2 mM EDTA and an appropriate concentration of KCl (determined as in Fig. 1A). Table I shows the preparation of the above phase system in three different volumes. Due to their viscosity, the Dextran and PEG solutions are weighed into the tubes. The 2-gm phase system is used for single-step partition. It is important to equilibrate the phase mixture at 4° before adding the chromatin, and to perform all subsequent steps at that temperature since formation of a two-phase system is highly dependent on temperature (27). The tubes are inverted mechanically (for 5 minutes at approximately 10 cycles/minute) and then centrifuged for 5 minutes (1000 g) to separate the phases. The upper phase is removed by pipette without disturbing the interphase, and the lower phase is recovered by puncturing the bottom of the tube. A negligible amount of chromatin is found in the interphase which is not collected; however, if chromatin is highly dependent on temperature (27). The tubes are inverted mechanically (for 5 minutes at approximately 10 cycles/minute) and then centrifuged mined by measuring the radioactivity in TCA-insoluble precipitates after the addition of TCA (5% final) and bovine serum albumin (50 μg). TCA concentrations higher than 5% should not be used since precipitation of the polymers may occur. The chromatin precipitates are collected on Whatman GF/B glass-fiber filters, which are washed 3 times with 5% TCA, then 95% ethanol, dried, and the radioactivity determined in a scintillation counter. When chromatin is labeled with [^{14}C]thymidine and [^{3}H]uridine, a correction is made for spillover of ^{14}C counts into the ^{3}H channel.

C.   Multistep Partition

To effect separation of the template-active regions (see Sections III,B and III,C) from the bulk of the chromatin, a countercurrent distribution is carried out. A stock of pure upper and lower phases, at a KCl concentration to give a chromatin K value of 1.0, is prepared in a separatory funnel prior to the fractionation. This phase mixture (Table I, 200 gm) is equilibrated at 4°, mixed vigorously, and the phases allowed to separate overnight. The pure upper phase is removed by pipette while the pure lower phase is recovered from the bottom.

For the first tube of a multistep partition, a phase mixture (Table I, 10 gm) is prepared containing up to 10 mg chromatin. Partition of chromatin between the phases is independent of its concentration to at least 0.9 mg DNA/ml phase mixture. The tube is inverted mechanically and centrifuged as before. The upper phase of this tube is transferred to a new tube, number 2. At this point it is important to equalize the upper and lower phase volumes

TABLE I

PREPARATION OF THE DEXTRAN–POLYETHYLENE GLYCOL
PHASE SYSTEM

	Phase system		
Stock solution	2 gm	10 gm	200 gm
20.5% (w/w) Dextran	0.49 gm	2.44 gm	48.80 gm
40.0% (w/w) PEG	0.20 gm	1.00 gm	20.00 gm
10 mM Tris-HCl (pH 7.4 at 4°)	0.02 ml	0.10 ml	2.00 ml
2 mM EDTA (pH 7.4)	0.20 ml	1.00 ml	20.00 ml
50 mM KCl[a]	0.008 ml	0.04 ml	0.80 ml
Bidistilled H$_2$O	1.032 ml	0.42 ml	108.40 ml
Chromatin	0.05 ml	5.00 ml	—
(approx. 600 μg DNA/ml)			

[a] For the chromatin preparation used in this example, 0.2 mM KCl in
the phase mixture gives a partition coefficient of 1.0. The Dextran and
PEG are weighed into the reaction mixture while the remaining components are added by pipette.

and if some adjustment is necessary, a small amount of upper phase is removed from the second tube. An equal volume of pure upper phase is then
added to tube 1, while an equal volume of pure lower phase is added to tube
2. The mechanical inversion, centrifugation, and phase separation steps are
repeated with the addition of a new tube at the end of each sequence until
a total of 10 phase separations have been performed. Samples containing
50–100 μg of chromatin are taken from each tube (homogeneously mixed)
and assayed for TCA-insoluble radioactivity.

D. Recovery of Chromatin

Fractions from the multistep distribution, representing the template-
active and template-inactive chromatin (see Sections III,B and III,C), are
pooled separately and freed of the polymers by gel filtration through Sepharose 4B (Pharmacia, Uppsala, Sweden). To separate Dextran T500 (MW
500,000) from the chromatin fragments, a ratio of column bed volume/sam-
ple volume of at least 5 is required when the column is eluted at low ionic
strength [0.2 mM EDTA (pH 7.4)]. More recently, we prefer to use a ratio
of 8 to insure complete separation of the Dextran and chromatin. Thus if one
pools three tubes from the multistep distribution, a Sepharose 4B column
with a bed volume of 200–250 cm^3 (2.5 × 40 cm to 2.5 × 50 cm) will be sufficient. The Dextran concentration in the column fractions can be monitored
by polarimetry or refractive index, although the latter method is less accurate.
PEG (MW 6000) is retained within the column and elutes much later than

the Dextran. After determining the TCA-insoluble radioactivity in the eluted fractions (see Section II,B), those containing chromatin are pooled, dialyzed against 0.2 mM EDTA (pH 7.4), and either used immediately or dialyzed against bidistilled water and lyophilized.

III.   Results

A.   Chromatin Fractionation

A typical distribution pattern of sheared chromatin labeled with [14C] thymidine and pulse-labeled for 3 minutes with [3H]uridine is shown in Fig. 1B. The bulk of the chromatin, as estimated by DNA content, is located in fractions 8, 9, and 10. A small amount of pulse-labeled RNA was associated with this material. On the other hand, most of the pulse-labeled RNA is found in the early fractions of the distribution profile. Three regions of high, intermediate, and low [3H]RNA/[14C]DNA ratios can be observed. The region with lower ratios may include fragments which contain both template-active and template-inactive chromatin in the same molecule. As shown later (Fig. 3A), more intensive shearing can reduce the amount of this material. The fractions indicated by a bar are pooled and applied to Sepharose 4B columns to separate the chromatin fragments from the polymers. The template-active chromatin isolated in such a manner usually represents 5–15% of the total chromatin based on DNA content.

Figure 2 shows typical elution profiles of the template-active and template-

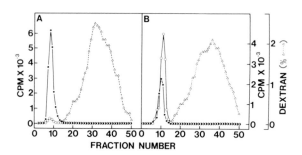

FIG.2.   Separation of chromatin fractions, after partition, from the phase mixture on Sepharose 4B. (A) Template-active fraction; (B) template-inactive fraction. The appropriate tubes from the countercurrent distribution shown in Fig. 1B are pooled and applied to Sepharose 4B columns (2.5 × 30 cm) equilibrated with 0.2 mM EDTA (pH 7.4). The columns are eluted with 0.2 mM EDTA and the fractions monitored for TCA-precipitable radioactivity and for Dextran concentration (Section II,D). (O) 14C-labeled DNA; (●) 3H-labeled RNA.

inactive chromatin fractions from Sepharose 4B columns. In both cases chromatin elutes at the void volume and is completely free of the polymers. The $[^3H]RNA/[^{14}C]DNA$ ratio for the template-active and template-inactive chromatin was 19.2 and 0.5, respectively, thus giving approximately a 38-fold enrichment of pulse-labeled RNA in the template-active fraction relative to the template-inactive material. The chromatin peaks from the column are then collected, dialyzed against 0.2 mM EDTA (pH 7.4), and used immediately.

B.   Distribution of RNA Polymerase

Alpha-amanitin and other amatoxins bind specifically to RNA polymerase B (29–31), the nucleoplasmic enzyme believed responsible for the synthesis of RNA destined to become messenger RNA (32–35). Cochet-Meilhac et al. (31–36) have developed a Millipore filtration assay using labeled amatoxins which allows the quantitation of RNA polymerase B present in cell homogenates. We have used this assay to examine the distribution of RNA polymerase in fractionated chromatin, utilizing O-[^3H]methyl-demethyl-γ-amanitin.

Figure 3B shows a countercurrent distribution of sheared chromatin labeled with [^{14}C]thymidine at low specific activity. The distribution profile is similar to that of the preparation used in Fig. 3A, which also contains pulse-labeled RNA. Samples, containing the same amount of DNA, are taken from each fraction and assayed for RNA polymerase B content by the Millipore filter technique. The major peak for RNA polymerase detected by amanitin binding (Fig. 3B) is located in the same region of the distribution profile as that which contains pulse-labeled RNA (Fig. 3A).

C.   Distribution of Nascent RNA

More than 70% of the total labeled RNA is associated with chromatin when cells are pulse-labeled with [^3H]uridine for 3 minutes. Evidence to support the conclusion that the majority of this newly synthesized RNA in chromatin is still attached to RNA polymerase is presented in Fig. 4. When chromatin from cells pulse-labeled with [^3H]uridine is analyzed by equilibrium density centrifugation in Cs_2SO_4, approximately 60% of the labeled RNA is found to band in the region of DNA (on the low-density side of the peak) at a density indicating that protein is also present (Fig. 4A). The remaining 40% of the labeled RNA, presumably derived from RNP particles or from nascent RNA fragmented during manipulation, is recovered in the pellet as free RNA ($\rho = 1.630$ gm/cm³). We have found that RNP particles extracted from purified nuclei by the method of Samarina et al. (37) disso-

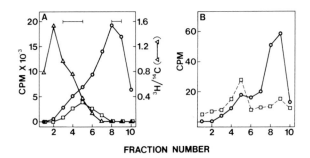

FRACTION NUMBER

FIG. 3. (A) Countercurrent distribution of chromatin containing pulse-labeled RNA. Chromatin prepared from cells labeled with [^{14}C]thymidine and pulse-labeled for 3 minutes with [^{3}H]uridine was sheared more intensively (average 2000 base pairs) than that used in Fig. 1B, and was fractionated by countercurrent distribution. The phase system was as described in Fig. 1A, with 0.5 mM KCl. The chromatin sample (containing 4 mg DNA) was loaded in tube 1. Fractions indicated by a bar were pooled for other studies. (○) ^{14}C-labeled DNA; (□) ^{3}H-labeled RNA. (B) Distribution of RNA polymerase, determined by the binding of [^{3}H]amanitin, in chromatin fractions after countercurrent distribution. Chromatin prepared from cells labeled with [^{14}C]thymidine (0.01 μCi/liter) was sheared and fractionated as described in Fig. 3A. From each fraction a sample, containing an equivalent amount of DNA, was taken and assayed for binding of [^{3}H]amanitin. Binding of O-[^{3}H]methyl-demethyl-γ-amanitin (0.02 μg/ml reaction mixture) to the chromatin fractions was performed for 2 hours at 0° in 100 mM Tris-HCl (pH 7.5)–0.1 mM EDTA–0.1 mM dithiothreitol–140 mM (NH$_4$)$_2$SO$_4$–30% glycerol (binding buffer) (30). The samples were then filtered slowly through Millipore HAWP filters (0.45 μm), which were washed 3 times with 10 ml of binding buffer containing 15% glycerol. The filters were dissolved in scintillation fluid (30) and counted. Blank values, obtained by preincubating the chromatin fractions for 10 minutes at 0° with unlabeled α-amanitin (2 μg/ml reaction mixture) before the addition of labeled amanitin, were subtracted. (○) ^{14}C-labeled DNA; (□)[^{3}H]amanitin retained on filters.

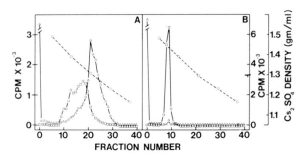

FRACTION NUMBER

FIG. 4. Analysis of chromatin containing pulse-labeled RNA by equilibrium density centrifugation in Cs$_2$SO$_4$ gradients. Chromatin was prepared from cells labeled for two generations with [^{14}C]thymidine and pulse-labeled for 3 minutes with [^{3}H]uridine. One portion of the chromatin was treated with proteinase K [Merck, chromatographically pure; 100 μg/ml in 5 mM Tris-HCl (pH 8.0)–5 mM EDTA–5 mM NaCl] at 37° for 1 hour. The control and treated chromatin (containing 50 μg DNA) were mixed with Cs$_2$SO$_4$ [in 10 mM Tris-HCl (pH 7.4) in a volume of 5.0 ml] to give a final density of 1.310 gm/cm^3 and centrifuged in a Spinco SW65 rotor at 40,000 rpm for 4 days at 20°. Fractions were collected from the bottom of the gradients, and the TCA-insoluble radioactivity was determined. (A) Original chromatin. (B) Chromatin treated with proteinase K. (□) ^{14}C-labeled DNA. (△) ^{3}H-labeled RNA. (○)

ciate completely into RNA and protein in Cs_2SO_4 gradients. In contrast, RNA being transcribed from chromatin (38) or DNA (39) remains attached to the template when centrifuged through 4 M CsCl.

When chromatin containing pulse-labeled RNA is treated with proteinase K, a powerful proteolytic enzyme with a broad action spectrum, more than 97% of the ³H-labeled RNA previously associated with the DNA in Cs_2SO_4 gradients is released and sediments as free RNA (Fig. 4B). The DNA bands in a sharp peak at its normal density (1.430 gm/cm³). No RNase or DNase activity could be detected in our proteinase K; this enzyme has been used by others to extract messenger RNA (40) and high-molecular-weight DNA (41).

D. Other Properties of the Chromatin Fractions

The chromatin fraction containing pulse-labeled RNA obtained by countercurrent distribution is enriched in tryptophan-labeled nonhistone proteins relative to the remaining chromatin (22). Sodium dodecyl sulphate (SDS)–polyacrylamide gel analysis of the nonhistone proteins in these two fractions is in agreement with this conclusion (unpublished results).

Fractionation of chromatin in this aqueous polymer system is not dependent of differences in DNA fragment size between the template-active and template-inactive regions, since no appreciable difference in DNA size is observed on alkaline sucrose gradients. The template-active chromatin, however, has a higher buoyant density than the inactive or nonfractionated chromatin when analyzed, after HCHO-fixation, by equilibrium density centrifugation in Cs_2SO_4. The difference (1.315 versus 1.290 gm/cm³) is probably due to the enrichment of nascent RNA in the active fraction, although a lower protein content would give a similar result.

Visualization of the template-active and template-inactive fractions by electron microscopy [Dubochet method (42)] reveals as yet no significant difference in structural organization (A. J. Faber, P. Oudet, and P. Chambon, unpublished observations); conditions for visualization of nascent RNA are at present being developed. Both fractions appear similar to the nonfractionated chromatin (25) for they are in a well-dispersed, nonaggregated state and show the characteristic nucleosomal structure described by Olins and Olins (43) and Oudet et al. (44).

IV. Comments

An advantage of this aqueous polymer system for chromatin fractionation is the mildness of the experimental conditions. Fractionation takes place at neutral pH, in a low ionic strength environment and at a low interfacial tension (0.0001–0.1 dyn/cm) (27). The chromatin is never exposed to ionic conditions causing intermolecular interactions (e.g., high concentrations of monovalent or divalent cations), and the fractions are easily recovered from the phase mixture by gel chromatography.

Partition of a substance in a two-phase aqueous polymer system is dependent on the ionic charge, structure, conformation, and nature of the groups in the substance (27). Hence the partition of chromatin in a two-phase system may have the potential for higher resolution than either centrifugation or column fractionation methods. Better resolution, or enrichment of other chromatin subfractions, may be possible by using other polymer systems or conditions. For analytical purposes, an automatic countercurrent distribution apparatus designed by Albertsson (27) has been shown to give high resolution (22).

The properties of transcribing chromatin responsible for its separation in this system are not yet known. When free RNA or DNA are fractionated under conditions used for chromatin, they remain in the first tube. Using chromatin reconstituted at different protein/DNA ratios one observes a greater displacement of the bulk of the chromatin from the origin with increasing protein/DNA ratios, although a portion of the chromatin remains in the early part of the distribution profile. Our working hypothesis is that the partition mechanism may be dependent on the presence of nascent RNA in the template-active chromatin, and/or on differences in the types of proteins and in the conformation of the protein–DNA structure which determines its affinity for a particular phase.

Acknowledgments

I am grateful for the advice and encouragement of Dr. Ron Hancock, in whose laboratory this work was done. I also thank Professor Pierre Chambon for his generous gift of labeled and unlabeled amanitin, and Dr. Peter Beard for reading the manuscript. This study was supported by the Swiss National Science Foundation.

References

1. Paul, J., and Gilmour, R. S., *J. Mol. Biol.* **34**, 305 (1968).
2. Bonner, J., Dahmus, M., Fambrough, D., Huang, R. C. C., Marushige, K., and Tuan, D., *Science* **159**, 47 (1968).
3. Smith, K. D., Church, R. B., and McCarthy, B. J., *Biochemistry* **8**, 4271 (1969).
4. Huang, R. C. C., and Huang, P. C., *J. Mol. Biol.* **39**, 365 (1969).
5. Frenster, J. H., Allfrey, V. G., and Mirsky, A. E., *Proc. Natl. Acad. Sci. U.S.A.* **50**, 1026 (1963).
6. Chalkley, R., and Jensen, R. H., *Biochemistry* **7**, 4380 (1968).
7. Yasmineh, W. G., and Yunis, J. J., *Biochem. Biophys. Res. Commun.* **35**, 779 (1969).
8. Duerksen, J. D., and McCarthy, B. J., *Biochemistry* **10**, 1471 (1971).
9. Murphy, E. C., Hall, S. H., Shepherd, J. H., and Weiser, R. S., *Biochemistry* **12**, 3843 (1973).

10. Berkowitz, E. M., and Doty, P., *Proc. Natl. Acad. Sci. U.S.A.* **72**, 3328 (1975).

11. Reeck, G. R., Simpson, R. T., and Sober, H. A., *Proc. Natl. Acad. Sci. U.S.A.* **69**, 2317 (1972).

12. Simpson, R. T., and Reeck, G. R., *Biochemistry* **12**, 3853 (1973).

13. Janowski, M., Nasser, D. S., and McCarthy, B. J., *Gene Transcription Reprod. Tissue, Trans. Karolinska Symp. Res. Methods Reprod. Endocrinol., 5th, 1972,* p. 112.

14. Duerksen, J. D., and Smith, R. J., *Int. J. Biochem.* **5**, 827 (1974).

15. Littau, V. C., Allfrey, V. G., Frenster, J. H., and Mirsky, A. E., *Proc. Natl. Acad. Sci. U.S.A.* **52**, 93 (1964).

16. Littau, V. C., Burdick, C. J., Allfrey, V. G., and Mirsky, A. E., *Proc. Natl. Acad. Sci. U.S.A.* **54**, 1204 (1965).

17. Unuma, T., Arendell, J. P., and Busch, H., *Exp. Cell Res.* **52**, 429 (1968).

18. Fakan, S., Puvion, E., and Spohr, G. *Exp. Cell Res.* **99**, 155 (1976).

19. Marushige, K., and Bonner, J., *Proc. Natl. Acad. Sci. U.S.A.* **68**, 2941 (1971).

20. Gottesfeld, J. M., Garrard, W. T., Bogi, G., Wilson, R. F., and Bonner, J., *Proc. Natl. Acad. Sci. U.S.A.* **71**, 2193 (1974).

21. Arnold, E. A., and Young, K. E., *Arch. Biochem. Biophys.* **164**, 73 (1974).

22. Turner, G., and Hancock, R., *Biochem. Biophys. Res. Commun.* **58**, 437 (1974).

23. Hancock, R., *J. Mol. Biol.* **86**, 649 (1974).

24. Schindler, R., Day, M., and Fisher, G. A., *Cancer Res.* **19**, 47 (1959).

25. Hancock, R., Faber, A. J., and Fakan, S.. *Methods Cell Biol.* **15**, 127 (1976).

26. Studier, F. W., *J. Mol. Biol.* **11**, 373 (1965).

27. Albertsson, P. Å., "Partition of Cell Particles and Macromolecules," 2nd ed. Wiley (Interscience), New York, 1971.

28. Walter, H., *Methods Cell Biol.* **9**, 25 (1975).

29. Kedinger, C., Gniazdowski, M., Mandel, J. L., Gissinger, F., and Chambon, P., *Biochem. Biophys. Res. Commun.* **38**, 165 (1970).

30. Lindell, T. J., Weinberg, F., Morris, P. W., Roeder, R. G., and Rutter, W. J., *Science* **170**, 447 (1970).

31. Cochet-Meilhac, M., and Chambon, P., *Biochim. Biophys. Acta* **353**, 160 (1974).

32. Roeder, R. G., and Rutter, W. J., *Nature (London)* **224**, 234 (1969).

33. Zylber, E. A., and Penman, S., *Proc. Natl. Acad. Sci. U.S.A.* **68**, 2861 (1971).

34. Cox, R. F., *Eur. J. Biochem.* **39**, 49 (1973).

35. Chambon, P., *Ann. Rev. Biochem.* **44**, 613 (1975).

36. Cochet-Meilhac, M., Nuret, P., Courvalin, J. C., and Chambon, P., *Biochim. Biophys. Acta* **353**, 185 (1974).

37. Samarina, O. P., Lukanidin, E. M., Molnar, J., and Georgiev, G. P., *J. Mol. Biol.* **33**, 251 (1968).

38. Bonner, J., and Widholm, J., *Proc. Natl. Acad. Sci. U.S.A.* **57**, 1379 (1967).

39. Chambon, P., personal communication.

40. Wiegers, V., and Hilz, H., *Biochem. Biophys. Res. Commun.* **44**, 531 (1971).

41. Gross-Bellard, M., Oudet, P., and Chambon, P., *Eur. J. Biochem.* **36**, 32 (1973).

42. Dubochet, J., Ducommun, M., Zollinger, M., and Kellenberger, E., *J. Ultrastruct. Res.* **35**, 147 (1971).

43. Olins, D. E., and Olins, A. L., *J. Cell. Biol.* **53**, 715 (1972).

44. Oudet, P., Gross-Bellard, M., and Chambon, P., *Cell* **4**, 281 (1975).

Chapter 31

Biochemical Approaches to Chromatin Assembly

R. HANCOCK

Swiss Institute for Experimental Cancer Research,
Lausanne, Switzerland

I.   Introduction

The DNA and histones from which chromatin is assembled are synthesized in a generally synchronous manner during S phase (*1*). Once assembled into chromatin, new histones H2A, H2B, H3, and H4 are conserved there for many cell generations (*2–4*); in contrast, histone H1 shows a turnover of about 15% per generation (*4*). Nonhistone proteins become associated with, and dissociate from, chromatin throughout the cell cycle (*5, 6*). A number of aspects of the assembly *in vivo* of the basic chromatin structure can be distinguished, for example: assembly of new histones to form the protein core of the nucleosome; assembly of new nucleosomes with DNA; events occuring to nucleosomes at the site of DNA replication; and the distribution of histones of the parental chromatin between the two daughter molecules during chromatin replication.

The experimental analysis of these processes is to some extent interdependent with developments in understanding of the primary and higher-order structures of chromatin. We have explored the possibility of density-labeling the components of chromatin, combined with buoyant density separation, to approach these questions.

II.   Methods

A.   Density-Labeling of DNA

The only density label extensively studied (*7,8*) is 5-iododeoxyuridine (IdU) which is incorporated into newly synthesized DNA in place of thymidine. Complete replacement in one strand, using IdU at 0.2 μM/ml in the growth medium, increases the buoyant density of mouse P815 cell (3, 9) DNA from 1.70 to 1.79 (9). The rate of DNA synthesis in P815 cells is normal under these conditions during one complete cycle of replication with IdU (10), but these cells are not able to complete a second cycle (9). The newly made DNA strands containing IdU may be simultaneously labeled with [³H]thymidine (Fig. 1A); alternatively, labeling with [¹²⁵I]IdU, which may be specifically measured in a gamma counter, allows simultaneous use of ³H and ¹⁴C to label other macromolecules (Fig. 4) if the levels of precursors are chosen so that a precise correction for β radioactivity from ¹²⁵I may be made.

The buoyant density of hybrid DNA could theoretically be increased to approximately 1.82 by simultaneous incorporation of the three other deoxynucleosides labeled with ¹³C, ¹⁵N, and ²H.

B. Chromatin Preparation

Isolation of chromatin under ionic conditions in which intermolecular interactions occur can lead to formation of intermolecular aggregates (11, 12) and to transfer of histone 1 (and possibly others) (12, 13) between chromatin fragments. An alternative procedure, using low ionic strength conditions, has therefore been developed for isolation of chromatin from cultured cells; interaction and histone 1 transfer do not occur during this procedure (14). Up to six samples containing 0.5–2 × 10⁷ cells (representing about 50–200 μg of chromatin DNA) may be processed in parallel. Two cycles of centrifuging through sucrose (solution A2) (14) yield chromatin which is sufficiently pure for the present purpose; the chromatin is centrifuged the second time in a Dounce homogenizer, suspended at a concentration of 50–100 μg DNA/ml in 0.2 mM EDTA–0.2 mM dithiothreitol (pH 7.2), and fragmented using the tight piston. The length of the DNA in the chromatin fragments produced is of the order of 3000 base pairs (14). Low-speed centrifugation to remove debris is not necessary.

C. Formaldehyde Fixation

The only procedures available at present for buoyant density analysis of density-labeled chromatin involve the covalent binding of proteins to DNA using formaldehyde (7, 8, 15, 16), followed by analysis in CsCl gradients. The high hydration of unfixed chromatin in metrizamide gradients masks the small density differences currently obtainable (10).

Fragmented chromatin prepared as described above is fixed by addition, with stirring, of 0.25 volumes of 30% formaldehyde (Merck p.a.) in 25 mM phosphate (pH 7.0). After 20–24 hours at 4°C, excess formaldehyde is removed by dialysis (three changes) against 100 volumes of 0.2 mM EDTA–0.2 mM dithiothreitol (pH 7.2) at 4°C. Fixed samples may be stored for several weeks at 4°C without change in their buoyant density pattern. Non-chromatin proteins do not detectably attach to chromatin during this procedure (12); sensitive evidence that intermolecular complexes are not formed is provided by the unchanged sedimentation coefficient of SV40 virus chromatin after fixation (17).

D.   Buoyant Density Analysis in CsCl

Self-generating CsCl gradients have been predominantly used (7,8,16, 18), although preformed gradients allow economy of centrifuge time (15). Maximum resolution of density-labeled from normal chromatin is obtained using angle rotors; the steeper gradients generated in swing-out rotors allow analysis of the whole range of densities between those of chromatin and free DNA (18). Samples containing up to 50 μg DNA are mixed with CsCl solution [5 M in 5 mM phosphate (pH 7.2)] to give the required starting density. Some useful conditions are as follows.

For density-labeled chromatin (Fig. 1A): 5-ml sample, initial density 1.380 (tubes filled with paraffin oil). Centrifugation at 33,000 rpm, 20°, $\nmid$ 60 hours, Beckman rotor 40 or 50Ti. Density range generated: 1.35 to 1.42.

For density-labeled chromatin lacking histone 1 (Fig. 1B): Same conditions but initial density 1.445. Density range generated: 1.41 to 1.49.

For separation of DNA and chromatin: 5-ml sample, initial density 1.50. Centrifugation at 45,000 rpm, 20°, $\nmid$ 60 hours, Beckman rotor SW65 or SW 50.1. Density range generated: 1.30 to 1.75.

Polyallomer tubes should be used since fixed chromatin (especially after histone 1 removal) may attach to nitrocellulose. Gradients are conveniently fractionated by collecting drops from the tube bottom.

E.   Dissociation of Histone 1

The interpretation of experiments using simultaneous density labeling of newly synthesized DNA and radioactive labeling of new proteins is complicated by a number of factors. The density difference between normal chromatin and chromatin whose DNA is fully substituted with IdU in one strand is small (Fig. 1A). A specific label for histones is not available, although nonhistone proteins may be specifically labeled with radioactive tryptophan. If other labeled amino acids are used, the fraction of the total radioactivity incorporated into histones must be calculated indirectly (8). Histone 1, unlike the nucleosomal histones 2A, 2B, 3, and 4, turns over slowly during growth (4), and its extranucleosomal location may be reflected

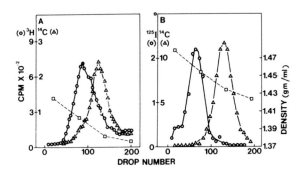

FIG. 1. Separation of chromatin containing newly synthesized density-labeled DNA (IdU replacing dT in one strand) from chromatin of normal density in CsCl gradients after formaldehyde fixation. (A) Total chromatin: (B) chromatin lacking HI and most nonhistone proteins. Normal chromatin was from mouse P815 cells, growing exponentially in suspension culture (3), labeled for 24 hours with [¹⁴C]dT (0.01 µCi/ml with unlabeled dT, 0.2 µg/ml). Density-labeled chromatin was from cells labeled for 60 minutes with IdU (0.2 µM/ml) together with [³H]dT (5 µCi/ml) (A) or together with ¹²⁵IdU (0.5 µCi/ml) (B). The appropriate two cell populations were mixed; chromatin was prepared, and in (B) histone 1 was removed as described in Section II.

in a different pattern of assembly. In order to circumvent these problems, we have employed a procedure based on that used by Oudet *et al.* (*19*) to remove HI and the major part of the nonhistone proteins from chromatin, leaving the basic nucleosomal structure.

Chromatin is prepared as described above but sedimented a second time through sucrose (solution A2) (*14*) in a 15-ml capacity centrifuge tube. The supernatant is pipetted off to leave 0.1 ml with the chromatin pellet, and 2 ml of a solution (buffer C) containing 0.6 M NaCl–5 mM phosphate, (pH 7.2)–0.2 mM EDTA–0.2 mM dithiothreitol are pipetted onto the pellet. The tube is closed and rocked slowly ($\sim$ 10 cycles/minute) in a horizontal position for 2 hours at 4°. During this period HI is dissociated, and the chromatin unfolds to give a clear, viscous solution which shows marked elastic recoil when gently mixed. This material is layered (using a 2–3 mm polystyrene pipette) on the surface of 1.5 ml of 30% sucrose in buffer C, itself layered on 1.5 ml of 60% sucrose in buffer C in a nitrocellulose tube for the Beckman SW65 rotor, which is then centrifuged at 35,000 rpm for 2 hours at 4°. To collect the chromatin band, the upper layer containing dissociated H1 and nonhistone proteins is pipetted off from the surface together with the upper 2–3 mm of the 30% sucrose layer. Using a wide-bore (5-mm) pipette, the remainder of the 30% sucrose layer and the upper one-half of the 60% sucrose layer, containing the viscous band of HI-less chromatin, are removed together. (Incomplete removal of HI, due usually to use of too large a quantity of chromatin, may result in pelleting of part of the sample to the bottom of the

60% sucrose layer.) The Hl-less chromatin is transferred (wide-bore pipette), to a dialysis tube, dialyzed against 100 volumes of 0.2 mM EDTA–0.2 mM dithiothreitol (pH 7.2) (3 changes), and fragmented and fixed as described above.

The structure of chromatin prepared by an essentially identical procedure (but employing centrifugation through glycerol in place of sucrose) has been studied in some detail (19), and the complete removal of Hl has been demonstrated (10,19). Over 80% of the chromatin nonhistone (tryptophan-labeled) proteins are also removed; over 90% of incorporated labeled arginine or lysine are present in the nucleosomal histones, which may thus be studied virtually specifically. Hl-less chromatin containing IdU-labeled DNA is almost completely separated in CsCl gradients from Hl-less chromatin of normal density (Figs. 1B and 4).

Reconstruction experiments show that no detectable migration of either global or pulse-labeled nucleosomal histones between chromatin molecules occurs during this procedure (Figs. 2 and 4B). However, if Hl-less chromatin taken from the sucrose step gradient is fragmented *before* dialysis, both intermolecular aggregation and displacement of histones occur (Fig. 2). Although this phenomenon deserves closer study, our working hypothesis is that fragmentation may allow the supertwisted configuration of DNA (20, 21) to relax to a less constrained structure to which nucleosomes are less tightly bound (19) and generate sites at which intermolecular interactions can occur. This model may also provide an explanation for the transfer of histones from sheared chromatin onto free DNA (an energetically much more favorable process than transfer between chromatin fragments) which has been observed to occur in 0.6 M NaCl (22,23).

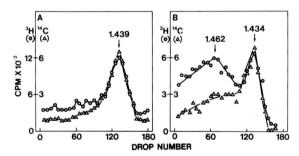

FIG. 2. Reconstruction experiment to show that intermolecular aggregation and histone migration can occur in sheared (B), but not in unsheared (A), Hl-less chromatin. Cells were grown for 24 hours with [^{14}C]dT (4 μCi/l) and [^{3}H]arginine (5 μCi/ml) and mixed with a 5-fold excess of cells whose chromatin had been density-labeled by growth for 15 hours with IdU (0.2 μM/ml). Hl-less chromatin was prepared as described in Section II for (A), but for (B) the chromatin taken from the sucrose step gradient was fragmented *before* dialyzing out sucrose and NaCl. Samples were fixed and analyzed in CsCl as described in Section II.

III.   Applications

A.   Association of New Nonhistone Proteins with Chromatin

Newly synthesized nonhistone proteins (specifically labeled with radio-active tryptophan) are incorporated into all density species of chromatin in exponentially growing cells (Fig. 3A). When newly replicated DNA is simultaneously density-labeled with IdU, the new nonhistones are not found uniquely in dense chromatin, but are distributed approximately equally between chromatin containing newly replicated DNA and that containing DNA which had not replicated during the experimental period (Fig. 3B). This result is compatible with the idea that some preexisting nonhistones migrate to regions of chromatin containing newly replicated DNA (for example, RNA polymerase to transcribed regions) and are replaced by newly synthesized nonhistones.

B.   Association of New Nucleosomal Histones with Chromatin

Experiments of a design similar to those just described, but using Hl-less chromatin, allow examination of the pattern of distribution of newly synthesized nucleosomal histones in chromatin. For example, after simultaneous density-labeling of new DNA and radioactive labeling of new histones during 30 minutes, the new nucleosomal histones are found predominantly

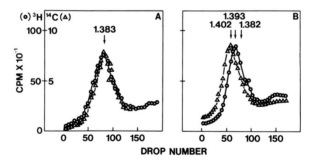

Fig. 3.   Distribution of newly synthesised nonhistone proteins in chromatin. In (A), cells were prelabeled by growth for 24 hours with [^{14}C]tryptophan (1 μCi/ml), then labeled for 3 hours with [^{3}H]tryptophan (20 μCi/ml). In (B), cells labeled in the same way with [^{3}H]tryptophan together with IdU (0.2 μM/ml) were mixed with cells grown for 12 hours with [^{14}C]tryptophan (3 μCi/ml) and IdU (0.2 μM/ml) to provide a marker of the density distribution of nonhistone proteins in fully dense chromatin. The density of normal chromation (1.382), measured relative to the same experimental sample in a second parallel chromatin preparation, is also indicated in (B).

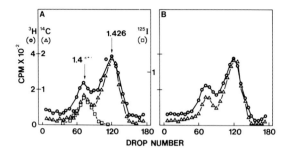

FIG. 4. Distribution of newly synthesized nucleosomal histones in Hl-less chromatin. Cells were prelabeled by growth for 24 hours with [^{14}C]dT (2 μCi/liter with unlabeled dT, 0.2 μg/ml). They were then centrifuged and resuspended in fresh medium with [^{3}H]arginine (25 μCi/ml), together with IdU (0.2 μM/ml) and ^{125}IdU (25 μCi/liter) to label newly synthesized DNA. After harvesting the cells were divided into two parts; one (A) was processed alone, and to the other (B) a 2-fold excess of cells containing fully dense chromatin (prepared as in Fig. 2) was added before chromatin isolation. Hl-less chromatin was prepared and analyzed as described in Section II.

in chromatin fragments containing DNA which had not replicated during the labeling period (Fig. 4A) *(24)*.

C. Density labeling of New Nucleosomal Histones

Density-labeling of proteins offers a potentially useful approach, complementary to density-labeling of DNA, to investigate the assembly of chromatin (and of other cellular macromolecules). Amino acids labeled with ^{15}N, ^{13}C, and ^{2}H are incorporated by P815 cells, and fragments of Hl-less

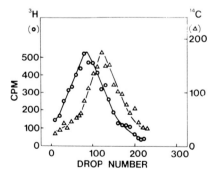

FIG. 5. Separation of chromatin containing [^{3}H]arginine- and density-labeled proteins from chromatin of normal density labeled with [^{14}C]arginine. Density-labeled chromatin was from P815 cells grown for 4 hours in medium containing [^{3}H]arginine (25 μCi/ml) in which 16 normal amino acids were replaced by their ^{15}N, ^{13}C-, ^{2}H-labeled forms. Chromatin of normal density was from cells grown for 24 hours in normal medium with [^{14}C]arginine (0.1 μCi/ml).

chromatin containing newly synthesized proteins labeled in this way may be separated by buoyant density analysis from chromatin containing proteins of normal density (Fig. 5).

REFERENCES

1. Prescott, D., *J. Cell Biol.* **31**, 1 (1966).
2. Byvoet, P., *J. Mol. Biol.* **17**, 311 (1966).
3. Hancock, R., *J. Mol. Biol.* **40**, 457 (1969).
4. Gurley, L. R., and Hardin, J. M., *Arch. Biochem. Biophys.* **130**, 1 (1969).
5. Stein, G. S., and Borun, T. W., *J. Cell Biol.* **52**, 292 (1972).
6. Gerner, E. W., and Humphrey, R. M., *Biochim. Biophys. Acta* **331**, 117 (1973).
7. Hancock, R., *J. Mol. Biol.* **48**, 357 (1970).
8. Jackson, V., Granner, D. K., and Chalkley, R., *Proc. Natl. Acad. Sci. U.S.A.* **72**, 4440 (1975).
9. Cramer, J. W., and Morris, N. R., *Mol. Pharmacol.* **2**, 363 (1966).
10. Hancock, R., unpublished material.
11. Chalkley, R., and Jensen, R. H., *Biochemistry* **7**, 4380 (1968).
12. Hancock, R. *J. Mol. Biol.* **86**, 649 (1974).
13. Jensen, R. H., and Chalkley, R., *Biochemistry* **7**, 4388 (1968).
14. Hancock, R., Faber, A. J., and Fakan, S., *Methods Cell Biol.* **15**, 127 (1976).
15. Ilyin, Yu. V., and Georgiev, G. P., *J. Mol. Biol.* **41**, 229 (1969).
16. Brutlag, D., Schlehuber, C., and Bonner, J., *Biochemistry* **8**, 3214 (1969).
17. Sen, A., Hancock, R., and Levine, A. J., *Virology* **61**, 11 (1974).
18. Varshavsky, A. J., Ilyin, Yu. V., and Georgiev, G. P., *Nature (London)* **250**, 602 (1974).
19. Oudet, P., Gross-Bellard, M., and Chambon, P., *Cell* **4**, 281 (1975).
20. Cooke, P. R., and Brazell, I. A., *J. Cell Sci.* **19**, 261 (1975).
21. Germond, J. E., Hirt, B., Oudet, P., Gross-Bellard, M., and Chambon, P., *Proc. Natl. Acad. Sci. U.S.A.* **72**, 1843 (1975).
22. Varshavsky, A. J., and Ilyin, Yu. V., *Biochim. Biophys. Acta* **340**, 207 (1974).
23. Clark, R. J., and Felsenfeld, G., *Nature (London) New Biol.* **229**, 101 (1971).
24. Hancock, R. *Br. Soc. Cell Biol., Symp. Chromatin, 1975*, p. A5.

Chapter 32

Nuclease Digestion of RNA in Chromatin

LEONARD H. AUGENLICHT

Memorial Sloan-Kettering Cancer Center,
New York, New York

I. Background

The RNA/DNA ratio is highest in chromatin isolated from tissues active in RNA synthesis (*1,2*). Several investigators have recently reported that some of this chromatin-associated RNA is rapidly labeled RNA of relatively high molecular weight. Monahan and Hall (*3*) isolated chromatin from mouse L cells on Metrizamide buoyant density gradients and found RNA of about 20–30 S in this fraction. The synthesis of this chromatin–RNA was for the most part inhibited by α-amanitin. Data of these investigators suggested that the chromatin–RNA turned over rapidly. We have also demonstrated the presence of 20–30 S RNA in chromatin isolated from cells of a human colon carcinoma line (HT-29) by sucrose gradient centrifugation (*4*). Following very brief pulses with [³H]uridine ($\leq$ 30 seconds), all of the nuclear radioactivity (following removal of nucleoli) was found in this chromatin fraction. However, label very rapidly appeared in the nuclear ribonucleo-protein fraction (nRNP) of these cells. With pulse times of 2 minutes to 1 hour, 60–70% of the radioactivity was found associated with chromatin and the balance in nRNP. This subnuclear distribution of pulse-labeled RNA is similar to that found by Tata and Baker (*5*), who reported that over 60% of the 1-hour [³H]adenosine-labeled RNA of rat liver was associated with euchromatin.

The presence of RNA as ribonucleoprotein associated with regions of chromatin active in transcription is well known. Electron microscopy has shown this to be true for the lampbrush chromosomes of amphibian oocytes (*6–9*), the puffs of insect polytene chromosomes (*10–12*), and in mammalian cells (*9*). Scott and Sommerville (*13*) have reported that the proteins of isolated nRNP are antigenically similar to protein associated with the loops of lampbrush chromosomes, further suggesting that RNA is associated with chromatin as ribonucleoprotein. However, the nature of the interaction between the RNA–protein complex and the chromatin is not clear. It has been suggested that processing of RNA (addition of poly A) takes place while the RNA is associated with chromatin (5). RNA of certain temperature-induced puffs of *Drosophila* may be stored in chromatin (14), and, during metaphase arrest, ribosomal precursor RNA remains associated with chromatin as ribonucleoprotein (15).

Several investigators have noted that a portion of the RNA in nRNP of fixed tissue is protected from nuclease digestion, presumably by the protein of the complex (7,11,16). Similar observations have also been made in isolated nRNP (17–20).

We describe below a procedure for investigation of the chromatin-associated RNA by digestion with staphylococcal nuclease (EC 3.1.4.7). Reddi (21,22) first showed that this enzyme would hydrolyze RNA as well as DNA. It will also hydrolyze homopolymers (23). The distinct advantage afforded by use of this nuclease is that all of the protected chromatin DNA and chromatin protein precipitates from the limit digest (24,25), and they can then be easily removed by low-speed centrifugation leaving the protected RNA–protein complex as the only macromolecular material in the supernate.

II. Method

A. Preparation of Chromatin

Bhorjee and Pederson (26) have published a method of chromatin isolation from HeLa cells that minimizes the contamination of chromatin by RNA and nRNP. We have adapted this method for the isolation of chromatin from an epithelioid colonic carcinoma cell line (HT-29) (4). The cultures, which grow as a monolayer, are rinsed once with Hanks' balanced salt solution, scraped into Hanks', and the cells pelleted by centrifugation at 1000 g for 5 minutes. Nuclei are prepared by resuspending about 2×10^7 cells in 4 ml 0.01 M NaCl–0.0015 M MgCl$_2$–0.01 M Tris (pH 7.0) (buffer A) to which is added 1 ml of the same buffer containing 0.5% Triton X-100 (final concentration 0.1%). The suspension is vigorously vortexed and the nuclei pelleted as above. The nuclei are washed once in buffer A without Triton. Clean nuclei, completely free of cytoplasmic contamination as judged by electron microscopy, are obtained quite easily by this method. This method is not, however, suitable for all cell types. Human fibroblast nuclei, for instance, usually break in the hypotonic buffer A, and nuclei must be prepared by other means (see Augenlicht and Baserga, 27).

Nuclei are resuspended in 2 ml of buffer A and sonicated briefly to break the nuclei. We have found two 5-second pulses with a Branson W-350 Sonifier at its lowest setting (output approximately 50 W) sufficient to break 80–90% of the nuclei.

Morphologically intact nucleoli are removed by layering the sonicate over 30% sucrose in buffer A and centrifuging at 1000 g in a clinical centrifuge. It is our experience that at higher speeds, as originally suggested (26), a great deal of chromatin is removed with the pelleted nucleoli. At the lower speed, the nucleoli enter the sucrose but are not pelleted, while any unbroken nuclei and some clumps of chromatin are pelleted.

The material remaining above the 30% sucrose is gently removed using a Pasteur pipette and in turn is layered over 2.5 ml of 60% sucrose in 0.01 M NaCl–0.024 M EDTA–0.0025 M Tris (pH 7.2). The upper two-thirds of this material is thoroughly mixed, leaving a 1-ml cushion of 60% sucrose. The chromatin is pelleted by centrifugation at 140,000 g in the Beckman SW50.1 rotor for 90 minutes. Recovery of DNA layered over 60% sucrose is between 55 and 100%. Highest recoveries are generally obtained with larger amounts of material, although for the volumes mentioned, we never use more than 3–4 × 10⁷ cells for preparation.

The RNA/DNA ratio of the chromatin preparation for HT-29 cells is 0.05. As mentioned above, if the cells have been labeled for 1 hour with [³H]uridine, the chromatin fraction contains 60% of the radioactivity layered on 60% sucrose. This same value is found if the chromatin is isolated by buoyant density centrifugation in Metrizamide gradients.

B. Nuclease Digestion

For digestion with staphylococcal nuclease, all ethylenediaminetetraacetic acid (EDTA) must be removed from the chromatin. The preparation is therefore resuspended in 1 ml of 0.005 M phosphate (pH 6.8)–25 μM CaCl₂ and is dialyzed overnight against two changes of 500 volumes of the same buffer (buffer B).

The chromatin is digested by the method of Clark and Felsenfeld (24). The enzyme is dissolved in buffer B at a concentration of 50 μg/ml (10,000 units/mg, Worthington) and 0.1 ml (5 μg) is added to 1 ml of the chromatin suspension which is then incubated at 37°. The progress of digestion can be followed most efficiently by incorporation of radioactivity into the cellular nucleic acid. DNA is labeled by growing the cells in medium containing 0.01 μCi/ml [¹⁴C]thymidine, and RNA is labeled by pulsing the cells for 1 hour before the experiment with 2–10 μCi/ml [³H]uridine. Carrier DNA (200 μg calf thymus DNA) is added to a sample (50–100 λ) of the digest, and the total is precipitated by the addition of 2 ml of cold 10% trichloroacetic acid (TCA). After removal of the precipitate by centrifugation at 1000 g for 10 minutes, the radioactivity in both DNA and RNA which is rendered acid-soluble can be readily determined by assay of the supernatant. There is a progressive hydrolysis of both the DNA and RNA in the chromatin to acid-soluble oligo (deoxy) nucleotides. At the limit of digestion (1–2 hours incubation), 55% of the $[^{14}C]$thymidine-labeled DNA and 85–90% of the $[^3H]$uridine pulse-labeled RNA are acid-soluble. The protected RNA from HT-29 chromatin is about 26 nucleotides in length as determined by electrophoresis in 5% formamide–polyacrylamide gels.

C. Isolation of a Structure Containing the Protected RNA Fragment

At the limit of digestion (generally after 1–2 hours), all of the protected DNA and chromatin protein are insoluble and precipitate (23,24). The reaction can be stopped by the addition of EDTA to 10 mM, and this precipitate can be removed by centrifugation at 3000 g for 5 minutes. About 15% of the $[^3H]$uridine label is also pelleted by this low-speed centrifugation, but this represents both acid-soluble and acid-precipitable radioactivity. Only one-half of the pelleted ^3H is acid-precipitable (therefore 7–8% of the total), and 9% of the $[^3H]$uridine-labeled RNA in the supernatant (or again 7–8% of the total) is acid-precipitable. Therefore, of the 10–15% of the $[^3H]$uridine pulse-labeled RNA in chromatin protected from digestion, about one-half is lost when the protected DNA–chromatin protein precipitate is removed.

Following removal of the precipitate, acid-soluble oligo (deoxy) nucleotides can be removed from the supernatant by passing the material through a 13 × 1 cm column of Sephadex G-150. The column is equilibrated and run with 0.005 M sodium phosphate (pH 7.4)–0.0002 M EDTA (buffer C). The protected fragment comes off in the void volume (2–3 ml) and the acid-soluble material is eluted in a second, major peak. The digested oligo (deoxy) nucleotides, although acid-soluble, cannot be removed by dialysis, suggesting they are 5 to 10 bases in length (28).

We have analyzed the protected fragment from HT-29 cells on 5–20% sucrose gradients in buffer C, spun at 167,000 g in the Beckman SW41 rotor at 4° for 21 hours. Under these conditions, the material sediments uniformly about one-quarter the distance from the top of the gradient. In the analytical ultracentrifuge, it has a sedimentation coefficient of no greater than 2 S. This material, when analyzed by electrophoresis in 10% sodium dodecyl sulfate (SDS)–polyacrylamide gels, is found to be associated with a major protein band of 40,000 molecular weight and a minor band of 66,000. It should be noted that neither of these bands corresponds to the major protein of nRNP isolated from HT-29 cells, which has a molecular weight of 34,000 (4).

We have not met with success in analyzing the undigested fragment by isopycnic centrifugation in CsCl. We have tried to fix the material with formaldehyde by the method of Spirin et al. (29), but upon subsequent centrifugation in CsCl, most of the RNA has the buoyant density of free RNA. However, we know that the RNA must be associated with protein, because under identical conditions protein-free RNA isolated from chromatin is completely digested (even in the presence of chromatin). This same difficulty in achieving covalent links between oligonucleotides and protein by aldehyde treatment has been reported by others (30).

An alternative method of demonstrating that the proteins are indeed bound to the small RNA fragment and are not simply coisolating with it is to digest the protected RNA and demonstrate that no protein is then isolated by the above procedures. This can be done by adding 5 μg of pancreatic A and 12.5 units of T_1 ribonuclease to the staphlococcal nuclease digest during the final hour of incubation. Kish and Pederson (31) have reported that these enzymes render >99% of uridine pulse-labeled RNA in nRNP acid-soluble. We have observed that the same is true for chromatin-associated RNA. When the RNA fragment is terminally digested with these ribonucleases, no protein is detected on SDS–polyacrylamide gels of material from any region of the sucrose gradient used in the normal isolation scheme.

The protection from digestion of an RNA fragment is apparently due to some of the proteins that complex with the transcript to form nuclear ribonucleoprotein (32). Since these proteins are tissue-specific (33), different results (as to the extent of protection and protein species involved) may be anticipated with other cell types.

REFERENCES

1. Dingman, C. W., and Sporn, M. S., J. Biol. Chem. 239, 3483 (1964).
2. Hjelm, R. P., and Huang, R. C., Biochemistry 14, 1682 (1975).
3. Monahan, J. J., and Hall, R. H., Biochim. Biophys. Acta 383, 40 (1975).
4. Augenlicht, L. H., and Lipkin, M., J. Biol. Chem. 251, 2592 (1976).
5. Tata, J. R., and Baker, B., Exp. Cell Res. 93, 191 (1975).
6. Gall, J. G., and Callan, H. G., Proc. Natl. Acad. Sci. U.S.A. 48, 562 (1962).
7. Snow, M. H. L., and Callan, H. G., J. Cell Sci. 5, 1 (1969).
8. Miller, O., and Hamkalo, B. A., Int. Rev. Cytol. 33, 1 (1972).
9. Miller, O., and Bakken, A. H., Gene Transcription Reprod. Tissue, Trans. Karolinska Symp. Res. Methods Reprod. Endocrinol., 5th, 1972, p. 155.
10. Beermann, W., and Bahr, G. F., Exp. Cell Res. 6, 195 (1954).
11. Swift, H., Brookhaven Symp. Biol. 12, 134 (1959).
12. Stevens, B. J., and Swift, H. J., J. Cell Biol. 31, 55 (1966).
13. Scott, S. E. M., and Sommerville, J., Nature (London) 250, 680 (1974).

14. Berendes, H. D., *Chromosoma* **24**, 418 (1968).

15. Fan, H., and Penman, S., *J. Mol. Biol.* **59**, 27 (1971).

16. Monneron, A., and Bernhard, W., *J. Ultrastruct. Res.* **27**, 266 (1969).

17. Georgiev, G. P., and Samarina, O. P., *Adv. Cell. Biol.* **2**, 47 (1971).

18. Stevenin, J., and Jacob, M., *Eur. J. Biochem.* **47**, 129 (1974).

19. Sekeris, C. E., and Neissing, J., *Biochem. Biophys. Res. Commun.* **62**, 642 (1975).

20. Faiferman, I., and Pogo, A. O., *Biochemistry* **14**, 3808 (1975).

21. Reddi, K. K., *Nature (London)* **182**, 1308 (1958).

22. Reddi, K. K., *Biochim. Biophys. Acta* **36**, 132 (1959).

23. Alexander, M., Heppel, L. A., and Hurwitz, J., *J. Biol. Chem.* **236**, 3014 (1961).

24. Clark, R. J., and Felsenfeld, G., *Nature (London), New Biol.* **229**, 101 (1971).

25. Clark, R. J., and Felsenfeld, G., *Biochemistry* **13**, 3622 (1974).

26. Bhorjee, J. S., and Pederson, T., *Biochemistry* **12**, 2766 (1973).

27. Augenlicht, L. H., and Baserga, R., *Arch. Biochem. Biophys.* **158**, 89 (1973).

28. Cleaver, J. E., and Boyer, H. W., *Biochim. Biophys. Acta* **262**, 116 (1972).

29. Spirin, A. S., Belitsina, N. V., and Lerman, M. I., *J. Mol. Biol.* **14**, 611 (1965).

30. Blobel, G., *Biochem. Biophys. Res. Commun.* **47**, 88 (1972).

31. Kish, V. M., and Pederson, T., *J. Mol. Biol.* **95**, 227 (1975).

32. Augenlicht, L. H., McCormick, M., and Lipkin, M., *Biochemistry* **15**, 3818 (1976).

33. Pederson, T., *J. Mol. Biol.* **83**, 163 (1974).

SUBJECT INDEX

A

12 A3 cl 10 cells, nuclei isolation from, 47, 60
Actin
  in cytoplasm, 64
  M.W. of, 272–273
  N-methylhistidine in, 272–273
  in nuclear protein fractions, 271–276
Actomyosin, formation of, 273–274
Acufine developing agent, 71
Adenylate kinase
  as cytoplasmic marker enzyme, 14
  in nuclei, 81
  in nucleoli, 32
Agla syringe, 132
Alkaline fast green (AFG), in histone
    identification, 241–255
Amanitin, RNA polymerase binding of, 453
Amaryllis, nuclei from, 89
Amatoxins, RNA polymerase binding of, 453
Amblystoma mexicana
  developmental stages in, 151
  enucleation of, 129
Amino acid analysis, of histones, 199
Amino acids, radioactive, vessel for, 147
Amoeba, nuclear transplantation in, 125
Amphibia, nuclear transplantation in, 125–139
Amphibian Ringer's medium, 147
  composition of, 148
Arginine-rich histones, 179–180
Ascites cells, nuclei isolation from, 21–23
ATP
  amount of, in whole cells and nuclei,
    64–65
  measurement of concentrations of, 56
Autoradiography
  of histone synthesis, 245–248
  of nuclear proteins, 311–312
  of nuclear transplanted embryos, 161–162
Azure stain
  for nuclei, 14
  for nucleoli, 27

B

BALB 3T3 cells, nuclei isolation from, 47, 52,
    53
Basic proteins, of low M.W., in spermatids,
    297–316

Bio-Gel P-60, in gel filtration chromatography
    of histones, 230–231
Biosol liquid, in frog medium, 149
Bisulfite, as protease inhibitor, 122
Bluegrass, nuclei from, 89
Brain cells, nuclei isolation from, 75
Branson Sonifier, 25, 27, 28, 33, 439
Broad beans, nuclei from, 89
BSC-1 cells, nuclei isolation from, 47
Buoyant density centrifugation, of chromatin,
    432–433, 461

C

Calf thymus, histone isolation from, 184–202
Cannon medium, modified, composition of, 348
Carbowax, 333, 382
Carver hydraulic press, in nuclei disruption, 34
Catalase, in nucleoli, 32
Centrifugation, of tissue homogenates, 12, 19
Chaetopterus oocytes, mitotic apparatus
    isolation from, 365
3-Chloro-1,2-propanediol, in nuclei isolation,
    54
CHO cells, nuclei isolation from, 70
Chromatin
  assembly of, biochemical studies on,
    459–466
  buoyant density centrifugation of, 432–433,
    461
  composition of, 113
  differential centrifugation of, 427–428
  differential solubility of, 433
  ECTHAM-cellulose chromatography of,
    430–431, 437–446
  formaldehyde fixation of, 460–461
  fractionation of, 421–472
    criteria for, 422–423
    in two-phase aqueous polymer system,
    447–457
  fractions of, properties, 443–445
  gel filtration chromatography of, 428–429
  histone association with, 464–466
  nonhistone proteins stimulating transcription
    of, 318–321
  partition chromatography of, 431
  preparation of, 1, 285–286, 330, 468–469
    from plant tissue, 87–95

473

2-Methyl-2,4-pentanediol, *see* Hexylene glycol
Metrizamide
    chromatin centrifugation in, 432–433
    in spermatid nuclei isolation, 307
Mg^{2+}-ATPase, 83
    of myosin, 276
Micrococcal nuclease, in chromosomal DNA
        fractionation, 425
Micrococcus luteus, RNA polymerase of, 320
Microdissector, 132
Microinjection pipettes, construction of, 170
Microneedles for nucleocytoplasmic exchange
        studies, 145
Micronuclei, isolation of, 103–105
Micropipette
    for transplantation, 132–133
        construction, 134
    volume calibration in, 145–146
Microplasmodia, nuclei isolation from, 117
Microscopes, for nucleocytoplasmic exchange
        studies, 144–147
Mitotic apparatus
    isolation methods for, 361–379
        DMSO-glycerol method
        ethanol, 375
        glycerol isolations, 365
        hexylene glycol method, 374–375
        MTM method, 375–376
        tubulin-polymerizing medium, 364–365
        using Weisenberg buffers, 376, 377
Modified Barth Saline (MBS), composition of,
        126
Molecular sieving, of histones, 205–225
Multicool MC-4-40 apparatus, 46
Myosin
    in actin precipitation, 273–274
    in nuclear protein fractions, 277–279
Myxamebas, nuclei isolation from, 116

N

Na^+, K^+-ATPase, 83
NAD pyrophosphorylase, in nuclei, 6
NADH-cytochrome c reductase, as cytoplasmic
        marker enzyme, 14
Needle puller, 145
Neomycin, in frog medium, 149
Neurospora crassa, nuclei isolation from,
        119–122
NHP-I protein, isolation and characterization of,
        318–321

NHP-II protein, isolation and characterization
        of, 321–325
NHP-III protein, isolation and characterization
        of, 325–327
Nonaqueous isolation, of nuclei, from cultured
        cells, 45–68
Nonhistone chromosomal proteins
    contractile proteins, 269–281
    fractionation and characterization of,
        257–420
        by hydroxyapatite chromatography,
        329–336
    high mobility group, isolation of, 257–267
    isoelectric focusing in fractionation of,
        381–386
    mitotic apparatus, 361–379
    in nuclear pore complex and nuclear
        peripheral lamina, 337–342
    phosphoproteins, 317–328
    of salivary gland cells, 343
    selective extraction methods for, 283–
        296
    two-dimensional polyacrylamide gel
        electrophoresis of, 407–420
Nonidet P-40, use in nuclear isolation, 21–22,
        31
Novikoff ascites tumor cells, nuclei isolation
        from, 73–74
N-terminal amino acid analysis, of histones,
        199
Nuclear-cytoplasmic exchange, 125–177
    of macromolecules, 167–177
    of nonhistone proteins in embryos, 141–165
        procedure, 156–160
Nuclear envelopes, preparation of, 339
Nuclear membrane, nuclear microscopy of, 3
Nuclear peripheral lamina, nuclear protein
        isolation from, 337–342
Nuclear pore complex, nuclear protein isolation
        from, 337–342
"Nuclear pores," 3
Nuclear proteins
    electrophoretic characterization and
        purification of, 387–406
    isolation of, 206–208
    phosphorylation of, in *S. coprophila,*
        351–354
Nucleases
    in chromosomal DNA fractionation, 424–426
    digestion of RNA in chromatin by, 467–472

CONTENTS OF PREVIOUS VOLUMES

Volume I

Volume IV

Volume V

Volume VI

Volume VIII

Volume IX

Volume XIII

Volume XIV

Volume XV